FOURTH EDITION

Mathematics of the SHOP

With hands like these and patient loving skill
Man wrought the wonders of the antique world,
Creates the marvels of the new — and I
Have hands like these

BLANCHE TURRELL

FOURTH EDITION

Mathematics
of the SHOP

FRANK J. McMACKIN
JOHN H. SHAVER
RALPH E. WEBER
ROBERT D. SMITH

DELMAR PUBLISHERS INC.
2 Computer Drive, West
Box 15-015
Albany, New York 12212

LIBRARY OF CONGRESS CATALOG CARD NUMBER: 76-6726
ISBN: 0-8273-1297-0

Printed in the United States of America
Published simultaneously in Canada
by Nelson Canada,
A division of The Thomson Corporation

Preface

In the present text, special consideration has been given to the increased use of the metric system, and the necessity of its use in vocational occupations. Chapter Five now presents the fundamentals of Algebra in such a way that the student can apply these basic principles in many different situations.

The arrangement of material is intended to increase the understanding for the student and increase the convenience for the instructor. Thinking precedes execution of any problem or job. Therefore, things to be thought are on the left side of the page; the things to be done appear opposite them on the right side of the page.

Examples have been presented in sufficient number to clarify each of the steps in a procedure or problem and to enable the student to move ahead steadily and with assurance. In the more immediate technical work of the shop, nomenclature, practices and procedures have been made to conform to practices accepted by industry.

The text lays the groundwork for the application of mathematics to problems by explaining in simple language the fundamentals of arithmetic algebra, geometry, and trigonometry. Emphasis is then given to basic problems arising in the building trades, the electrical shop, and the machine shop. Every opportunity has been taken to make the mathematics grow out of shop situations, and to have the student use his mathematics in connection with vocational activities. A multitude of standard trade and technical terms and symbols are introduced and used, and over 700 drawings, sketches, and photographs are included. Drawings and sketches are presented in orthographic projection, and in isometric and cabinet drawings to give the student wide experience in reading and interpreting these different forms.

The skilled worker is called upon to use tables and materials from handbooks. Some practice in the use of tables is given in the text. To further provide practice, several important and useful tables are included in the Appendix.

Acknowledgments

The authors gratefully acknowledge photographs and reference materials generously provided by

Atlas Supply Company, Springfield, New Jersey

American Standards Association, New York, New York

American Welding Society, New York, New York

Charles C. Bond Company, Philadelphia, Pennsylvania

Brown & Sharpe Manufacturing Company, Providence, Rhode Island

City Colleges of Chicago, Chicago, Illinois

Cullman Wheel Company, Chicago, Illinois

General Electric Company, Schenectady, New York

General Motors Corporation, Oldsmobile Division, Lansing, Michigan

Keuffel & Esser Company, Hoboken, New Jersey

Link Belt Company, Chicago, Illinois

Lufkin Rule Company, Saginaw, Michigan

Morse Twist Drill & Machine Company, New Bedford, Massachusetts

Tinius Olsen Testing Machine Company, Willow Grove, Pennsylvania

Rockwell Manufacturing Company, Power Tool Division, Pittsburgh, Pennsylvania

South Bend Lathe Inc., South Bend, Indiana

Stanley Tools, Division of the Stanley Works, New Britain, Connecticut

L.S. Starrett Company, Athol, Massachusetts

Ward Leonard Electric Company, Mount Vernon, New York

Weston Electrical Instrument Corporation, Newark, New Jersey

Contents

Other Related Books in the Applied Mathematics Series

1 Measurement

No matter what industry you plan to work in, one of the first things you must know is how to measure accurately, according to the requirements of that industry. Measurements and their representation in sketches and drawings are of great importance in every industrial occupation.

To measure accurately, you must have good measuring tools, know what is meant by accurate measurement, and acquire the necessary skill to use your tools properly.

This chapter is concerned with these fundamentals.

Common tools

Some of the measuring instruments frequently used are:

(a) *The common folding rule,* shown in Fig. 1-1, is 2 feet long, folds to a length of about 6 inches, is easily carried, and is used mostly by workers in wood.

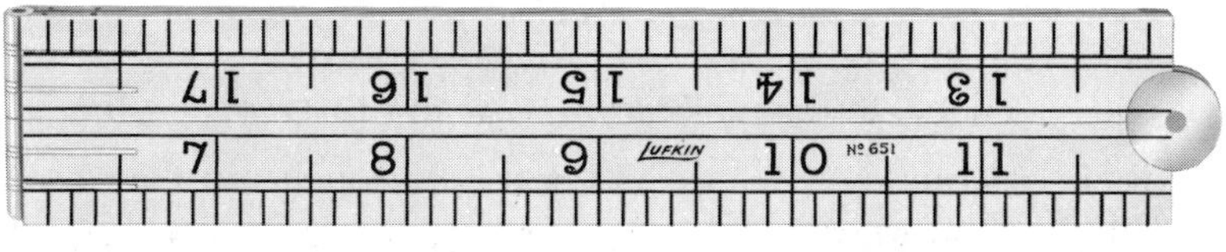

Fig. 1-1. Two-foot, four-fold boxwood rule. *(Courtesy Lufkin Rule Company.)*

(b) *The zigzag rule,* shown in Fig. 1-2, is usually 4, 6, or 8 feet long and folds to about 6 inches, which makes it easy to carry. It is used by carpenters, electricians, concrete workers, and other skilled mechanics.

(c) *The steel tape,* shown in Fig. 1-3, comes in 25-foot, 50-foot, and longer lengths. Contractors and surveyors frequently use steel tapes.

(d) *The machinist's rule or scale,* shown in Fig. 1-4, is a 6-inch steel rule with a carefully drawn scale. Machinists prefer it for its convenience in size and for the way in which each inch is divided into smaller (fractional) parts. The smallest division on this scale is 1/64 of an inch. Some of these scales are divided so carefully that the smallest division is 1/100 of an inch. Such small divisions are not practicable with wood rules.

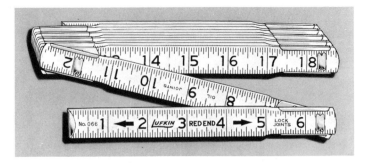

Fig. 1-2. Zigzag rule. *(Courtesy Lufkin Rule Company.)*

These are a few of the tools used to make accurate measurements in industry. You will meet others elsewhere in this book. Successful use of these measuring devices depends upon the accuracy and skill with which

Fig. 1-3. Steel tape. *(Courtesy Lufkin Rule Company.)*

you use them. Your eye, hand, and sense of touch must be trained to work together. You must be able to apply these devices in many forms.

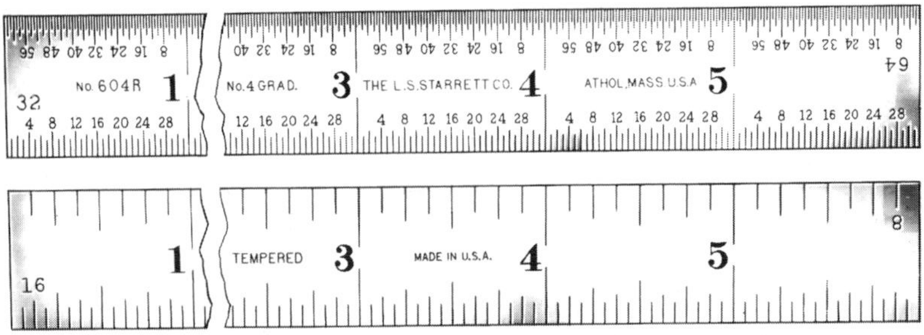

Fig. 1-4. Machinist's steel rule or scale. *(Courtesy L.S. Starrett Company.)*

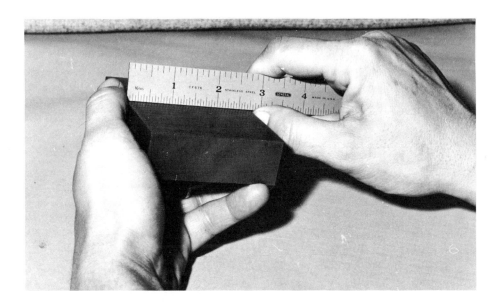

Fig. 1-5. Machinist using the steel rule.

You must be able to convert measurements and dimensions from one unit to another and to add, multiply, subtract, and divide them. It is the purpose of this chapter to provide practice in these operations and to help you develop skill in working with measurements and dimensions. We shall use the units of length common in American shop practice.

Units of length

The length of a line or the length of an object is measured by comparing its length with a line segment of unit length. To be mathematically correct, a "line" is indefinitely long and cannot be measured; hence we should use the term "line segment." In actual practice, we often use the term "line" when we mean "line segment," because it is a shorter and more convenient expression.

In American industry the *English system* of measurement is in general use. The standard unit of length in English measure is the *yard*. This is the distance between two parallel marks on a platinum bar kept at a certain constant temperature at the Bureau of Standards in Washington, D.C.

The yard is divided, for convenience, into 36 equal parts, each called an *inch*. If the yard is divided into 3 equal parts, each part forms a unit called the *foot*.

The steel bar shown in Fig. 1-6 is 3 inches long, because the unit of length, the inch, is contained in the length of the bar 3 times. Or we say, 3 inches are contained in the length of the bar.

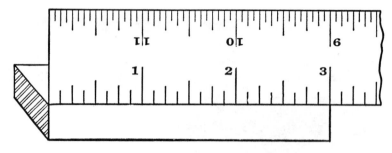

Fig. 1-6*

Common units of length

1 yard = 36 inches	1 yd. = 36 in. (1 yd. = 36″)
1 yard = 3 feet	1 yd. = 3 ft. (1 yd. = 3′)
1 foot = 12 inches	1 ft. = 12 in. (1′ = 12″)
1 rod = 5 1/2 yards	1 rd. = 5 1/2 yds.
1 mile = 1,760 yards	1 mi. = 1,760 yds.
or	
1 mile = 5,280 feet	1 mi. = 5,280 ft. (1 mi. = 5,280′)

EXERCISES

1. How tall are you? Express the result in inches. In feet and inches. In yards, feet and inches.
2. The Australian, Kevin Berry, captured the 100-yard butterfly world swimming record in 1963. He covered the distance in 59 seconds. How many feet did he average in one second?
3. List the following in order of size, the shortest first and the longest last: foot, rod, inch, mile, yard.
4. Complete the following table:

Number of Miles	Number of Rods	Number of Yards	Number of Feet	Number of Inches
1	320	1,760	5,280	?
	1	?	?	?
		1	?	?
			1	?

Changing units

A measurement that is expressed in one unit may be changed to another unit. The following examples will show you how.

 *The lines drawn across the end of the steel bar are cross-hatching lines and indicate that the bar is made of steel.

Fig. 1-7. A draftsman uses a drafting machine for accurate measurement.

Example 1. A piece of lumber is 1 in. thick, 6 in. wide, and 96 in. long. What is the length of the board in feet?

THINK	DO THIS
1. Since 12 in. equal 1 ft., there will be as many feet in this length as the number of times 12 is contained in 96.	1. 12 in. = 1 ft.
2. Divide 96 by 12.	2. $\dfrac{8}{12\,)\,\overline{96}}$ $\underline{96}$

Ans. 8 ft.

In Example 1, note that the thickness and width of the board have nothing to do with the question asked. Always read the problem carefully to determine exactly what is wanted and whether any extra information that does not apply to the problem has been given.

Example 2. Find the number of feet in a piece of oak 48 in. long.

THINK	DO THIS
1. Since 12 in. equal 1 ft., there are as many feet in 48 in. as the number of times 12 is contained in 48.	1. 12 in. = 1 ft.
2. Divide 48 by 12.	2. $\dfrac{4}{12\,)\,48}$ $\underline{48}$

Ans. 4 ft.

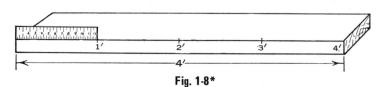

Fig. 1-8*

(Note in Fig. 1-8 above that the foot rule covers just 12 in. The foot rule is contained in the entire length, 48 in., as many times as 12 is contained in 48.)

Example 3. How many yards are there in a piece of tracing cloth 57 ft. long?

THINK	DO THIS
1. Since there are 3 ft. in 1 yd., there are as many yards in 57 ft. as the number of times 3 is contained in 57.	1. 3 ft. = 1 yd.
2. Divide 57 by 3.	2. $\dfrac{19}{3\,)\,57}$ $\underline{57}$

Ans. 19 yds.

Whenever you wish to find how many times one number is contained in another, *divide* the *second* by the *first.*

Summary	EXAMPLE
To change a length to a larger unit, divide the given length by the number of smaller units contained in one of the larger units. The result is the number of larger units.	Find the number of miles in 3,520 yds. 1 mile = 1,760 yds. $\dfrac{2}{1{,}760\,)\,3{,}520}$ $\underline{3\,520}$

Ans. 2 mi.

*This is the symbol for wood. Note also how dimensions (here length) are shown on a drawing.

Number of	Divided by		Number of
Inches	÷	12	=	Feet
Feet	÷	3	=	Yards
Inches	÷	36	=	Yards
Feet	÷	5,280	=	Miles
Yards	÷	1,760	=	Miles
Yards	÷	5 1/2	=	Rods

EXERCISES

Copy this chart, then put the correct numerical values in place of the question marks.

	Number of Miles	Number of Yards	Number of Feet	Number of Inches
1		?	?	36
2		?	45	?
3	?	880	?	
4		?	?	156
5		?	?	18
6		?	?	180
7			?	42
8	?	?	13,200	?
9		?	?	9

Were you able to do Numbers 2, 3, and 8?

Example 4. A room is 8 ft. high. What is the height of this room when the measurement is expressed in inches?

THINK	DO THIS
1. Each foot contains 12 in.	1. 1 ft. = 12 in.
2. Multiply 8 by 12.	2. 8 x 12 = 96
	Ans. 96 in.

Example 5. How many yards are there in 3 1/2 mi.?

THINK	DO THIS
1. In each mile there are 1,760 yds.	1. 1 mi. = 1,760 yds.
2. Multiply 3 1/2 by 1,760.	2. 3 1/2 x 1,760 = 5,280 + 880 = 6,160
	Ans. 6,160 yds.

Summary

To express in terms of one unit a length given in terms of a larger unit, multiply the given length by the number of smaller units in one of the larger units. The result is the number of smaller units.

EXAMPLE

Find the number of inches in 2 1/4 yd.

1 yd. = 36 in.

2 1/4 x 36 = 72 + 9 = 81

Ans. 81 in.

Number of		Multiplied by		Number of
Feet	x	12	=	Inches
Yards	x	36	=	Inches
Yards	x	3	=	Feet
Miles	x	5,280	=	Feet
Miles	x	1,760	=	Yards

EXERCISES

Copy this chart, then put the correct numerical values in place of the question marks.

	Number of Miles	Number of Yards	Number of Feet	Number of Inches
1		8	?	?
2		?	4	?
3		?	6 1/2	?
4	3	?	?	
5	6 1/2	?	?	
6		3	?	?
7		5 1/2	?	?
8	?	220	?	
9		5 3/4	?	?
10			7 1/4	?

Example. How many inches are there in a piece of conduit that is 3 ft. 4 in. long?

THINK

1. Since each foot contains 12 in., multiply 3 by 12 to find the number of inches in 3 ft.

2. Add 4 to the product to find the total number of inches.

DO THIS

1. 3 x 12 = 36

2. 36 + 4 = 40

Ans. 40 in.

Summary	EXAMPLE
When a length is given in terms of two units, to find the length in terms of the smaller unit:	Change 2 ft. 3 in. to inches.
1. Multiply the number of large units in the length by the number of small units in one large unit.	1. 2 x 12 = 24
2. Add this result to the number of smaller units.	2. 24 + 3 = 27
	Ans. 27 in.

EXERCISES

Express the lengths given in Exercises 1 to 12 in terms of the unit indicated.

1. 3 ft. 8 in. in inches
2. 7 yd. 1 ft. in feet
3. 12 ft. 6 in. in inches
4. 3 yd. 8 in. in inches
5. 8 ft. 3 in. in inches
6. 4 yd. 2 ft. in feet
7. 3 mi. 70 yd. in yards
8. 2 ft. 11 in. in inches
9. 6 yd. 9 in. in inches
10. 6 yd. 9 ft. in feet
11. 5 ft. 8 in. in inches
12. 9 yd. 1 ft. in inches

PROBLEMS

1. The dimensions of the concrete foundation for the garage shown in Fig. 1-9 are expressed in feet. Express the dimensions of the length and width of the foundation in yards.

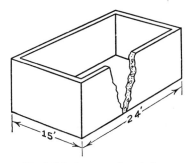

Fig. 1-9.* Concrete foundation.

2. In erecting a transmission line (Fig. 1-10) consisting of six conductors, 68,640 ft. of wire were used for each conductor. How many miles of wire were used altogether?

*This is the symbol for concrete.

Fig. 1-10. Transmission line.

3. A draftsman finds that he needs 147 ft. of blueprint paper. Will a stock roll of 50 yds. be sufficient for the job? Explain.

4. The door shown in Fig. 1-11 has a stile 6 ft. 9 in. long. What is the length of the stile in inches?

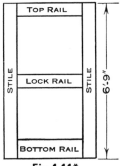

Fig. 1-11*.

5. An electrical contractor estimated that 2 1/3 mi. of 1″ conduit would have to be run in order to install the electrical wiring in a new building. How many feet of conduit would be required? How many 10′ lengths of conduit should he order?

6. A concrete driveway leading to a garage is 2 1/3 yds. wide and 19 1/2 yds. long. What are the dimensions of the driveway expressed in feet?

7. A cylindrical steel tank designed for use with an oil burner has a diameter of 3′ 8″ and length of 8′ 6″. What size tank, measured in inches, should be ordered?

8. A distance was measured as 18 1/2 yds. Express this distance in feet.

9. A plane travels at an altitude of 25,000 ft. Express this altitude in miles.

The linear scale

Figure 1-12 shows a portion of a 6-in. machinist's steel rule with the usual scales. Notice that the numbers 1, 2, 3, etc., indicate the whole

---

*Note how dimensions in feet and inches are shown on the drawing. Dimensions up to 72 in. should preferably be shown in inches, and those 72 in. or greater should be shown in feet and inches. Omit the symbol (″) when *all* dimensions are given in inches.

number of inches from the left end of the scale. On the lower scale, each inch is divided into 8 equal parts. Each of these smaller divisions is 1/8 in. Two such small divisions, 2/8 in., equal 1/4 in. Four such small divisions, 4/8 in., equal 2/4 or 1/2 in.

Fig. 1-12.

To make the division lines on a scale easy to read, each division representing a given fractional part is shown by a line of a particular length. In general, the inch division lines are the longest, the half-inch division lines are next to the longest, the quarter-inch divisions are somewhat shorter, and the smaller division lines are still shorter (Fig. 1-12). If the inch is divided into ten equal parts, usually the half-inch division line is shorter, than the inch division line, and the other division lines are the shortest, but are all equal in length.

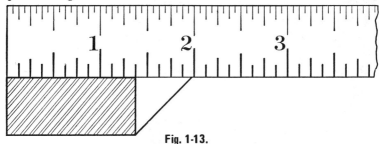

Fig. 1-13.

The width of the steel bar in Fig. 1-13 is one inch plus 3/8 of one inch, or 1 3/8 in., because it contains one full inch and, in addition to that, three small divisions of 1/8 in. each.

The length of the stud bolt in Fig. 1-14 is 3 in. plus 7/8 in., or 3 7/8 in.

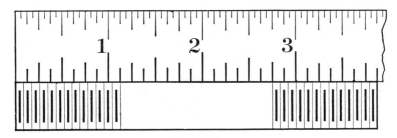

Fig. 1-14*.

*This is one method of representing a thread.

Fig. 1-15.

In Fig. 1-15 note that on the upper edge of the rule the scale is graduated to sixteenths of an inch. That means that each inch space is divided into sixteen equal parts, each of the smallest divisions being 1/16 in.

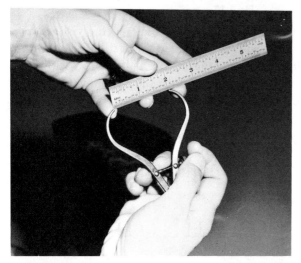

Fig. 1-16.* Setting outside calipers to a scale measurement.

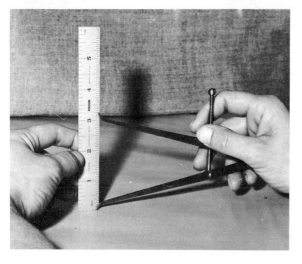

Fig. 1-17. Setting inside calipers to a scale measurement.

*Calipers are instruments used for determining the thickness or diameter of objects or the distance between surfaces, etc. They are not measuring tools but are used when a scale cannot be applied directly.

Two such small divisions, 2/16 in., equal 1/8 in.
Four such small divisions, 4/16 in., equal 2/8 in., or 1/4 in.
Eight such small divisions, 8/16 in., equal 4/8 in., or 2/4 in., or 1/2 in.

Measuring with the linear scale

When the end of the line segment or object being measured falls exactly at a subdivision on the scale, as in Fig. 1-18, it is easy to take the measurement.

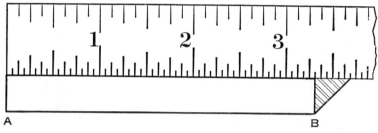

Fig. 1-18.

Summary

To measure a length with a given scale:

1. First determine what part of the inch the smallest subdivision represents. If the inch is divided into 8 equal parts, each is 1/8 in. If the inch is divided into 16 equal parts, each part is 1/16 in., etc.

2. Place the zero (or any whole division) of the scale at one end of the object to be measured and the scale alongside of, and as close as possible to, the object to be measured.

3. Read from the scale the number of whole inches and the number of additional subdivisions included within the length to be measured.

4. Write the measurement as the number of whole inches and the number of smallest divisions. Express the fraction in lowest terms.

EXAMPLE

Measure the length of the cast steel bar *AB* (Fig. 1-18).

1. The inch is divided into 16 equal parts; each is 1/16 in.

2. See Fig. 1-18.

3. There are three whole inches, plus five of the smallest subdivisions, sixteenths.

4. Length of the bar is 3 in. plus 5/16 in., or 3 5/16 in.

Ans. 3 5/16 in.

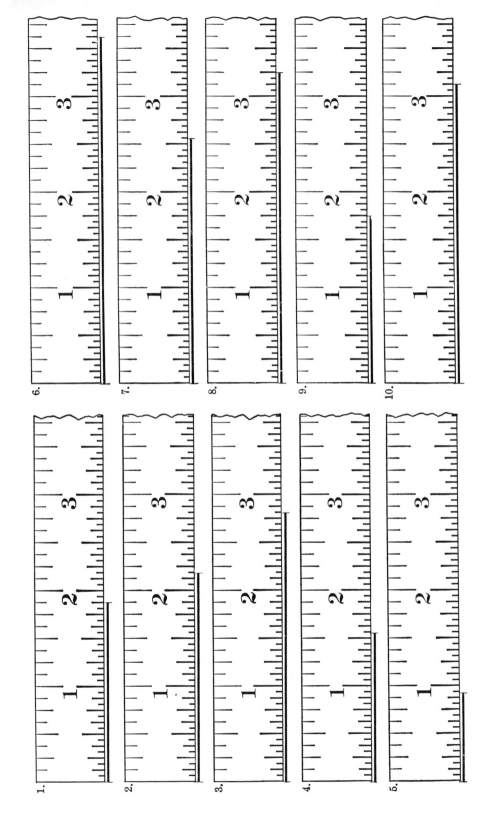

14

EXERCISES

Read each of the measurements shown in Exercises 1 to 10 on page 14.

Sometimes the end of the line segment or object does not fall exactly on a subdivision of the scale. The following examples show you how to read the measurements to the accuracy desired.

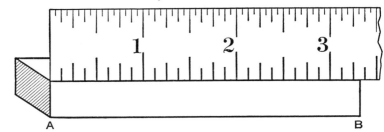

Fig. 1-19*.

Example 1. Read the length of *AB* (Fig. 1-19) to the nearest 1/8 in.

THINK

1. To measure the nearest eighth of an inch, use the scale divided into eight equal parts per inch.

2. Read the scale division nearest the end of the object. Note that this is not an *exact* measurement. It is read here to the *nearest eighth of an inch*.

DO THIS

1. Put the zero of the scale at one end of the object to be measured.

2. The right end of *AB* falls between 3 2/8 in. (or 3 1/4 in.) and 3 3/8 in., but nearer to 3 3/8 in. We call this length 3 3/8 in.
    Ans. 3 3/8 in.

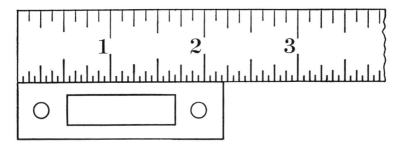

Fig. 1-20.

Example 2. Read the length of the brass lock strike (Fig. 1-20) to the nearest sixteenth of an inch.

THINK

1. Use the scale divided into sixteen equal parts per inch.

DO THIS

1. Put the zero of the scale at one end of the lock strike.

 *These cross-hatching lines are used to represent cast iron.

2. Read the division nearest the end of the object.

2. The right end of the lock strike falls between 2 3/16 in. and 2 4/16 in. (or 2 1/4 in.). It is nearer to 2 3/16 in., and is thus read 2 3/16 in.

3. Check the measurement by moving the scale so a point other than the zero point is at the left end of the length to be measured (Fig. 1-21).

3. Place the left end at 1 in. The right end now falls between 3 3/16 in. and 3 4/16 in. (or 3 1/4 in.) but is nearer 3 3/16 in.

4. Subtract the reading at the left end of the scale from that at the right end.

4. The measurement is 3 3/16 in. – 1 in., or 2 3/16 in.

Ans. 2 3/16 in.

Fig. 1-21.

Summary

To measure any length with a linear scale:

1. Choose the scale divided into the desired number of parts per inch.

2. Put the zero point of the scale (or any other whole inch point) at the left end of the length.

3. Make the reading, choosing the line on the scale that is nearest the end of the length being measured.

4. Subtract the reading at the left end of the length to obtain the measurement.

5. Check the measurement by placing some other point of the scale at the left end of the object to be measured.

EXAMPLE

Find the length of the block in Fig. 1-22.

1. To measure to the nearest sixteenth, choose the scale divided into sixteen parts per inch.

2. Line up scale and block carefully. The 1-in. point is often convenient for beginning a measurement.

3. The right end of the block falls at 3 5/16 in.

4. 3 5/16 in. – 1 in. = 2 5/16 in.

Ans. 2 5/16 in.

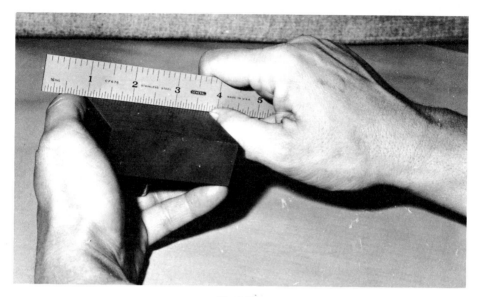

Fig. 1-22.

Note: For purposes of this text, if the end falls exactly in the center of a subdivision, choose the longer length.

EXERCISES

Make the following measurements to the nearest 1/32 in. and check your results.

1. _____

2. _____

3. _____

4. _____   5. _____

6. _____

7. _____

8. _____

9. _____

10. _____

REVIEW PROBLEMS

For the following problems, show your dimensions in a table like this:

PROBLEM	DIMENSIONS				
	A	B	C	D	E
1					
2					

1. The dimensions of certain parts of the "V" block shown in Fig. 1-23 are indicated by letters. Use a scale to determine these dimensions to the nearest 1/64 in. Check each measurement.

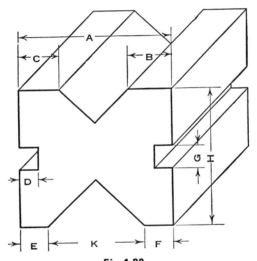

Fig. 1-23.

2. Find each of the dimensions of the dado shown in Fig. 1-24. Measure to the nearest 1/32 in.

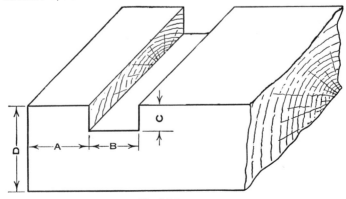

Fig. 1-24.

3. Figures 1-25 and 1-26 show a mortise and tenon, respectively.
   (a) Find each of the dimensions indicated by letters. Make all measurements to the nearest 1/32 in.
   (b) Will the tenon in Fig. 1-26 fit the mortise in Fig. 1-25? Explain.

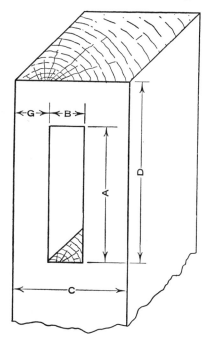

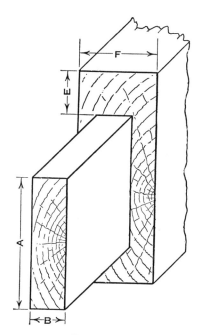

Fig. 1-25. Mortise.   Fig. 1-26. Tenon.

4. Using your scale, determine the dimensions of the template in Fig. 1-27 to the nearest thirty-second of an inch. Make a full-size copy of the drawing and record the dimensions on your copy.

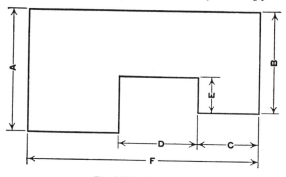

Fig. 1-27. Template.

5. Figure 1-28 is an orthographic projection in two views of a tool-post screw. Measure the dimensions represented by letters to the nearest 1/64 in.

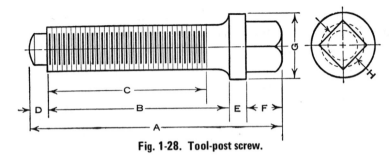

Fig. 1-28. Tool-post screw.

6. A piece of No. 16 gage sheet steel is to be used to cover the top of a work bench that is to be 74″ long and 27″ wide. Express in feet and inches the dimensions of the steel sheet to be ordered.

7. The over-all length of the lathe mandrel shown in Fig. 1-29 is 7 1/4″. How many mandrels can be made out of a piece of tool steel 12′ long? Allow 3/16″ for cutting off and facing each piece.

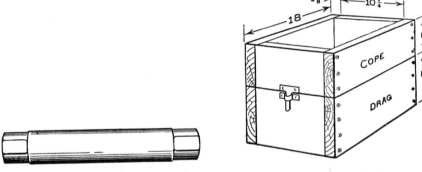

Fig. 1-29. Lathe mandrel.   Fig. 1-30. Foundry flask.

8. Twelve foundry flasks having the dimensions indicated in Fig. 1-30 are to be made of yellow pine. If this lumber is sold in 8′, 10′, 12′, 14′ and 16′ lengths, which would be the most economical lengths to buy, and how many lengths of each width would be required, if an allowance of 1/8″ is made for each cut?

9. A carpenter orders two pieces of yellow pine 7/8″ x 2″ x 8′ to make the sink drainboard shown in Fig. 1-31. If 1/8 in. per piece is allowed for cutting, will he have sufficient lumber to do the job? How much lumber, if any, will he have left?

10. How many 3″ corner braces (Fig. 1-32) can be made out of a piece of steel 12′ long if 1/8″ per piece is allowed for cutting? Will there be any stock left over? If so, how much?

11. A surveyor measured the distance between an electric power station and a substation and found it to be 4 3/8 mi. The linemen, installing

a transmission line between these two points, used 5 1/4 mi. of cable for each conductor. How many extra feet of cable were used for the "sag" of each conductor?

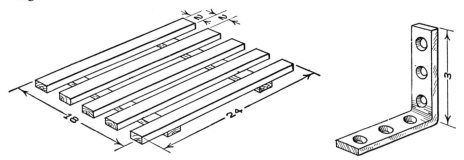

Fig. 1-31. Drain board.   Fig. 1-32. Corner brace.

12. Round head brass machine screws (Fig. 1-33) are to be manufacured in an automatic screw cutting machine. If 1/8" is allowed for cutting off each, how many feet of brass are required to make a gross (144) of screws?

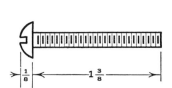

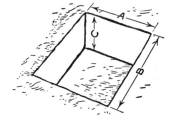

Fig. 1-33. Round head
brass machine screw.   Fig. 1-34. Excavation.

13. In measuring the excavation (Fig. 1-34) for the foundation of a building with a 6-ft. zigzag rule, the dimension A was found to be 4 1/2 lengths; the dimension B, 6 lengths plus 2 ft.; and the dimension C, 1 length plus 2 1/2 ft. Express the length, width, and depth of the excavation in feet and inches.

14. A machine shop measures 8 1/4 yds. wide and 17 3/4 yds. long. A bench 24" wide is to be built along two adjacent sides. How many lineal feet of lumber will be required for the job? Assume the lumber to be 8" wide D 4 S (dressed on four sides). Make no allowance for cutting.

Working with fractions

The jobs done in these last few pages should make you realize that in order to be a good mechanic or technician it is absolutely necessary to be thoroughly familiar with fractions and how to make computations with fractions.

We, therefore, take up next the rules and practices in making computations with common fractions. Study and practice these operations on fractions until you become thoroughly familiar with them.

2 Common Fractions

Some facts about fractions

You have often talked about a half of a thing, such as a half apple or half dollar. You know that one-half is written as 1/2 and that one-third is written 1/3. When we talk about one-third of a thing, for example, one-third of a sheet of paper, we mean one of the three equal parts into which the sheet of paper might be cut.

A number such as 1/8 means that the number 1 is divided by another number, 8. It is a *common fraction,* or simply a *fraction.* The fraction 1/8 means 1 ÷ 8.

The number above the fraction bar is called the *numerator* of the fraction.

The number below the bar is the *denominator* of the fraction.

Note: It is necessary to learn these names in order to study intelligently that which follows.

A number such as 3 5/8 is a *mixed number.* It consists of a whole number, 3, and a fraction, 5/8. It means 3 + 5/8.

If the numerator of a fraction is smaller than the denominator, as in 3/8, 7/12, 1/10, the fraction is a *proper fraction.*

If the numerator is equal to or larger than the denominator, the fraction is an *improper fraction.* Numbers like 19/16, 8/8, 65/64, are improper fractions.

EXERCISES

In each of the following, name the numerator and the denominator and tell whether the number is a proper fraction, an improper fraction, or a mixed number.

1. (a) 3/8   (c) 35/64   (e) 9/8
   (b) 7/16   (d) 11/32   (f) 3 5/32

2. (a) 7 5/16   (c) 19/16   (e) 5 11/64
   (b) 5/4   (d) 7/8   (f) 35/32

Fig. 2-1 Machinist's rule, or scale, showing divisions to 64ths of an inch.

Multiplying fractions

Use the machinist's rule to help in answering the questions in these examples.

Example 1. What does 1/2 of 3/4″ equal?

THINK	DO THIS
1. On the machinist's rule 3/4″ contains 6/8″. 1/2 of 3/4″ then would be 1/2 of 6/8″ or 3/8″.	1. $\dfrac{1}{2}$ of $\dfrac{3''}{4} = \dfrac{3''}{8}$
2. But 3/8″ is also 1/2 x 6/8″.	2. 1/2 x 3/4″ = 3/8″
	Ans. 1/2 of 3/4″ = 1/2 x 3/4″ = 3/8″

Example 2. What does 1/4 of 5/8 mean?

THINK	DO THIS
1. 1/4 of 5/8 means 1/4 x 5/8.	1. 1/4 of 5/8 means 1/4 x 5/8
2. Multiply 1/4 by 5/8.	2. $\dfrac{1}{4}$ x $\dfrac{5}{8} = \dfrac{1 \times 5}{4 \times 8} = \dfrac{5}{32}$
	Ans. 1/4 of 5/8 = 1/4 x 5/8 = 5/32

Example 3. What does 3/4 of 5/8 equal?

THINK	DO THIS
3/4 of 5/8 = 3/4 x 5/8 Verify the result on the machinist's rule.	$\dfrac{3}{4}$ x $\dfrac{5}{8} = \dfrac{3 \times 5}{4 \times 8} = \dfrac{15}{32}$ *Ans.* 3/4 of 5/8 = 15/32

Example 4. What is 5/8 of 7/8?

THINK	DO THIS
5/8 of 7/8 = 5/8 x 7/8 Verify the result on the machinist's rule.	$\dfrac{5}{8}$ x $\dfrac{7}{8} = \dfrac{5 \times 7}{8 \times 8} = \dfrac{35}{64}$ *Ans.* 5/8 of 7/8 = 35/64

Example 5. 5/12 of 3/4 = ? x ?

THINK

What did previous examples show that will help solve this problem?

DO THIS

$$\frac{5}{12} \times \frac{3}{4} = \frac{5 \times 3}{12 \times 4} = \frac{15}{48}$$

Ans. 5/12 of 3/4 = 5/12 x 3/4
= 15/48

Summary

To multiply two proper fractions:

1. Multiply the numerators to obtain the numerator of the product.

2. Multiply the denominators to obtain the denominator of the product.

EXAMPLE

Multiply $\frac{5}{8}$ x $\frac{3}{4}$

1. 5 x 3 = 15

2. 8 x 4 = 32

Ans. 5/8 x 3/4 = 15/32

EXERCISES

Find the products in the following:

1. 8/11 x 3/4 = ?
2. 9/16 x 5/8 = ?
3. 3/4 x 1/2 = ?
4. 5/6 x 7/8 = ?
5. 9/10 x 1/5 = ?

6. 16 x 1/2 = 16/1 x 1/2 = ?
7. 24 x 5/6 = ?
8. 32 x 3/8 = ?
9. 4/5 x 20 = ?
10. 2/3 x 5/12 = ?

In each of the following find the fraction that is needed:

11. 16/1 x ? = 8
12. 24/1 x ? = 8

13. 32 x ? = 1/2
14. 16 x ? = 2

Solve the following:

15. What is 1/2 of 18?
 (1/2 x 18/1 = ?)
16. What is 1/3 of 21?
17. What is 1/2 of 3/4?
18. 2/3 of 12 = ?
19. 1/5 of 10 = ?

20. 9/10 of 7/8 = ?
21. What is 7/8 of 12?
22. 7/8 x 1 = ?
23. 5 x 1 = ?
24. 1 x 3/4 = ?

The principle of one

A basic principle in mathematics that can greatly assist in understanding and performing the operations with fractions is the "principle of 1." This principle states that *any number may be multiplied by 1 without changing the number.*

Examples: 18 x 1 = 18; 5/8 x 1 = 5/8; 1 x 7/8 = 7/8

We can also use this principle of 1 to obtain another important principle. This new one states that *any number may be divided by 1 without changing the number.*

Examples: 8 ÷ 1 = 8; 8/1 = 8; 3/2 ÷ 1 = 3/2; 10/1 = 10

Forms of 1

To make good use of these principles it is necessary to recognize the forms in which the number 1 may be written.

Example 1. 36/36 is a form of 1. It is another name for the number one. The principle of 1 then says 8 x 36/36 = 8 x 1 = 8.

Example 2. 100/100 = 1. 100/100 is another name for the number 1. The principle of 1 then says 100/100 x 15 is the same as 1 x 15, or 15.

Example 3. Some other forms of the number 1 are: 5/5 = 1, ½/½ = 1; ¼/¼ = 1; ¾/¾ = 1; etc. In fact *any number divided by itself is a form of 1.*

Summary

A number may be multiplied or divided by the number 1 in any of its forms, and the result is the original number.

EXERCISES

Write the product in each of the following:

1. 18 x 1 = ?
2. 1 x 12 = ?
3. 1/2 x 1 = ?
4. 3/4 x 1 = ?
5. 1 x 8/9 = ?
6. 12 x 3/3 = ?
7. 15 x 8/8 = ?
8. 4/4 x 3/8 = ?
9. 10/10 x 6 = ?
10. 16/16 x 5/8 = ?
11. 11/48 x 8/8 = ?
12. 10/9 x 3/3 = ?
13. 5/5 x 1/12 = ?
14. 8/8 x 3 1/2 = ?
15. 16/16 x 1 1/2 = ?

To express a fraction with a smaller or larger denominator

To compare, add or subtract fractions it often is necessary to express each fraction with a new denominator.

Example 1. Express 3/4 as sixteenths $\left(\dfrac{3}{4} = \dfrac{?}{16}\right)$.

THINK

By the principle of 1 we may multiply 3/4 by 1 in any of its forms without changing 3/4.

1. Since 4 x 4 = 16, we choose a form of 1 that will give the new denominator 16. We use 1 = 4/4.

2. Multiply 3/4 by 4/4.

DO THIS

1. 4 x 4 = 16

2. $\dfrac{3}{4}$ x $\dfrac{4}{4}$ = $\dfrac{12}{16}$

$Ans. \dfrac{3}{4} = \dfrac{12}{16}$

Example 2. Change 5/16 to 64ths $\left(\dfrac{5}{16} = \dfrac{?}{64}\right)$.

THINK

1. 16 x ? = 64.

2. If we multiply by 1 in the form 4/4 we obtain the desired fraction.

DO THIS

1. 16 x 4 = 64

2. $\dfrac{5}{16}$ x $\dfrac{4}{4}$ = $\dfrac{20}{64}$

$Ans. \dfrac{5}{16} = \dfrac{20}{64}$

Example 3. Change 32/64 to 16ths $\left(\dfrac{32}{64} = \dfrac{?}{16}\right)$.

THINK

1. 64 x ? = 16.

2. Multiply by 1 in the form ¼/¼ to obtain the desired denominator.

Note: Multiplying 32 by 1/4 is equivalent to dividing 32 by 4.

DO THIS

1. 64 x 1/4 = 16

2. $\dfrac{32}{64}$ x $\dfrac{\frac{1}{4}}{\frac{1}{4}}$ = $\dfrac{8}{16}$

or $\dfrac{32}{64}$ = $\dfrac{32 \div 4}{64 \div 4}$ = $\dfrac{8}{16}$

$Ans. \dfrac{32}{64} = \dfrac{8}{16}$

Example 4. Reduce 32/64 to lowest terms* $\left(\dfrac{32}{64} = \dfrac{?}{?}\right)$.

THINK

1. We seek the largest number that will divide 32 *and* 64 exactly, and will also give a smaller denominator. This number is 32.

2. Multiply the fraction by 1 in the form $1/32/1/32$ or divide 32 and 64 each by 32.

DO THIS

1. The largest factor of 32 and 64 is 32.

2. $\dfrac{32}{64} \times \dfrac{\frac{1}{32}}{\frac{1}{32}} = \dfrac{1}{2}$, or,

$$\dfrac{32 \div 32}{64 \div 32} = \dfrac{1}{2}$$

Ans. 32/64 in lowest terms is 1/2

Example 5. Reduce 18/32 to lowest terms $\left(\dfrac{18}{32} = \dfrac{?}{?}\right)$.

THINK

1. The largest number that divides both 18 and 32 exactly is 2.

2. Divide numerator (18) and denominator (32) by 2.

DO THIS

1. Largest factor of 18 and 32 is 2.

2. $\dfrac{18 \div 2}{32 \div 2} = \dfrac{9}{16}$

Ans. $\dfrac{18}{32}$ in lowest terms is $\dfrac{9}{16}$

Can you do this example in another way? Explain.

Summary

To express a fraction with a different denominator:

1. Find the number which, multiplied by the old denominator, will equal the required denominator. This gives the form of 1 to use.

2. Multiply the given fraction by this form of 1.

EXAMPLE

Change 3/4 to 32nds.

1. 4 x ? = 32
   8/8 is the form of 1 to use.

2. $\dfrac{3}{4} \times \dfrac{8}{8} = \dfrac{3 \times 8}{4 \times 8} = \dfrac{24}{32}$

Ans. 3/4 = 24/32

*This expression means to write the name for this number using the smallest possible denominator.

To change a fraction to its simplest form:

1. Find the *largest* number that divides both the numerator and denominator of the fraction exactly.

2. Divide the numerator and the denominator of the given fraction by this number.

Change 21/36 to its simplest form.

1. The largest number that divides both 21 and 36 exactly is 3.

2. $\dfrac{21 \div 3}{36 \div 3} = \dfrac{7}{12}$

Ans. 21/36 in simplest form is 7/12

EXERCISES

Express each of the following fractions with the specified denominator:

1. 10/20 as halves
2. 15/24 as 8ths
3. 16/48 as 12ths
4. 2/5 as 10ths
5. 3/8 is how many 64ths?
6. 5/8 is how many 32nds?
7. 48/64 = ?/4
8. 48/64 = ?/8

Express each of the following fractions with the smallest possible denominator:

9. 10/16
10. 28/96
11. 40/64
12. 8/32
13. 72/144
14. 26/32
15. 20/16
16. 22/32
17. 20/40
18. 24/64
19. 5 48/64
20. 14 10/12
21. 16 28/32
22. 4 18/64
23. 8 8/12
24. 9 16/48
25. 52 27/60

Expressing improper fractions as mixed numbers

Example 1. Express 8/5 as a mixed number.

THINK

Since 8/5 means 8 divided by 5, divide 8 by 5 and write the remainder over 5.

DO THIS

$5\overline{)8}$ $\;1 + \dfrac{3}{5} = 1\dfrac{3}{5}$

Ans. 8/5 = 1 3/5

Example 2. Express 56/16 as a mixed number.

THINK	DO THIS
1. Since 56/16 means 56 divided by 16, we divide 56 by 16 and write the remainder over 16.	1. $16\overline{)56}$ $\dfrac{3 + \frac{8}{16}}{} = 3\frac{8}{16}$ $\dfrac{48}{8}$
2. Write the fraction in simplest form.	2. $3\,8/16 = 3\,1/2$

Ans. 56/16 = 3 1/2

Summary	EXAMPLE
	Express 98/48 as a mixed number.
1. To express an improper fraction as a mixed number, divide as the improper fraction indicates and write the remainder over the denominator.	1. $\dfrac{98}{48}$ means 98 ÷ 48. $48\overline{)98}$ $\dfrac{2 + \frac{2}{48} \text{ or } 2\frac{2}{48}}{}$ $\dfrac{96}{2}$
2. Write the resulting fraction in simplest form.	2. $2\dfrac{2}{48} = 2\dfrac{1}{24}$

Ans. 98/48 = 2 1/24

EXERCISES

Express each of the following as a mixed number:

1. 36/8	4. 56/12	7. 126/16	10. 152/16
2. 42/16	5. 84/8	8. 58/4	11. 248/32
3. 5/3	6. 96/16	9. 184/12	12. 562/64

Changing a mixed number to an improper fraction

Example 1. Express 3 1/2 as an improper fraction.

THINK	DO THIS
1. 3 1/2 means 3 and 1/2 or 3 + 1/2.	1. $3\dfrac{1}{2} = 3 + \dfrac{1}{2}$
2. Using the principle of 1, express 3 as halves. Multiplying 3 x 2/2 will not change its value.	2. $3 \times \dfrac{2}{2} = \dfrac{6}{2}$
3. Add the numerators and write the sum over the denominator.	3. $\dfrac{6}{2} + \dfrac{1}{2} = \dfrac{7}{2}$

Ans. 3 1/2 = 7/2

Example 2. Express 4 3/8 as an improper fraction.

THINK	DO THIS
1. 4 3/8 equals 4 plus 3/8.	1. $4\frac{3}{8} = 4 + \frac{3}{8}$
2. Multiply 4 by 8/8 to express it as 8ths.	2. $4 \times \frac{8}{8} = \frac{32}{8}$
3. Add 32/8 and 3/8.	3. $\frac{32}{8} + \frac{3}{8} = \frac{35}{8}$
	Ans. 4 3/8 = 35/8

Summary

To express a mixed number as an improper fraction:

1. Separate the mixed number into the whole number plus the fraction.
2. Using the principle of 1, express the whole number part as a fraction having the same denominator as the fractional part of the mixed number.
3. Add the numerator of the new fraction to the numerator of the original fraction. Write this sum over the denominator. The result is the improper fraction that is another name for the mixed number.

EXAMPLE

Express 5 1/16 as an improper fraction.

1. $5\frac{1}{16} = 5 + \frac{1}{16}$

2. $5 \times \frac{16}{16} = \frac{80}{16}$

3. $\frac{80}{16} + \frac{1}{16} = \frac{81}{16}$

Ans. 5 1/16 = 81/16

EXERCISES

Change each of the following mixed numbers to improper fractions:

1. 3 7/8	10. 26 1/2	19. 18 62/64
2. 9 1/2	11. 9 1/8	20. 26 11/12
3. 12 3/4	12. 8 3/4	21. 42 3/32
4. 14 7/12	13. 5 7/16	22. 15 463/1728
5. 7 1/2	14. 7 1/12	23. 2 9/16
6. 12 5/32	15. 4 9/16	24. 17 11/32
7. 8 1/2	16. 12 5/8	25. 1 29/64
8. 16 2/3	17. 9 35/64	26. 3 1/7
9. 4 5/8	18. 23 73/144	

Multiplying fractions and mixed numbers

Example 1. Multiply 7/8 x 1 1/2.

THINK	DO THIS
1. Express 1 1/2 as an improper fraction.	1. $1\frac{1}{2} = \frac{2}{2} + \frac{1}{2} = \frac{3}{2}$
2. Multiply this result by 7/8.	2. $\frac{7}{8} \times \frac{3}{2} = \frac{21}{16}$
3. Express the product in simplest form.	3. $16\overline{)21}$ $\frac{1 + \frac{5}{16} = 1\frac{5}{16}}{}$ $\frac{16}{5}$

Ans. 7/8 x 1 1/2 = 1 5/16

Example 2. Multiply 6 1/2 x 5 7/8.

THINK	DO THIS
1. Express each mixed number as an improper fraction.	1. $6\frac{1}{2} = \frac{12}{2} + \frac{1}{2} = \frac{13}{2}$ $5\frac{7}{8} = \frac{40}{8} + \frac{7}{8} = \frac{47}{8}$
2. Multiply the resulting fractions.	2. $\frac{13}{2} \times \frac{47}{8} = \frac{13 \times 47}{2 \times 8} = \frac{611}{16}$
3. Reduce this result to an improper fraction in lowest terms.	3. $16\overline{)611}$ $\frac{38 + \frac{3}{16} = 38\frac{3}{16}}{}$ $\frac{48}{131}$ $\frac{128}{3}$

Ans. 6 1/2 x 5 7/8 = 38 3/16

Example 3. Multiply 16 1/4 x 12.

THINK	DO THIS
1. Express each mixed number as an improper fraction.	1. $16\frac{1}{4} = \frac{64}{4} + \frac{1}{4} = \frac{65}{4}$; $12 = \frac{12}{1}$
2. Multiply the resulting fractions.	2. $\frac{65}{4} \times \frac{12}{1} = \frac{65}{4} \times \frac{3 \times 4}{1} =$ $\frac{65}{1} \times \frac{3 \times 4}{4}$

3. Here, and in many problems, we may use the principle of one to shorten the computation by first writing $\frac{12}{1}$ as $\frac{3 \times 4}{1}$ and then arranging the factors so that a pair of like factors equals 1, as, for example, 4/4.

3. $\frac{4}{4} = 1$, hence $\frac{65}{1} \times 3 \times 1 = 195$

Ans. 16 1/4 x 12 = 195

Example 4. 16 x 1 15/32 x 7/16 = ?

THINK

1. Express any mixed numbers as improper fractions.

2. Place the resulting fractions in order for multiplication.

3. Rearrange the factors to obtain a pair of like factors to equal 1; then multiply.

4. Reduce the result to simplest form.

DO THIS

1. $1\frac{15}{32} = \frac{32 + 15}{32} = \frac{47}{32}$; $16 = \frac{16}{1}$

2. $\frac{16}{1} \times \frac{47}{32} \times \frac{7}{16}$

3. $\frac{47}{32} \times \frac{7}{1} \times \frac{16}{16} = \frac{47 \times 7}{32} \times 1$

$\frac{47 \times 7}{32} = \frac{329}{32}$

4. $32\overline{)329}$ $\begin{array}{r} 10 \\ \underline{32} \\ 9 \end{array}$ $\frac{10 + \frac{9}{32}}{} = 10\frac{9}{32}$

Ans. 16 x 1 15/32 x 7/16 = 10 9/32

Summary

To multiply fractions and mixed numbers:

1. Express each mixed number or whole number as an improper fraction and place in order for multiplication.

2. Write the factors in a form in which some fractions equal 1. Then multiply the resulting fractions.

3. Express the product as a mixed number.

EXAMPLE

3 5/8 x 10 2/3 x 4 = ?

1. $\frac{29}{8} \times \frac{32}{3} \times \frac{4}{1} = ?$

2. $\frac{29}{2} \times \frac{16 \times 2}{3} \times \frac{4}{4} = \frac{29}{1} \times$

$\frac{16}{3} \times \frac{2}{2} \times \frac{4}{4}$; $\frac{29}{1} \times \frac{16}{3} = \frac{464}{3}$

3. $3\overline{)464}$ $\begin{array}{r} 154 \end{array}$ $\frac{154 + \frac{2}{3}}{} = 154\frac{2}{3}$

Ans. 3 5/8 x 10 2/3 x 4 = 154 2/3

EXERCISES

Perform the multiplications indicated.

1. 3/4 x 8	13. 3 7/8 x 8 5/16
2. 7/8 x 12	14. 4 3/8 x 5 1/2
3. 1 5/8 x 9	15. 48 x 3 1/7
4. 3 7/16 x 12	16. 7 3/8 x 3 1/7
5. 3/4 x 15	17. 18 5/16 x 9 1/2
6. 4 11/12 x 18	18. 12 31/32 x 20 3/8
7. 3/4 x 5/8	19. 5/8 x 3/4 x 2
8. 3/8 x 11/16	20. 7/8 x 5/8 x 48
9. 5/12 x 8/15	21. 3/8 x 5/12 x 1/2
10. 3 1/2 x 5/8	22. 16 x 15/32 x 9/16
11. 29/32 x 27/144	23. 3 1/2 x 6 5/8 x 3 3/4
12. 7/12 x 5 1/2	24. 7 1/12 x 3 3/8 x 8 3/16

PROBLEMS

1. The riser of the main stairway (Fig. 2-2) of a one-family house measures 7 1/8″. What is the rise if the stairway has seventeen risers?
2. The resistance of No. 18 annunciator wire (bell wire) is approximately 6 3/8 ohms per 1,000 ft. What is the total resistance of 6,000 ft. of such wire?
3. There are 7 1/2 gallons in a cubic foot. How many gallons of lard oil are required to fill a steel storage tank having a capacity of 9 2/3 cu. ft.?
4. The circumference of a circle may be found by multiplying the diameter of the circle by 3 1/7. Find the circumference of a bar of

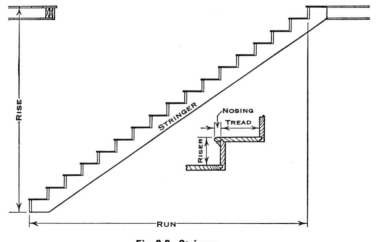

Fig. 2-2. Stairway.

steel 1 3/16 in. in diameter. If this steel weighs 1 3/4 lb. per ft., how much will a 12-ft. bar weigh?

5. In stone masonry a perch is equal to 24 3/4 cu. ft. How many cubic feet of stone masonry are there in 17 2/3 perches?

6. A 7″ steel channel beam used as a lintel (Fig. 2-3) in the construction of a brick doorway opening is 5 2/3′ long and weighs 9 3/4 lb. per ft. What is the weight of the beam?

Fig. 2-3 Lintel.

Dividing fractions and mixed numbers

Example 1. Divide 3/4 x 5/8.

THINK	DO THIS
1. First write the problem in the fractional form.	1. $\dfrac{\frac{3}{4}}{\frac{5}{8}}$
2. Use the principle of 1 to reduce the denominator, 5/8, to 1. Since 5/8 x 8/5 = 1, we multiply by 8/5 /8/5 .	2. $\dfrac{\frac{3}{4}}{\frac{5}{8}}$ x $\dfrac{\frac{8}{5}}{\frac{8}{5}}$ = $\dfrac{\frac{3}{4} \times \frac{8}{5}}{1}$ = $\dfrac{3}{4}$ x $\dfrac{8}{5}$
3. Multiply the resulting fractions.	3. $\dfrac{3}{4}$ x $\dfrac{8}{5}$ = $\dfrac{3}{1}$ x $\dfrac{2}{5}$ x $\dfrac{4}{4}$ = $\dfrac{6}{5}$
4. Write the result in simplest form.	4. $\dfrac{6}{5}$ = $1\dfrac{1}{5}$

Note: In Step 2, 3/4 /5/8 actually became 3/4 x 8/5.

Ans. 3/4 ÷ 5/8 = 1 1/5

Example 2. Divide 9/16 by 2/3.

THINK	DO THIS

1. Write this problem in fractional form.

1. $\dfrac{\dfrac{9}{16}}{\dfrac{2}{3}}$

2. Use the principle of 1 to reduce the denominator, 2/3, to 1. 2/3 x 3/2 = 1. Multiply by 3/2 / 3/2 .

2. $\dfrac{\dfrac{9}{16}}{\dfrac{2}{3}}$ x $\dfrac{\dfrac{3}{2}}{\dfrac{3}{2}}$ = $\dfrac{\dfrac{9}{16} \text{ x } \dfrac{3}{2}}{1}$ = $\dfrac{9}{16}$ x $\dfrac{3}{2}$

3. Multiply the resulting fractions.

3. $\dfrac{9}{16}$ x $\dfrac{3}{2}$ = $\dfrac{27}{32}$

Note: In both examples 1 and 2, to reduce the divisor to 1 we multiplied by the reciprocal of the divisor.

Ans. 9/16 ÷ 2/3 = 27/32

Example 3. Divide 3 1/2 by 7 5/12.

THINK	DO THIS

1. Express each mixed number as an improper fraction.

1. $3\dfrac{1}{2}$ = $\dfrac{7}{2}$; $7\dfrac{5}{12}$ = $\dfrac{89}{12}$

2. Write the problem in fractional form.

2. $\dfrac{3\dfrac{1}{2}}{7\dfrac{5}{12}}$ = $\dfrac{\dfrac{7}{2}}{\dfrac{89}{12}}$; $\dfrac{\dfrac{7}{2}}{\dfrac{89}{12}}$ x $\dfrac{\dfrac{12}{89}}{\dfrac{12}{89}}$ =

3. *Since the principle of 1 enables use to obtain the result by multiplying the numerator by the reciprocal of the denominator, multiply 7/2 by 12/89.

3. $\dfrac{7}{2}$ x $\dfrac{12}{89}$

4. To simplify the work, rewrite the problem with factors that give 1.

4. $\dfrac{7}{2}$ x $\dfrac{6 \text{ x } 2}{89}$ = $\dfrac{7}{1}$ x $\dfrac{6}{89}$ x $\dfrac{2}{2}$ = $\dfrac{42}{89}$

Ans. 3 1/2 ÷ 7 5/12 = 42/89

Note: From this point on, instead of the second part of step 2 we may simply say, "By the principle of 1, 7/2/89/12 = 7/2 x 12/89."

Example 4. Divide 2 3/4 by 5 1/2.

THINK	DO THIS
1. Write the mixed numbers as improper fractions.	1. $2\frac{3}{4} = \frac{11}{4}$; $5\frac{1}{2} = \frac{11}{2}$
2. Write the problem in fractional form.	2. $\dfrac{\frac{11}{4}}{\frac{11}{2}}$
3. Using the principle of 1, multiply the numerator by the reciprocal of the divisor.	3. $\frac{11}{4}$ x $\frac{2}{11}$
4. Simplify the problem by using factors that give 1.	4. $\frac{11}{11}$ x $\frac{2}{2}$ x $\frac{1}{2} = \frac{1}{2}$
	Ans. 2 3/4 ÷ 5 1/2 = 1/2

Summary	EXAMPLE
To divide fractions or mixed numbers:	$4\frac{1}{8} \div \frac{3}{7} = ?$
1. Express mixed numbers as improper fractions.	1. $4\frac{1}{8} = \frac{33}{8}$
2. Write the problem in fractional form.	2. $\dfrac{\frac{33}{8}}{\frac{3}{7}}$
3. Multiply the numerator by the reciprocal of the divisor, simplifying the problem by using factors that give 1.	3. $\frac{33}{8}$ x $\frac{7}{3} = \frac{11}{8}$ x $\frac{7}{1}$ x $\frac{3}{3} = \frac{77}{8}$
4. Write the result in simplest form.	4. $\frac{77}{8} = 9\frac{5}{8}$
	Ans. 4 1/8 ÷ 3/7 = 9 5/8

EXERCISES

Divide as indicated.

1. 7/8 ÷ 3/4
2. 3/16 ÷ 7/12
3. 5/8 ÷ 4
4. 9/16 ÷ 8
5. 15/16 ÷ 12
6. 3 1/2 ÷ 12
7. 9 3/4 ÷ 36
8. 72 ÷ 3 5/8
9. 144 ÷ 9 1/16
10. 84 5/7 ÷ 3 1/7
11. 154 3/7 ÷ 3 1/7

12. 14 3/4 ÷ 5 1/2
13. 54 ÷ 3 1/3
14. 48 ÷ 5 1/2
15. 15 7/12 ÷ 16 1/2
16. 462 ÷ 12 1/2
17. 12 ÷ 1/2
18. 15 ÷ 1/2
19. 12 1/2 ÷ 3/4
20. 96 ÷ 3 1/7
21. 3 1/4 ÷ 1· 2/5

PROBLEMS

7. A carpenter estimates 32 courses of shingles for a roof 14 ft. wide (Fig. 2-4). The shingles are to be laid 5 1/4 in. to the weather. Is his estimate correct?

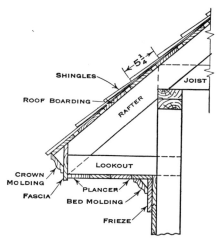

Fig. 2-4*. Section of a roof.

8. Armature cores are built up of thin layers of insulated iron plates. The core shown in Fig. 2-5 is 4 3/16 in. long. How many laminations 3/64 in. thick are required to build up this core?

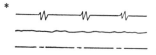

* This line represents a long break.

This line represents a short break.

This line is a center line.

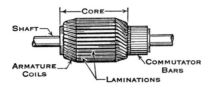

Fig. 2-5. D.C. armature.

9. A radio repair man has a strip of Bakelite 8″ long, which he wishes to use to make the terminal strip shown in Fig. 2-6. What will be the distance between the centers of the holes if there are to be nine terminals on the strip?

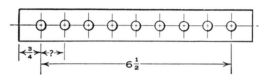

Fig. 2-6. Terminal strip.

10. It requires 17 rivets to make the lap joint shown in Fig. 2-7. What is the pitch of the rivets? (The pitch is the distance between the centers of the rivets.)

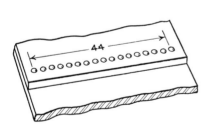

Fig. 2-7. Riveted lap joint.

Fig. 2-8. Plain milling cutter.

11. In estimating the stock required to make the high-speed, steel plain milling cutter shown in Fig. 2-8, a machinist must allow 9/16″ for each cutter. This includes 1/8″ for cutting off and facing. How many cutters can be made from a piece of stock 8 7/16″ long?

12. How many 2 1/2″ bright steel mending plates (Fig. 2-9) can be obtained from a piece of stock 72″ long if the length of stock per plate including allowance for cutting is 2 5/8″?

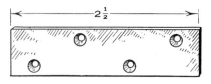

Fig. 2-9. Mending plate.

Finding the common denominator of two or more fractions

To add or subtract fractions, there are two things that must be done: (1) Find a *common denominator* for the fractions. (When two or more fractions have the same denominator, they are said to have a common denominator.) (2) Express each fraction in terms of that common denominator. This section shows how to find the common denominator.

Example. Find the common denominator of 3/4, 5/6 and 7/8.

THINK	DO THIS
1. To change all the denominators to the same number, we try to find factors that will divide exactly into all of the denominators. Write the denominators one beside the other in a division box.	1.) 4, 6, 8
2. Divide *all* the denominators by the same number if that is possible, here 2. Then 2 is a factor of the denominator.	2. 2) 4, 6, 8 2, 3, 4
3. Repeat this process if possible. If it is not possible to divide *all* results by the same factor, then divide *as many as possible* by a factor and bring down those which do not divide exactly, as is done with the 3 in the second division by 2.	3, 4. 2) 4, 6, 8 2) 2, 3, 4 2) 1, 3, 2 3) 1, 3, 1 1, 1, 1
4. Repeat this process until all results are 1.	
5. Multiply all the divisors (factors). This number, 24, is the lowest common denominator of the three given fractions. It is the smallest number into which 4, 6 and 8 will divide exactly.	5. 2 x 2 x 2 x 3 = 24 *Ans.* A common denominator for 3/4, 5/6 and 7/8 is 24.

Summary	EXAMPLE
To find a common denominator for several fractions:	Find the common denominator of 5/16, 7/12 and 3/10.
1. Write all the denominators in a division box.	1.) 16, 12, 10
2. Select as a factor a number that will divide exactly into all or nearly all the denominators, here 2.	2. 2) 16, 12, 10

3. Divide by this number. Any denominator that cannot be divided exactly by this number is written below itself.	3, 4. 2) 16, 12, 10
	      2) 8, 6, 5
	      4) 4, 3, 5
	      3) 1, 3, 5
	      5) 1, 1, 5
	        1, 1, 1

4. Repeat this process until all results are unity, 1.	
5. Multiply all the divisors. This result is the common denominator.	5. 2 x 2 x 4 x 3 x 5 = 240 *Ans.* A common denominator for 5/16, 7/12 and 3/10 is 240.

EXERCISES

Find the common denominator of the fractions in each of the following exercises:

1. 3/8, 11/12, 1/4
2. 7/8, 5/16, 3/8
3. 13/16, 5/6, 11/12
4. 3/10, 7/8, 33/40
5. 5/16, 21/40, 1/5
6. 17/24, 3/8, 5/6
7. 9/10, 7/40, 3/5
8. 5/8, 23/24, 87/144

9. 7/18, 5/12, 3/4
10. 15/16, 7/32, 9/24
11. 1/8, 9/16, 3/64
12. 11/12, 13/40, 5/16
13. 5/27, 2/9, 1/6
14. 51/72, 107/144, 5/8
15. 3/8, 31/64, 7/24
16. 3/10, 19/40, 7/16

Changing fractions to equivalent fractions having a common denominator
Example. Change 5/8 and 3/4 to equivalent fractions having a common denominator.

THINK

1. First find the common denominator of the two fractions. It is 8.

2. Express each fraction with the common denominator 8 by dividing 8 by the denominator of the fraction and then multiplying both numerator and denominator of the fraction by this result. (Use the principle of 1.) To express 3/4 with denominator 8, remember that 8 ÷ 4 = 2, then multiply 3/4 x 2/2, thus changing the form, but not the value of the fraction.

DO THIS

1. 4) 8, 4
    2) 2, 1
       1, 1

    4 x 2 = 8

2. $\frac{5}{8}$ x $\frac{1}{1}$ = $\frac{5}{8}$

    $\frac{3}{4}$ x $\frac{2}{2}$ = $\frac{6}{8}$

Ans. 5/8 = 5/8, 3/4 = 6/8

Summary

To change fractions to equivalent ones having a common denominator:

1. Find the lowest common denominator of all the fractions.

2. For each fraction, divide the common denominator by the denominator of the fraction, and multiply both numerator and denominator of the fraction by this result.

EXAMPLE

Change 5/6, 7/12 and 3/4 to equivalent fractions with a common denominator.

1. 2) 6, 12, 4
    2) 3, 6, 2
    3) 3, 3, 1
       1, 1, 1

    2 x 2 x 3 = 12

2. $\frac{5}{6}$ x $\frac{2}{2}$ = $\frac{10}{12}$

    $\frac{7}{12}$ x $\frac{1}{1}$ = $\frac{7}{12}$

    $\frac{3}{4}$ x $\frac{3}{3}$ = $\frac{9}{12}$

Ans. 5/6 = 10/12, 7/12 = 7/12, 3/4 = 9/12

EXERCISES

Change the fractions in each of the following to equivalent fractions with a common denominator.

1. 3/8, 5/6
2. 3/5, 7/10
3. 7/16, 7/8, 3/4
4. 3/5, 9/10, 11/20
5. 5/8, 31/64, 13/16
6. 3/8, 7/12, 5/6
7. 9/20, 4/5, 3/8
8. 23/24, 7/8, 5/12

9. 15/16, 3/64, 11/32
10. 61/72, 5/12, 123/144
11. 3/4, 5/6, 7/8
12. 2/3, 3/4, 4/5
13. 7/8, 51/64, 13/24
14. 7/12, 37/40, 11/16
15. 5/9, 5/6, 11/24

Adding fractions

Example. Add 7/8, 5/6 and 11/12.

THINK

1. Find the common denominator of all the fractions. This is always the first step in adding fractions.

2. Next, change each fraction to an equal fraction with this common denominator.

3. Add the numerators of the resulting fractions. Write the sum over the common denominator.

4. If possible, reduce this fraction to simpler form, or express it as a mixed number.

DO THIS

1. 2) 8, 6, 12
    2) 4, 3, 6
    2) 2, 3, 3
    3) 1, 3, 3
        1, 1, 1

    2 x 2 x 2 x 3 = 24

2. $\frac{7}{8}$ x $\frac{3}{3}$ = $\frac{21}{24}$

    $\frac{5}{6}$ x $\frac{4}{4}$ = $\frac{20}{24}$

    $\frac{11}{12}$ x $\frac{2}{2}$ = $\frac{22}{24}$

3. $\frac{63}{24}$

4. $\frac{63}{24}$ = $\frac{63 \div 3}{24 \div 3}$ = $\frac{21}{8}$

    $\frac{21}{8}$ = $8 \overline{) 21} \; 2\frac{5}{8}$

    $\frac{16}{5}$

Ans. 7/8 + 5/6 + 11/12 = 2 5/8

Summary

To add fractions:

1. Find the lowest common denominator of all the fractions.

2. Express each fraction as an equivalent fraction with this denominator. (Use the principle of 1.)

3. Add the numerators of the resulting fractions and write the total over the common denominator.

4. Express this result in simplest form, or as a mixed number if possible.

EXAMPLE

Add 1/3, 4/5 and 5/8.

1. 3) 3, 5, 8
   5) 1, 5, 8
   8) 1, 1, 8
      1, 1, 1

$3 \times 5 \times 8 = 120$

2. $\dfrac{1}{3} \times \dfrac{40}{40} = \dfrac{40}{120}$

$\dfrac{4}{5} \times \dfrac{24}{24} = \dfrac{96}{120}$

$\dfrac{5}{8} \times \dfrac{15}{15} = \dfrac{75}{120}$

3. $\dfrac{211}{120}$

4. $\dfrac{211}{120} = 120 \overline{)211} \quad 1\dfrac{91}{120}$
              $\dfrac{120}{91}$

Ans. $1/3 + 4/5 + 5/8 = 1\ 91/120$

Note: Those students who have studied algebra may want to use the formula $\dfrac{a}{b} + \dfrac{c}{d} = \dfrac{a \times d + c \times b}{b \times d}$ to find the sum of two fractions.

Example. Add 7/8 and 5/12, using the formula for finding the sum of two fractions.

THINK

1. Write the formula.

2. Write the values of a, b, c, and d for the present problem.

3. Write each number in place of its letter in the formula.

4. Perform the operations indicated on the right side of this equation. Perform the multiplications first, then the addition.

5. Express this result in the simplest form.

DO THIS

1. $\dfrac{a}{b} + \dfrac{c}{d} = \dfrac{a \times d + c \times b}{b \times d}$

2. $a = 7$, $b = 8$, $c = 5$, $d = 12$

3. $\dfrac{7}{8} + \dfrac{5}{12} = \dfrac{7 \times 12 + 5 \times 8}{8 \times 12}$

4. $\dfrac{7}{8} + \dfrac{5}{12} = \dfrac{84 + 40}{96} = \dfrac{124}{96}$

5. $\dfrac{124}{96} = \dfrac{124 \div 4}{96 \div 4} = \dfrac{31}{24} = 1\dfrac{7}{24}$

Ans. $7/8 + 5/12 = 1\ 7/24$

EXERCISES

Add the fractions in Exercises 1-9.

1. 1/2, 5/8
2. 3/4, 11/16
3. 5/6, 7/12
4. 5/8, 3/4, 3/8
5. 3/16, 5/12, 7/16

6. 3/4, 11/20, 4/5
7. 3/4, 5/6, 4/5
8. 11/12, 9/16, 1/6
9. 3/8, 7/64, 3/32

Find the following sums:

10. 2/3 + 3/4 + 7/8
11. 9/16 + 5/10 + 3/20
12. 5/12 + 3/4 + 11/16
13. 31/64 + 5/8 + 1/4
14. 2/3 + 3/4 + 11/12
15. 5/8 + 9/10 + 4/5
16. 5/24 + 5/12 + 9/16

17. 1/2 + 1/3 + 1/4 + 1/5
18. 3/40 + 5/8 + 11/32
19. 15/16 + 1/12 + 5/8
20. 9/20 + 33/40 + 3/4
21. 7/40 + 1/2 + 31/32
22. 1/4 + 5/16 + 13/32, + 3/4 + 7/16 + 19/32 + 3/8

PROBLEMS

13. Find the length of the eye bolt shown in Fig. 2-10.

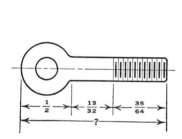

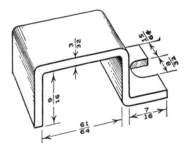

Fig. 2-10. Eye bolt.   Fig. 2-11. Clamp.

14. Calculate the dimensions of the flat brass blank required to make the clamp in Fig. 2-11. Make no allowance for bending.

15. What is the over-all length of the pin in Fig. 2-12?

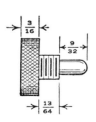

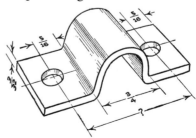

Fig. 2-12. Pin.   Fig. 2-13. Pipe clamp.

16. What is the distance between the centers of the holes of the pipe clamp shown in. Fig. 2-13?
17. Find the length of the rod end pin shown in Fig. 2-14.

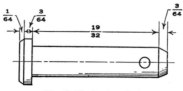

Fig. 2-14 Rod end pin.

Adding mixed numbers

Each mixed number consists of a whole number and a fraction. Adding mixed numbers involves adding fractions, adding whole numbers, and combining the results.

Example. Find the sum: 3 1/2 + 2 7/8 + 1 3/4

THINK	DO THIS
1. Find the common denominator of the fractions.	1. $\quad$ 2)2, 8, 4 $\qquad$ 2)1, 4, 2 $\qquad$ 2) 1, 2, 1 $\qquad$ 1, 1, 1 $\quad$ 2 x 2 x 2 = 8
2. Write the numbers in a column with the fractions under one another.	2, 3. $\quad 3\frac{1}{2} = 3\frac{4}{8}$ $\qquad 2\frac{7}{8} = 2\frac{7}{8}$ $\qquad 1\frac{3}{4} = 1\frac{6}{8}$
3. Rewrite the column, expressing all fractions with the common denominator, here 8.	4, 5. $\qquad\qquad 6\frac{17}{8}$
4. Find the sum of fractions, 17/8.	
5. Find the sum of the whole numbers, 6.	
6. Express the improper fraction as a mixed number, 17/8 to 2 1/8, and add this result to the whole number, 6, to find the total.	6. $\frac{17}{8} =$ $\quad\begin{array}{r} 2\frac{1}{8} \\ 8\overline{)17} \\ \underline{16} \\ 1 \end{array}$ 6 + 2 1/8 = 8 1/8 *Ans.* 3 1/2 + 2 7/8 + 1 3/4 = 8 1/8

Summary

To add mixed numbers:

1. Find the common denominator of all the fractions, then write the numbers in a column with the fractions under one another.

2. Rewrite the column, expressing all fractions with the common denominator.

3. Add the fractions.

4. Add the whole numbers.

5. Express the sum of the fractions as a mixed number in simplest form, and add this result to the sum of the whole numbers.

EXAMPLE

Add the following:
$$4\ 1/2 + 3\ 5/8 + 1\ 1/3$$

1, 2.
$$4\frac{1}{2} = 4\frac{12}{24}$$
$$3\frac{5}{8} = 3\frac{15}{24}$$
$$1\frac{1}{3} = 1\frac{8}{24}$$

3, 4.
$$8\frac{35}{24}$$

5.
$$\frac{35}{24} = 24\overline{)35}^{1\frac{11}{24}}$$
$$\underline{24}$$
$$11$$

$$8 + 1\ 11/24 = 9\ 11/24$$

Ans. $4\ 1/2 + 3\ 5/8 + 1\ 1/3 = 9\ 11/24$

EXERCISES

Add the numbers in each exercise.

1. $3\ 1/2 + 5\ 5/8 + 6\ 1/2$
2. $5\ 1/4 + 7\ 3/4 + 2\ 7/8$
3. $1\ 11/16 + 2\ 5/8 + 3\ 1/2$
4. $7\ 11/16 + 3\ 3/4 + 2\ 5/6$
5. $1\ 7/8 + 3\ 11/32 + 2\ 3/8$
6. $2 + 3\ 1/4 + 1\ 7/8$
7. $5\ 3/10 + 3\ 7/20 + 1\ 1/2$
8. $9\ 5/16 + 3\ 3/4 + 7\ 5/8$
9. $12\ 1/2 + 16\ 2/3 + 15$
10. $8\ 5/32 + 1\ 3/7 + 11/16$
11. $23/24 + 1\ 1/2 + 4\ 3/8$
12. $3\ 11/20 + 7\ 1/4 + 3\ 4/5$
13. $15\ 1/2 + 22\ 19/32 + 10\ 13/16$
14. $3\ 1/12 + 13/24 + 11\ 5/32$
15. $9\ 1/8 + 12\ 37/64 + 3\ 11/12$

PROBLEMS

18. How much 1 1/2″ stock is required to make the soldering-iron holder illustrated in Fig. 2-15? Total allowance for bending is 3/16″ over dimensions shown.

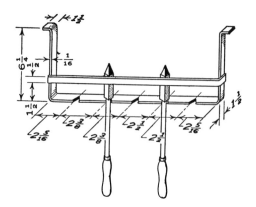

Fig. 2-15. Soldering-iron holder.

19. How many lineal feet of oak are required to make four table legs similar to that shown in Fig. 2-16? Allow 1 1/2 in. waste for each of the four legs.

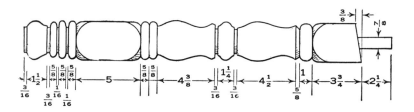

Fig. 2-16. Table leg.

20. Find the value of the missing dimensions in Fig. 2-17.

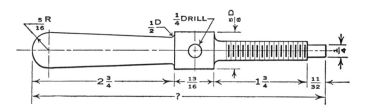

Fig. 2-17.* Adjustable handle of tap wrench.

21. How long is the tool post shown in Fig. 2-18?

22. Find the over-all length of the arbor in Fig. 2-19.

*D after a dimension indicates diameter; R indicates radius.

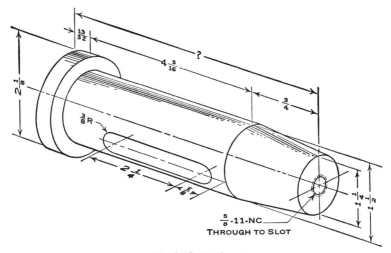

Fig. 2-18. Tool post.

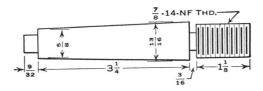

Fig. 2-19. Arbor.

Subtracting fractions

Subtraction is the *inverse* (opposite) of addition. Thus, almost everything we learned about addition of fractions also applies to subtraction of fractions. In the problem "9/16 – 5/32 = ?" we are really asking, "What number added to 5/32 gives 9/16?"

Example. Subtract 3/8 from 5/6.

THINK	DO THIS
We seek the number which, added to 3/8, results in 5/6. Therefore we call 5/6 the sum and 3/8 the addend.	
1. Write the fractions in order, with the sum (5/6) first and the addend (3/8, the number to be subtracted) second.	1. $\dfrac{5}{6} - \dfrac{3}{8}$

2. Find the common denominator for the fractions.

2. $2\,)\,\underline{8,\,6}$
   $4\,)\,\underline{4,\,3}$
   $3\,)\,\underline{1,\,3}$
   $1,\,1$

   $2 \times 4 \times 3 = 24$

3. Express the fractions with the common denominator.

3. $\dfrac{5}{6} \times \dfrac{4}{4} = \dfrac{20}{24}$

   $-\dfrac{3}{8} \times \dfrac{3}{3} = \dfrac{9}{24}$

4. Subtract the numerator of the addend, 9, from that of the sum, 20, and put the result over the common denominator.

4. $\dfrac{11}{24}$

   Ans. $5/6 - 3/8 = 11/24$

Summary

To subtract fractions:

EXAMPLE

Subtract 3/4 from 11/12.

1. Write the fractions in order, with the sum first and the addend (the number to be subtracted) second.

1. $\dfrac{11}{12} - \dfrac{3}{4}$

2. Find the common denominator of the fractions.

2. $4\,)\,\underline{12,\,4}$
   $3\,)\,\underline{3,\,1}$
   $1,\,1$

   $4 \times 3 = 12$

3. Using the principle of 1, express the fractions with a common denominator.

3. $\dfrac{11}{12} \times \dfrac{1}{1} = \dfrac{11}{12}$

   $-\dfrac{3}{4} \times \dfrac{3}{3} = \dfrac{9}{12}$

4. Subtract the numerators. Write this result over the common denominator.

4, 5. $\dfrac{2}{12} = \dfrac{1}{6}$

5. Change the result to simpler terms if possible.

Ans. $11/12 - 3/4 = 1/6$

Note: The formula for adding two fractions (page 44) may be used to subtract two fractions provided the "+" is changed to "–". That is,

$$\frac{a}{b} - \frac{c}{d} = \frac{a \times d - c \times b}{b \times d}.$$

EXERCISES

1. Subtract 3/5 from 9/10 (9/10 – 3/5).
2. Subtract 1/4 from 5/16.
3. Subtract 7/16 from 7/8.
4. From 19/20 take 4/5.
5. From 21/32 take 1/2.
6. From 1/2 take 15/32.

Subtract as indicated in the following problems:

7. 9/10 – 3/20
8. 39/64 – 3/8
9. 55/72 – 5/12
10. 17/24 – 5/8
11. 37/40 – 3/8
12. 53/64 – 9/32
13. 7/8 – 13/16
14. 5/12 – 9/32
15. 87/144 – 3/8

PROBLEMS

23. How long is the threaded section of the wing nut illustrated in Fig. 2-20?

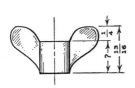

Fig. 2-20. Wing nut.

Fig. 2-21.* Knurled-head thumb screw.

24. Find the length of the threaded section of the knurled head thumb screw shown in Fig. 2-21.
25. Calculate the distance between the centers of the holes of the link shown in Fig. 2-22.

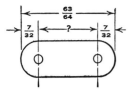

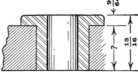
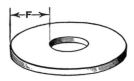

Fig. 2-22. Link.

Fig. 2-23. Jig bushing.

Fig. 2-24. Brass washer.

26. What is the thickness of the cast-iron plate into which the case-hardened steel jig bushing illustrated in Fig. 2-23 has been fitted?
27. This brass washer (Fig. 2-24) has an outside diameter of 3/4″ and an inside diameter of 11/32″. Find the value of the dimension F.

*This is an orthographic projection in two views.

Subtracting mixed numbers

Recalling that a mixed number consists of a whole number plus a fraction, it is easy to see that subtracting these numbers involves basic steps that we have already learned.

Example 1. From 7 5/8 take 2 7/16.

THINK	DO THIS

1. Find the common denominator of the fractions.	1. 8) 8, 16
	    2) 1, 2
	       1, 1
	
	    8 x 2 = 16

2. Write the numbers in order with the sum first and the addend (the number to be subtracted) second.	2, 3. $\quad 7\frac{5}{8} \quad = 7\frac{10}{16}$
3. Rewrite the problem. Express the fraction of each mixed number with the common denominator.	$\quad -2\frac{7}{16} = 2\frac{7}{16}$ 4. $\qquad\qquad\qquad 5\frac{3}{16}$
4. Subtract the fractions, then subtract the whole numbers.	*Ans.* 7 5/8 – 2 7/16 = 5 3/16

Example 2. Subtract 12 29/32 from 18 3/4.

THINK	DO THIS

1. Find the common denominator for the fractions.	1. 4) 32, 4
	    8) 8, 1
	       1, 1
	
	    4 x 8 = 32

2. Write the numbers in order with sum first, addend second.	2, 3. $\quad 18\frac{3}{4} \quad = 18\frac{24}{32} = 17\frac{56}{32}$
3. Rewrite the problem, expressing each fraction with the common denominator. Note that the fraction in the addend is the larger one. In this case take one unit, here 32/32, from the 18 and add it to the 24/32 (18 24/32 = 17 56/32).	$\quad -12\frac{29}{32} = 12\frac{29}{32} = 12\frac{29}{32}$ 4. $\qquad\qquad\qquad\qquad\qquad 5\frac{27}{32}$
4. Subtract the fractions, then subtract the whole numbers.	*Ans.* 18 3/4 – 12 29/32 = 5 27/32

Summary

To subtract mixed numbers:

1. Find the common denominator of the fractions.

2. Write the mixed numbers in order with sum first, addend second.

3. Express the fractions with the common denominator.

4. Subtract the fractions. If the fraction in the addend is the larger one, then, before subtracting take one unit from the whole number in the sum and combine it with the fraction of the sum. Then subtract the fractions.

5. Subtract the whole numbers.

6. Express this result in the simplest form.

EXAMPLE

Subtract: 5 2/21 - 2 3/7

1. $7 \,)\, 21, 7$
$3 \,)\, 3, 1$
$1, 1$

$7 \times 3 = 21$

2-5. $5\frac{2}{21} = 5\frac{2}{21} = 4\frac{23}{21}$

$-2\frac{3}{7} = 2\frac{9}{21} = 2\frac{9}{21}$

$2\frac{14}{21}$

6. $\frac{14}{21} = \frac{14 \div 7}{21 \div 7} = \frac{2}{3}$

$2\frac{14}{21} = 2\frac{2}{3}$

Ans. 5 2/21 - 2 3/7 = 2 2/3

EXERCISES

1. From 3 7/8 subtract 1 5/8.
2. From 6 3/4 subtract 2 15/32.
3. From 11 11/12 subtract 10 2/3.
4. From 53 1/4 subtract 2 1/3.
5. Take 6 5/8 from 9 49/64.
6. Take 3 11/24 from 15 7/8.
7. Take 1 23/32 from 10 55/64.
8. Take 13 15/16 from 23 29/32.

In Exercises 9 to 17 subtract as indicated.

9. 9 15/32 - 4 7/8
10. 72 5/12 - 35 23/24
11. 26 7/8 - 14 5/6
12. 14 19/36 - 9 2/3
13. 54 11/15 - 39 5/9
14. 17 1/16 - 12 7/8
15. 42 3/32 - 18 3/4
16. 11 17/24 - 9 63/144
17. 8 5/12 - 4 37/40

PROBLEMS

28. What is the width of the dado shown in Fig. 2-25? Find the value of *B* in this figure.
29. What is the value of the missing dimension in the drawing of the index pin in Fig. 2-26?
30. What is the length of the threaded section of the king bolt in Fig. 2-27?

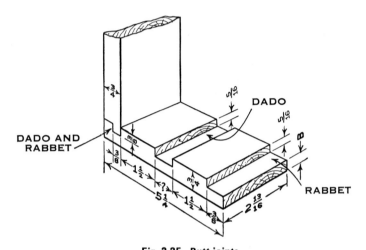

Fig. 2-25. Butt joints.

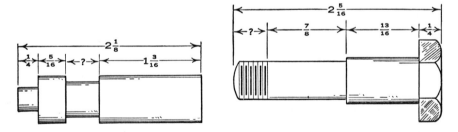

Fig. 2-26. Index pin.   **Fig. 2-27. King bolt.**

31. The insulator shown in Fig. 2-28 is used for electric power lines. How far away from the wall will the wire be?
32. When open, the turnbuckle shown in Fig. 2-29 measures 5 9/32" and when closed, measures 3 5/8". What is the amount of the take-up?

In the next several pages you are asked to solve problems especially intended to test your ability to use fractions. Review the rules given on page 55 and use them thoughtfully.

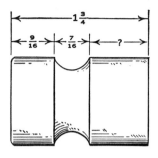

Fig. 2-28. Insulator.   Fig. 2-29. Turnbuckle.

Rules to follow in working with fractions

1. *In multiplication and division, change all mixed numbers to improper fractions, then perform the operations.*
2. *In division, invert the divisor, then multiply.*
3. *Before adding or subtracting, see that the fractions have a common denominator.*
4. *When fractions are to be multiplied or reduced, use the principle of 1 to make the work easier.*
5. *Always express the final answer as a proper fraction or as a mixed number.*

REVIEW PROBLEMS

1. Molten cast iron in solidifying will shrink 1/8″ per foot of length. If a patternmaker is to make a pattern for a bed plate 10″ wide and 50″ long, what will be the over-all dimensions of the completed pattern when measured with a standard rule? What will be the dimensions of the pattern when measured with a shrinkage rule?*
2. What is the thickness of the molding shown in Fig. 2-30?
3. The eccentric pin illustrated in Fig. 2-31 is made of two pieces of steel. What is the over-all length of the pin?

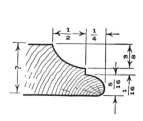

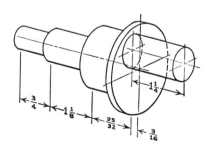

Fig. 2-30. Molding detail.   Fig. 2-31. Eccentric pin.

*See Appendix for table of shrinkage of castings.

4. What is the width of the board required to make the lap dovetail joint shown in Fig. 2-32?

5. A carpenter wishes to drill a hole 15/16″ deep in a piece of lumber with a brace and auger bit (Fig. 2-33). If one turn of the brace advances the bit 1/32″, how many turns will be required to drill the hole?

6. The outside length of a proof-coil chain link (Fig. 2-34) is 1 51/64″. It is made of steel 5/16″ in diameter.
   (a) What is the combined length of 2 links?
   (b) What is the combined length of 3 links?
   (c) A chain 12 ft. long will contain how many links?

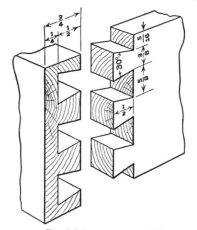

Fig. 2-32. Lap dovetail joint.

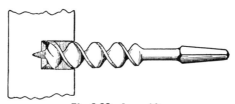

Fig. 2-33. Auger bit.

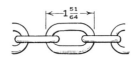

Fig. 2-34. Proof-coil chain links.

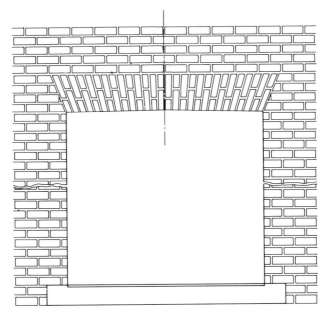

Fig. 2-35. Window opening in brick wall.

7. The opening provided for a window in a brick building (Fig. 2-35) has 24 courses of common brick between the sill and the lintel. If a standard size brick is 2 1/4" x 3 3/4" x 8" and the joints are 3/16", what is the length of the sash required for this opening?

8. A millman receives a piece of Idaho pine lumber, "B" select, 1 1/2" thick and 11" wide. In dressing this lumber, 1/8" is removed from each face by the planer and 1/16" from one edge by the jointer. What is the thickness of the dressed board? What is the width?

9. A carpenter's nail box is made of 3/4" lumber and is assembled as shown in Fig. 2-36. What is the size of each of the 8 equal compartments?

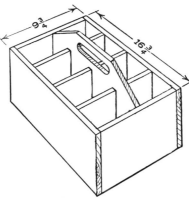

Fig. 2-36. Nail box.

10. Find the weight of each item and the total weight of the following bill of material:

Item No.	No. of Pieces	Stock Length	Section	Material	Wt. per Foot	Total Weight
1	3	60'		Structural I beams, 5"	14 3/4 lb.	
2	6	20'		Angles 1 x 1 x 1/4 bar size	1 1/2 lb.	
3	3 1/3	60'		Structural channels, 8"	13 3/4 lb.	
4	2 1/2	65'		8 x 8 Wide flange H section	34 1/3 lb.	

Total weight _____

11. Three hundred and seventy-seven 7" x 14" slate shingles, when exposed 5 1/2" to the weather, will cover one square of 100 sq. ft. Two and two-thirds pounds of nails are required to lay these shingles. How many slate shingles and how many pounds of nails are required for a roof containing 37 2/3 squares?

12. The radius of a circle is one-half the length of its diameter. What is the radius of each of the following circles?

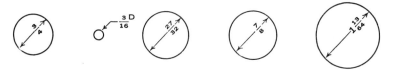

Fig. 2-37. Circles.

13. Twenty-four pieces of 3/4" round stock 3 5/16" long are required for a certain job. If 1/8" is to be allowed for waste in cutting each piece, how many feet of stock are required?

14. Determine the dimensions *A*, *B*, and *C* in Fig. 2-38.

15. A machinist uses 577 1/2 cu. in. of lubricating oil to fill a gear-case unit for a lathe. Then he takes 1/3 gallon of the same oil to fill several oil cans. If this oil is removed from a full 55-gallon drum, how many gallons of oil are left in the drum? (Assume there are 231 cu. in. to the gallon.)

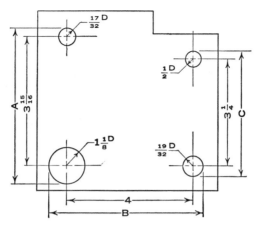

Fig. 2-38. Fixture plate.

16. If 1 3/4-in. diameter stock is used to make the machine part with shoulders shown in Fig. 2-39, what is the total depth of each cut?

17. (a) Find the dimension X in Fig. 2-40.

   (b) How many pieces can be obtained from a piece of stock 12' long? Allow 3/16" per piece for cutting.

   (c) How much stock will be left?

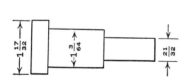

Fig. 2-39. Machine part with shoulders.

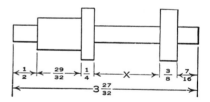

Fig. 2-40. Pin.

3 The Micrometer and Decimal Fractions

Many times our shop measurements must be more precise than those we can get with the rules described in Unit 1. Then we use instruments such as the micrometer or the vernier caliper, which regularly give us precision to 1/1000 in., or can be made even more precise, giving us measurements of 1/10,000 in. or finer. The micrometer operates on the principle of a screw moving in a fixed nut; hence to understand it, we must learn something about screws and screw threads. Our first few paragraphs deal with these fundamentals. Then we shall discuss the micrometer itself, and review the decimal numeration that lets us write fractions in tenths, hundredths, or thousandths without writing the denominator each time.

Screw threads

If you will observe carefully the construction of a screw, you will see that it is cut like a spiral of steel wound around a steel cylinder.

A screw that has just a single band wound around a cylinder is a single-threaded screw. The screw shown in Fig. 3-1 is a single-threaded screw.

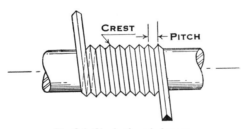

Fig. 3-1. Single-threaded screw.

If two bands, one beside the other, seem to be wound around a cylinder, the screw is a double-threaded screw (Fig. 3-2.).

Pitch and lead of a screw are terms met frequently in industrial work. *Pitch* is the distance from any point on a screw thread to a corresponding point on the next thread, measured parallel to the axis of the screw. It is usually measured from the crest of one thread to the crest of the next.

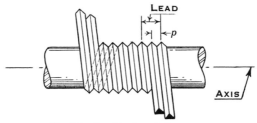

Fig. 3-2. Double-threaded screw.

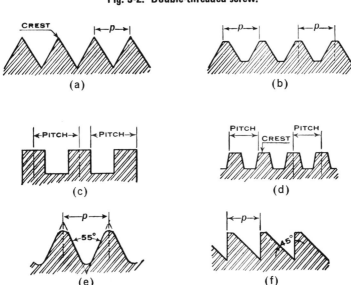

Fig. 3-3. Pitch of different types of screw threads. (a) "V" thread; (b) American Standard or National Form Thread; (c) Square thread; (d) Acme thread; (e) Whitworth thread; (f) Buttress thread.

In Figs. 3-1 and 3-2, the pitch is the distance p. The pitch for different screws might be $1/8''$, $1/10''$, $1/12''$, $1/14''$, $1/18''$, $1/40''$, etc. Fig. 3-3 shows the pitch for several different forms of threads. The *lead* of a screw is the distance the screw progresses into the work with one complete turn. The lead of a screw may be $1/8''$, $1/10''$, $1/12''$, $1/14''$, $1/18''$, $1/40''$, etc.

For a single-threaded screw, the pitch and lead are always the same. In Fig. 3-3 (c), the pitch and lead are $1/2$ in. In Fig. 3-3 (d), the pitch and lead are $5/16$ in.

We shall refer to single-thread screws in this text, unless otherwise noted.

Example 1. A flat-head brass machine screw has a pitch of $1/20''$. How many threads does such a screw have to one inch?

Fig. 3-4. Flat-head brass machine screw.

THINK

Since the pitch of each thread is 1/20 in., there will be as many threads per inch as the number of times 1/20 is contained in 1. Divide 1 by 1/20.

DO THIS

$$1 \div \frac{1}{20} = 1 \times \frac{20}{1} = 20$$

threads per inch

Ans. Screw has 20 threads per in.

Example 2. A certain round-head iron machine screw has 40 threads per in. What is the pitch of such a screw?

Fig. 3-5. Round-head iron machine screw.

THINK

Since there are 40 threads in 1 in., each thread will measure $1 \div 40$, or 1/40 in. This is the pitch of the screw. Divide 1 by 40.

DO THIS

$$1 \div 40 = \frac{1}{40} \text{ in.}$$

Ans. Pitch = 1/40 in.

EXERCISES

1. Using the machinist's scale, measure the pitch of each of the screws shown below. Find the number of threads per inch mathematically, then check by counting the number within an inch on the scale.

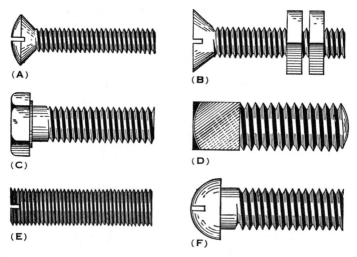

(A)

(B)

(C)

(D)

(E)

(F)

Fig. 3-6. (a) Oval-head machine screw; (b) sink bolt; (c) hexagon-head cap screw; (d) regular set screw; (e) slotted headless set screw; (f) button-head cap screw.

Fig. 3-7. Fillister-head machine screw.

2. A No. 8 fillister-head machine screw when driven into the work is found to advance 1/32" for each turn of the screw. What is the lead of this screw? What is the pitch? (See Fig. 3-7.)
3. What is the lead of each of the screws shown in Fig. 3-6? How far will each screw advance into the work in eight turns? In twelve turns?
4. A screw with a pitch of 1/20" is threaded for a distance of 1 3/4". How many turns will be necessary to drive it completely into the work?

The micrometer

One of the most important instruments for making shop measurements is the *micrometer* (Fig. 3-8). This consists of a frame with a fixed

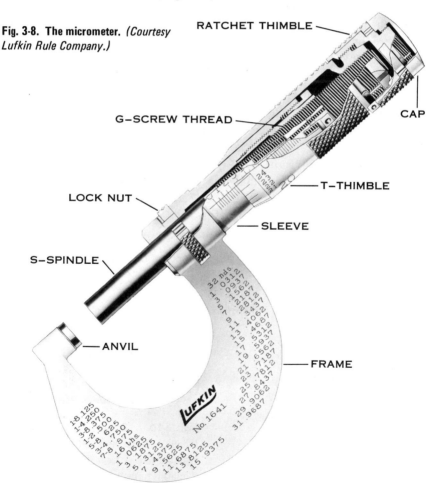

Fig. 3-8. The micrometer. *(Courtesy Lufkin Rule Company.)*

RATCHET THIMBLE

G—SCREW THREAD

CAP

T—THIMBLE

LOCK NUT

SLEEVE

S—SPINDLE

ANVIL

FRAME

barrel or sleeve, and a movable spindle and thimble. The spindle, **S**, and the thimble, **T**, are fastened together inside the sleeve so that turning **T** also turns **S**. A screw thread, **G**, joins the spindle and thimble with the sleeve and frame.

As **T** is turned, both spindle and thimble move forward or backward. The screw thread used in the micrometer has forty threads to the inch. Its pitch and lead are both 1/40 in. Consequently, giving the thimble one complete turn moves both thimble and spindle forward or backward 1/40 in. Some micrometers have caps with a ratchet thimble and/or a locknut. The ratchet thimble makes it possible to obtain consistent readings independent of "feel," and the locknut permits locking of spindle at any reading.

Figure 3-9 shows the scale on the fixed sleeve. Each large or major division, 0, 1, 2, 3, etc., is divided into four equal smaller or minor divisions. Each minor division represents the distance the spindle and thimble move when the thimble makes one complete turn — that is, 1/40 in. Each major division represents the distance the spindle and thimble move when the thimble makes four complete turns — that is, four times 1/40, or 4/40, or 1/10 in. *Each minor sleeve division is 1/40 in.; each major sleeve division is 1/10 in.*

The end of the movable thimble (Fig. 3-10) is divided into 25 equal parts. These divisions are marked 0, 5, 10, 15, 20, but on some micrometers (see Fig. 3-8) every graduation is numbered. *When you turn the thimble exactly one division, the spindle moves 1/25 of 1/40 in., or 1/1000 in.* (Written as a decimal this is 0.001 in.) A complete turn of the thimble is 25 divisions, or 25 x 0.001, or 0.025 in. Hence, 1/40 in. = 0.025 in., read "twenty-five thousandths of an inch."

When you measure with the micrometer, remember:

1. One large division on the sleeve is 1/10″ (0.1″ or 0.100″).

2. One small division on the sleeve is 1/40″ (0.025″).

3. One small division on the thimble is 1/1000″ (0.001″).

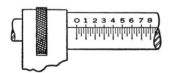

Fig. 3-9. Graduations on the sleeve.

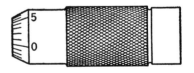

Fig. 3-10. Graduations on the thimble. Note the knurling to prevent the fingers from slipping as the thimble is turned.

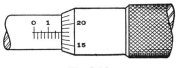

Fig. 3-11.

Example. What is the measurement represented by the reading indicated above?

THINK	DO THIS
1. There is one major division on the sleeve. This is 0.1″ or 0.100″.	1. 1 x 0.100″ = 0.100″
2. In addition, there are three minor divisions completely visible on the sleeve. Each is 0.025″. Multiply 3 x 0.025″.	2. 3 x 0.025″ = 0.075″
3. Eighteen divisions show on the thimble. Each is 0.001″. Multiply 18 x 0.001″.	3. 18 x 0.001″ = 0.018″
4. The sum of the results is the reading.	4. 0.100 + 0.075 + 0.018 = 0.193
	Ans. Reading is 0.193″

Summary

To measure with the micrometer:

1. Count the number of major divisions entirely visible on the sleeve. Multiply this by 0.1″ (or 0.100″).

2. Count the additional minor divisions entirely visible on the sleeve. Multiply this number by 0.025″.

3. Read the number of divisions on the thimble through which the thimble has been turned. Multiply this by 0.001″.

4. Add the results of 1, 2, and 3.

EXAMPLE

Read the setting in Fig. 3-8.

1. Two major divisions are visible.
   2 x 0.1″ = 0.2″ or 0.200″

2. Two additional minor divisions are visible.
   2 x 0.025″ = 0.050″

3. Thimble has not been turned through any further divisions.
   0 x 0.001″ = 0.000″

4. 0.2 + 0.050 + 0.000 = 0.250

Ans. Reading is 0.250 in.

EXERCISES

State as a decimal the length that each of the following distances indicates:

1. Five major on sleeve
2. Three minor on sleeve
3. Eight major on sleeve
4. Three major, two minor on sleeve
5. Five major, three minor on sleeve

6. Five on thimble
7. Twelve on thimble
8. Nineteen on thimble
9. One on thimble
10. Thirteen on thimble

11. Two minor on sleeve, and eight on thimble
12. Three minor on sleeve, and sixteen on thimble
13. Two major and two minor on sleeve, and six on thimble
14. Eight major, one minor on sleeve, and eleven on thimble
15. Four major on sleeve, nine on thimble

Make the following readings:

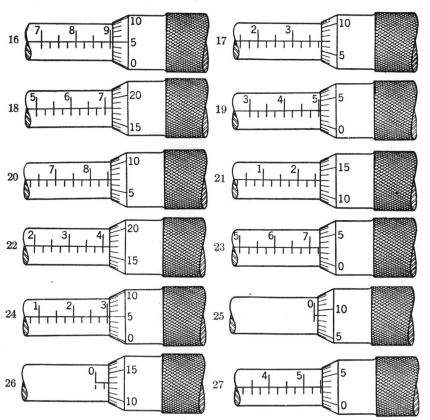

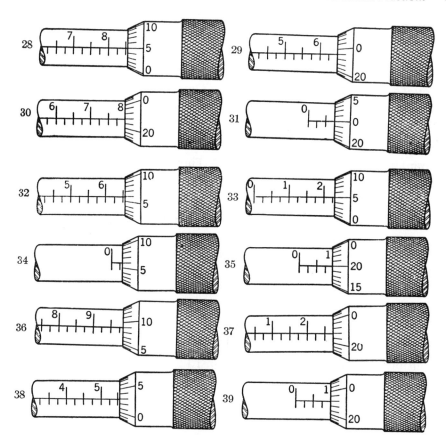

What to do when a line on the thimble is not exactly at the main line of the sleeve

How would you make the reading on the micrometer shown below? In this case, there are six major division points on the sleeve. This corresponds to a distance of 0.6 in. There are in addition three minor divisions showing on the sleeve. This corresponds to 3 x 0.025, or 0.075 in. The main line of the sleeve falls between 8 and 9 on the thimble. We choose the division point which is nearer the main line. Here the nearer one is 8. Hence the reading is 0.6 + 0.075 + 0.008, or 0.683 in. (In order to avoid errors in addition, it would be well to write the decimals in a column with decimal points in line.)

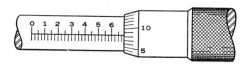

Fig. 3-12.

Summary

To make the reading from a micrometer:

1. Count the number of major divisions shown on the sleeve. Multiply this number by 0.1″ (or 0.100″).

2. Count the number of additional minor divisions on the sleeve. Multiply this by 0.025″.

3. Choose the division point on the thimble that is nearest to the main line on the sleeve. If the main line seems to be exactly half way between two lines on the thimble, choose the division point with the higher number.

4. Multiply this result by 0.001″.

5. Add the results of 1, 2, and 4.

EXAMPLE

Read the micrometer in Exercise 4 below.

1. One major division is visible.
   1 x 0.1″ = 0.1″

2. Three additional minor divisions are visible.
   3 x 0.025″ = 0.075″

3. Main line is half way between 12 and 13 on the thimble. Choose 13.

4. 13 x 0.001″ = 0.013″

5. 0.1 + 0.075 + 0.013 = 0.188

Ans. Reading is 0.188″

EXERCISES

Make the following readings to the nearest thousandth of an inch.

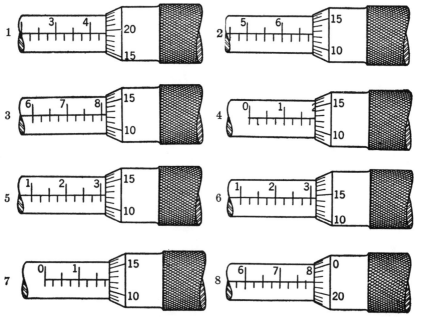

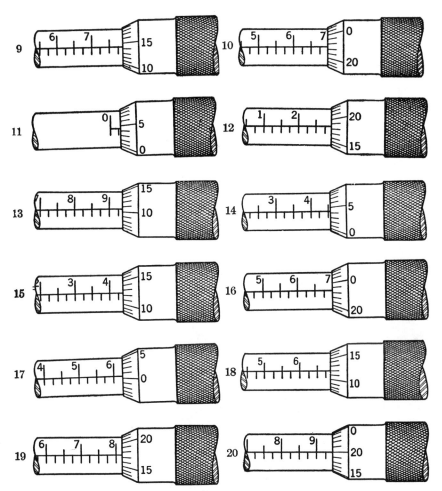

The ten-thousandths micrometer (vernier micrometer)

This instrument is similar to the regular micrometer, but in addition it has a 10-part vernier scale on the sleeve. The appended vernier permits measurements accurate to the nearest ten-thousandths of an inch. This micrometer is used when approximation to the nearest thousandth of an inch, as described above, is not accurate enough for the specific requirements of the job.

Ten spaces on the vernier scale match nine spaces on the thimble. If a thimble graduation is not exactly on the index line of the sleeve, one of the thimble graduations will coincide with some line on the vernier scale. The number of this line is added to the sleeve and thimble reading as a number of ten-thousandths. (For a more detailed discussion on the principle of operation of a vernier scale, see the section called "The vernier caliper," page 96.)

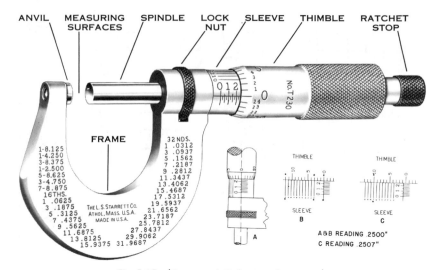

Fig. 3-13. *(Courtesy L.S. Starrett Company.)*

How to read a micrometer graduated in ten-thousandths of an inch

Example 1. Read the micrometer shown in **A** and **B** above.

THINK	DO THIS
1. Count the number of major divisions showing on the sleeve. Multiply this number by 0.1″ (or 0.100″) to get the first significant digit of the reading.	1. There are two major divisions visible on the sleeve. 2 x 0.1″ = 0.2″
2. Count the number of additional minor divisions on the sleeve. Multiply this by 0.025″.	2. There are two additional minor divisions visible. 2 x 0.025″ = 0.050″
3. Read the number of whole divisions on the thimble. Multiply this by 0.001″.	3. Line "0" on the thimble coincides with the main line on the sleeve. 0 x 0.001 = 0.000″
4. Read the vernier scale. See which of the lines on the vernier coincide with a line on the thimble. Multiply the number of this line by 0.0001″.	4. The "0" lines on the vernier coincide with lines on the thimble. 0 x 0.0001″ = 0.0000″
5. Add the results. Note that the final zeros in the reading 0.2500 are significant figures that must be included.	5. 0.2 + 0.505 + 0.000 + 0.0000 = 0.2500 *Ans.* Reading is 0.2500″

Example 2. Read the micrometer shown in C on page 70.

THINK	DO THIS
1. Count the number of major divisions showing on the sleeve. Multiply this number by 0.1″ (or 0.100″).	1. There are two major divisions visible on the sleeve. 2 x 0.1″ = 0.2″
2. Count the number of additional minor divisions on the sleeve. Multiply this by 0.025″.	2. There are two additional minor divisions visible. 2 x 0.025″ = 0.050″
3. Read the number of whole divisions on the thimble. Multiply this by 0.001″.	3. The main line on the sleeve lies between "0" and "1" on the thimble. 0 x 0.001″ = 0.000″
4. Read the vernier scale. Multiply the number of the line on the vernier coinciding with a line on the thimble by 0.0001″.	4. The "7" line on the vernier coincides with a line on the thimble. 7 x 0.0001″ = 0.0007″
5. Add the results.	5. 0.2 + 0.050 + 0.000 + 0.0007 = 0.2507 *Ans.* Reading is 0.2507″

EXERCISES

1. Read and record the setting of each vernier micrometer shown in the figures on page 72.
2. Set a vernier micrometer to each of the following measurements.
   (a) 0.1234   (b) 0.0216   (c) 0.9983   (d) 0.5050   (e) 0.4327
3. State as a decimal the length of each of the following vernier micrometer settings:

	Major Divisions on Sleeve	Minor Divisions on Sleeve	Whole Divisions on Thimble	Reading on Vernier
(a)	5	3	21	8
(b)	0	0	4	4
(c)	3	1	0	5
(d)	2	2	2	2
(e)	6	1	6	6

Setting a micrometer to a given measurement

Example. A machinist has a round steel rod that measures 3/4″ in diameter. He is to turn this in the lathe to make it 0.687″ in diameter. How should he set the micrometer to have exactly 0.687″?

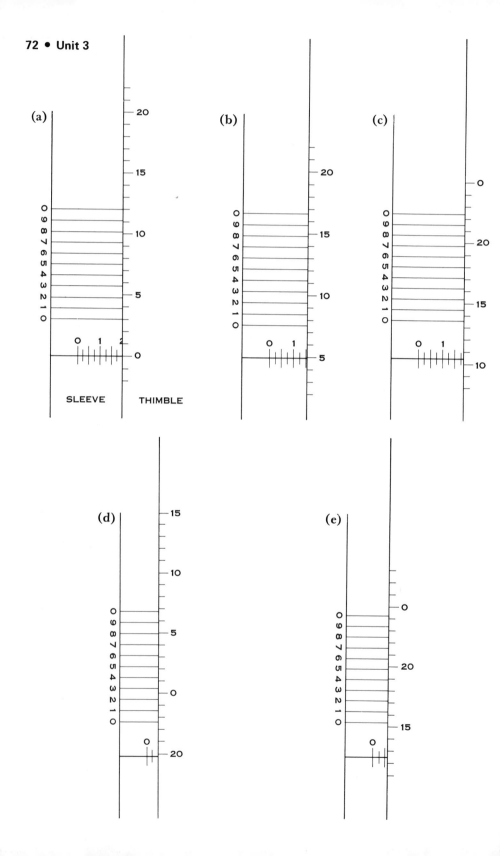

THINK	DO THIS
1. Since each major division on the sleeve is 0.1″, as many of these divisions must show as there are tenths in the measurement. In 0.687 there are 6 tenths. Six complete major sleeve divisions must show. This leaves 0.087 still to be represented.	1. $0.687 = 0.6 + 0.087$ 6 major divisions on the sleeve must show.
2. Since each minor division on the sleeve is 0.025″, we need as many minor divisions as the number of times 0.025 is contained in 0.087. We need three minor sleeve divisions.	2. $0.087 = (3 \times 0.025) + 0.012$ 3 minor divisions on the sleeve must show.
3. The remaining 0.012 must show on the thimble.	3. Exactly 12 small divisions on the thimble must show. *Ans.* The micrometer setting for 0.687″ shows 6 major and 3 minor divisions on the sleeve, plus 12 divisions on the thimble (Fig. 3-14).

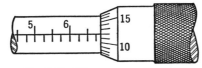

Fig. 3-14. Micrometer set for 0.687″.

Summary	EXAMPLE
To set the micrometer at a given length:	Set a micrometer to check measurements of 0.342″.
1. The number of major divisions on the sleeve is the number of tenths in the length.	1. $0.342 = 0.3 + 0.042$ 3 major sleeve divisions must show.
2. Subtract this number of tenths from the given length.	2. $0.342 - 0.3 = 0.042$
3. The number of minor divisions on the sleeve is the number of times 0.025 is contained in the difference found in Step 2.	3. $0.042 = (1 \times 0.025) + 0.017$ 1 minor division on the sleeve must show.

4. The number of divisions on the thimble equals the number of thousandths remaining in Step 3.

4. 0.017 requires 17 divisions on the thimble.

Ans. Setting for 0.342″ shows 3 major and 1 minor sleeve division, and 17 divisions on the thimble.

EXERCISES

Describe the setting on the 1″ micrometer to measure each of the following:

1. 0.025″
2. 0.075″
3. 0.175″
4. 0.300″

5. 0.500″
6. 0.450″
7. 0.515″
8. 0.892″

9. 0.466″
10. 0.832″
11. 0.965″
12. 0.141″

Decimals

Practical shop applications require more than a knowledge of reading and setting a micrometer to a given measurement. You must understand the meaning of the readings, and be able to work with them.

Micrometer measurements are read in decimals, which are fractions in tenths, hundredths, thousandths, or ten-thousandths, written without the denominator. The location of the decimal point indicates the denominator, and thus the value of the fraction, as explained more fully below. Working with these measurements requires a knowledge of adding, subtracting, multiplying and dividing numbers expressed in decimals. The following sections give you the necessary practice with these important operations.

Reading decimals

The decimal point is a dot placed in certain numerals to indicate their value. When a numeral ends one place to the right of the decimal point, the value is in *tenths.* Thus 0.2 is two tenths. This is the same as the common fraction 2/10. Two places to the right of the point indicates *hundredths.* Thus 0.32 is thirty-two hundredths, the same as 32/100. Three places to the right of the point indicates *thousandths.* Thus 0.432 is four hundred thirty-two thousandths. This is the same as 432/1000. Four places indicates *ten-thousandths.* Thus 0.7432 is seven thousand four hundred thirty-two ten-thousandths. This is the same as 7432/10000. A numeral such as 5.3 is read "five *and* three tenths," or 5 3/10. A numeral such as 543.82 is read "five hundred forty-three *and* eighty-two hundredths," or 543 82/100. Note that the whole number part is read first, then the word "and" is inserted to tell where the decimal point is located, and then the decimal is read.

The place names of the figures in a numeral are illustrated here:

5	5	4	6	8	2	3	2	7	9	1	8	4	3
millions	hundred thousands	ten thousands	thousands	hundreds	tens	units	tenths	hundredths	thousandths	ten thousandths	hundred thousandths	millionths	ten millionths

Summary

To read a numeral that contains a decimal point:

1. Read the whole number part, which is the part to the left of the decimal point.

2. Read the word "and" for the decimal point.

3. Read the decimal. One place of decimals is tenths, two places hundredths, three places thousandths, etc.

EXAMPLE

Read 62.638

1. Sixty-two

2. and

3. six hundred thirty-eight thousandths

Ans. 62.638 is read sixty-two and six hundred thirty-eight thousandths

EXERCISES

Read each of the following numbers:

1. 0.8	6. 5.0675	11. 0.7208	16. 0.8299
2. 1.75	7. 3.1285	12. 19.129	17. 10.01
3. 1.846	8. 75.125	13. 742.8	18. 1.0008
4. 3.92	9. 96.26	14. 9.6209	19. 48.276
5. 0.072	10. 0.005	15. 8762.45	20. 80.007

Adding and subtracting numbers expressed in decimal numerals

Example 1. Add 72.05, 9.638 and 0.00722.

THINK

1. Write the numerals in a column with the decimal points immediately below one another.

DO THIS

1.
$$
\begin{array}{r}
72.05 \\
9.638 \\
\underline{0.00722} \\
\end{array}
$$

2, 3.   81.69522

2. Add these as you would add whole numbers.

3. Place the decimal point in the sum immediately below the decimal points of the addends.

Ans. 81.69522

Example 2. Subtract 8.029 from 11.238.

THINK	DO THIS
1. Write the numerals in a column with decimal points immediately below each other and the numeral to be subtracted below the other one.	1. 11.238 8.029 2. 3.209
2. Subtract as you would whole numbers, placing the decimal point in the answer immediately below the decimal points in the problem.	*Ans.* 3.209

Example 3. From 18.92 subtract 11.175.

THINK	DO THIS
1. Write the numerals in a column with the number to be subtracted below the other one and with the decimal points immediately below each other.	1. 18.92 11.175
2. Since, in this case, there seems to be no way to subtract the 5, fill out the top numeral with zeros until it contains as many decimal places as the number to be subtracted.	2. 18.920 11.175 3. 7.745
3. Subtract, and place the decimal point in the answer immediately below the other decimal points.	*Ans.* 7.745

Summary

To add or subtract decimals:

EXAMPLES

A. Add 81.625, 1.054 and 52.608.

1. Write the numerals in a column with the decimal points immediately below one another.

2. Add or subtract as required.

```
1.        81.625
           1.054
          52.608
2, 3.    135.287
```

Ans. **A.** 135.287

3. Place the decimal point in the result immediately below the decimal points in the problem.

B. Subtract 8.16235 from 10.0924.

1'. In subtraction put the number to be subtracted below the other one.

```
1'.       10.09240
           8.16235
2', 3'.    1.93005
```

2'. Subtract as required.

3'. Place the decimal point in the result.

Ans. **B.** 1.93005

Note: As the numerals in Example B were given, there seemed to be no way to subtract the 5 of 8.16235. A zero was annexed to 10.0924 to make this subtraction possible.

As many zeros as are needed may be annexed to a decimal numeral without changing its value.

EXERCISES

Add the numerals given in each of the Exercises 1-5.

1. 8.625, 93.125 and 15.375.
2. 592.875, 8.063 and 0.250.
3. 184.388, 52.750 and 274.125.
4. 0.638, 1.038 and 2.688.
5. 5.375, 3.938 and 0.625.
6. Subtract 8.6719 from 12.9531.
7. Subtract 0.1094 from 5.3594.
8. Subtract 2.1406 from 8.4375.
9. Subtract 3.0781 from 5.5312.
10. From 17.375 subtract 8.938.
11. From 2.906 subtract 1.625.
12. From 8.0156 subtract 3.3906.
13. From 12 subtract 4.63 (write 12 as 12.00).
14. From 9 subtract 5.453.

15. From 24 subtract 6.281.
16. From the sum of 8.421, 9.625 and 0.750 subtract 15.156.
17. From the sum of 10.578, 3.859 and 11.031 subtract 14.984.

PROBLEMS

1. What is the over-all length (*E*) of the stud shown in Fig. 3-15?

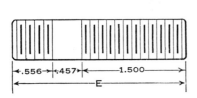

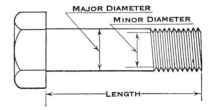

Fig. 3-15. Stud.   **Fig. 3-16. Standard hexagonal-head bolt.**

2. The major diameter of the standard hexagonal-head bolt shown in Fig. 3-16 is 1.125″. The minor diameter is 0.939″. What is the depth of the thread? *Hint:* Subtract minor diameter from major diameter. Divide difference by 2.
3. The inside diameter of a piece of 2″ seamless cold-drawn tubing is 1.939″. The thickness of the wall is 0.218″. What is the outside diameter of the pipe?
4. Find the over-all dimensions of the template shown in Fig. 3-17. What is the value of *R* ?

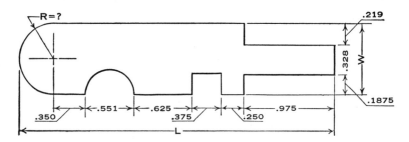

Fig. 3-17. Template.

5. The diameter of the large end of a No. 6 standard taper pin, 1″ long, is 0.341″. The diameter of the small end is 0.320″. What is the taper per inch?
6. The addendum of a spur gear (Fig. 3-19) is 0.1250″. The whole depth is 0.2696″, and the clearance is 0.0196″. How does the dedendum compare with the addendum?
7. In the drawing of the shifter cone (Fig. 3-20), two dimensions, *B* and *L*, are missing. Find the value of each.
8. In the adjusting screw shown in Fig. 3-21, find each of the following lengths to the nearest thousandth: *A* + *B* + *C*; *C* + *D*; *G*; *L*.

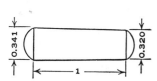

Fig. 3-18. Standard taper pin.

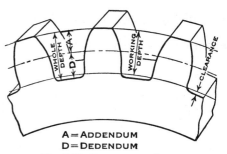

A=ADDENDUM
D=DEDENDUM

Fig. 3-19. Spur gear teeth.

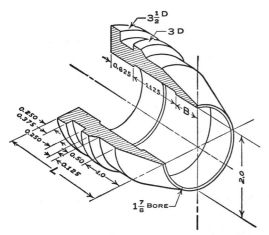

Fig. 3-20.* Shifter cone.

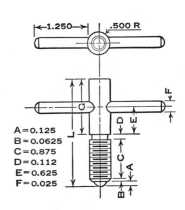

A=0.125
B=0.0625
C=0.875
D=0.112
E=0.625
F=0.025

Fig. 3-21. Adjusting screw of an acetylene regulator.

9. A machinist has received an order to make a link with two reamed holes (Fig. 3-22), each 0.117" in diameter. They are to be spaced 0.625" apart. What is the distance between centers?

10. A cored bar of bearing bronze has an outside diameter of 3" and an inside diameter of 1.675". In order to work this material down to size, it will be necessary to take a rough cut of 0.052" and a finish cut of 0.020" on both the outside and inside diameters. What will the diameters measure when the job is completed? What will be the thickness of the wall?

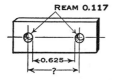

REAM 0.117

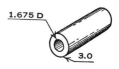

1.675 D

3.0

Fig. 3-23. Cored bearing bronze casting.

*This is an isometric drawing in half section.

Multiplying numbers expressed in decimal numerals

Example. Multiply 8.26 by 1.507.

THINK	DO THIS

1. Arrange the numerals as you would when multiplying whole numbers. It is preferable to use the one with fewer figures as the multiplier.	1. 1.507
	        8.26
	       9042
	      3014
	    12 056

2. Multiply as you would multiply whole numbers, starting from the right.	2, 3. 12.44782
3. Place the decimal point so that *there will be as many places to the right of the point in the product as there are places to the right of the point in the two factors combined.* Here 3 + 2, or 5 places.	Three places to the right of the decimal point in 1.507 and two places to the right of the decimal point in 8.26 require five (3 + 2) places to the right of the decimal point in the product. *Ans.* 12.44782

Note: Why do we require as many places to the right of the decimal point in the product as there are places to the right in the two factors combined? The following simple illustrations may show why.

0.5 = 5/10 and 0.7 = 7/10; then 0.5 x 0.7 is the same as 5/10 x 7/10, which equals 35/100; and 35/100 = 0.35. Also 0.03 x 0.12 is the same as 3/100 x 12/100 = 36/10,000 = 0.0036.

Notice that the products of the denominators contain as many zeros as there are zeros in the denominators of the two factors combined. Since a decimal is just a shorter way to write fractions whose denominators are powers of 10, the product of decimals will have as many decimal places as the factors combined.

Summary

To multiply decimal numerals:

1. Arrange the numerals as you would when multiplying whole numbers.

2. Multiply as you would multiply whole numbers.

EXAMPLE

Multiply 5.623 by 8.24

1.   5.6 23
       8.24
    22 4 92
  1 12 4 6
  44 98 4
2, 3. 46.33 3 52

3. To place the decimal point, start at the right and count off as many places as there are decimal places in the two factors combined. Here 3 + 2 = 5 places in the product.

Ans. 46.33352

Note: If the factors were measured numbers, the product would retain only as many of the digits to the right of the decimal point as there were in the factor with the lesser number of decimal places, here two.

EXERCISES

Multiply as indicated.

1. 8.45 by 7.2
2. 9.91 by 8.5
3. 4.7 by 2.5
4. 12.0 by 2.375
5. 10.7 by 3.004
6. 46.3 by 12.5
7. 7.2 by 0.3165
8. 6.725 by 3.117
9. 4.6922 by 4.5
10. 12.911 by 7.25
11. 1.045 by 7.25
12. 8.196 by 16.67

PROBLEMS

11. The depth of an American Standard screw thread (Fig. 3-24) may be found by multiplying the pitch by 0.6495. Find the depth of a thread having a pitch of 0.125".

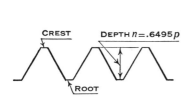

6.750 D

Fig. 3-24. American Standard
screw thread.

Fig. 3-25. Flywheel.

12. Find the circumference of the flywheel shown in Fig. 3-25. The diameter times π, or 3.1416, will give the circumference. For this type of work the value for π may be taken as 3.14.

13. A plumbing job requires 20 lengths of 1" brass pipe. If the standard length of pipe of this size is 12' and weighs 1.2 lbs. per ft., what will be the total weight of the shipment?

14. A patternmaker estimates that a casting to be made from a certain pattern will have a volume of 14.75 cu. in. If cast iron weighing 0.284 lb. per cu. in. were to be used to make the casting, what would be the estimated weight of 65 such castings? At $.14 per pound, what would be the cost of the cast iron for these castings?

15. A machinist orders a quantity of square and round steel bars in 12' lengths. The delivery slip reads as follows:

> 9 bars 7/8"   round S.A.E.* 1020 @ $.09
> 4 bars 1 3/16"   square S.A.E.* 1112 @ $.09
> 7 bars 2 1/8"   round S.A.E.* X1112 @ $.09

If the weight per foot of each of the items is 2.044, 4.795, and 12.058 lbs. respectively, what will be the amount of the bill if the base price of steel in these quantities and sizes is $.09 per pound?

16. In laying out a finished head for a hexagonal bolt (Fig. 3-26), a draftsman obtains the width across the flats (W) by multiplying the diameter of the bolt (D) by 1.500, and the height (H) by multiplying the diameter (D) by 0.666. Find the width across the flats and the height for each of the following finished hexagonal-head bolts:

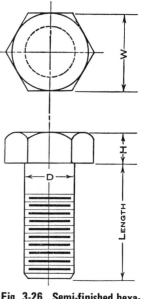

Fig. 3-26. Semi-finished hexa-gonal-head bolt.

Diameter of bolt	Width across flats	Height
0.02500	. . . .	. . . .
0.3750	. . . .	. . . .
0.7500	. . . .	. . . .
1.000	. . . .	. . . .
1.125	. . . .	. . . .

17. A work bench is to be covered with a piece of No. 12 gage sheet steel having an area of 32.75 sq. ft. If No. 12 gage steel weighs 4.375 lbs. per sq. ft., what is the total weight of the steel used to make the top?

18. If the operating cost of a job shop is $18.75 per hour, what is the operating cost of a job which takes 17.5 hours?

19. Turpentine weighs 0.031 lb. per cu. in. What will be the weight of the turpentine that will fill a 30-gallon drum? (1 gallon = 231 cu. in.)

*These symbols represent specific types of steel. S.A.E. stands for Society of Automotive Engineers.

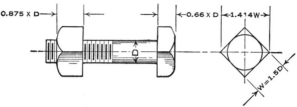

Fig. 3-27. Square-head bolt.

20. A piece of untempered steel when heated will change its length 0.00000689″ for each degree of change of temperature and for each inch of length. What will be the length of a 16″ piece of steel after it has been heated through a temperature change of 250° Fahrenheit?

21. Figure 3-27 shows a two-view orthographic projection of an American Standard unfinished regular square-head bolt. The dimensions on the drawing are given in terms of the diameter (D). Find the value of each of these dimensions for a bolt 0.875″ in diameter.

22. Free-cutting round brass rod is made in 8- to 12-ft. lengths and is sold by the pound. Assuming the cost per pound to be $.90, complete the following price list:

Brass Rods

Round-Free Cutting

Diameter	Weight, Lb. per Foot	8-Ft. Length		12-Ft. Length	
		Weight	Cost	Weight	Cost
1/16	0.01130	. . . . . .	. . . . . .	. . . . . .	. . . . . .
1/8	0.0452	. . . . . .	. . . . . .	. . . . . .	. . . . . .
1/4	0.1808	. . . . . .	. . . . . .	. . . . . .	. . . . . .
1/2	0.7234	. . . . . .	. . . . . .	. . . . . .	. . . . . .
1	2.893	. . . . . .	. . . . . .	. . . . . .	. . . . . .

Dividing numbers expressed in decimal numerals

Example 1. Divide 4.726 by 1.65.

THINK

1. To divide 4.726 by 1.65 means to find the number which when multiplied by 1.65 gives a product of 4.726. To do this, first arrange the numerals as you

DO THIS

1. $1.65 \overline{)4.726}$

would arrange whole numbers for division. Put the numeral that is the product in the box, here 4.726, and the factor you are dividing by, here 1.65, in front of the box.

2. Move the decimal point of the factor to the right-hand end. Here we moved the point two places, from 1.65 to 165. Also move the decimal point in the product, 4.726, the same number of places in the same direction. (See *note* below.)

2. $165 \overline{)472.6}$

3. Divide as with whole numbers; 165 into 472 goes 2 times. Place this 2 in the units place in the result. Then 165 into 1426 goes 8 times. Note that *the decimal point in the result is immediately above the point in the product.*

3, 4.
```
        2.86
165 ) 472.60
      330
      1426
      1320
      1060
       990
        70
```

4. By affixing a zero following the 6 in the product, we carry the division one more place. Then 165 into 1060 goes 6 times. Finally, we stop here and say that 4.726 divided by 1.65 carried to hundredths, or two decimal places, is 2.86.

Ans. 2.86

Note: If we write the problem in the form 4.726/1.65, we see that what we do here is multiply this fraction by 100/100, or 1, and obtain 472.6/165.

Example 2. Divide 5.29 by 12.345. Carry the result to the nearest hundredth.

THINK

DO THIS

1. Arrange the numerals for division.

1. $12.345 \overline{)5.29}$

2. Move the point in both product and factor *three* places to the right, to make the factor a whole number. Notice that to do this we annex a 0 to 529.

3. Divide as for whole numbers. To carry the result to the nearest hundredth we must annex *three* zeros to the right of the decimal point in the product.

4. Make sure the decimal point in the result is immediately above the decimal point in the product.

5. Since the problem requires the resulting factor to be carried to the nearest hundredth, that is, to the nearest digit in the second decimal place, we drop the 8 and add one to the 2. The result is 0.43.

2, 3, 4.

$$12345 \overline{\smash{)}\begin{array}{r} 0.428 \\ 5290.000 \end{array}}$$
$$\begin{array}{r} \underline{4938\ 0} \\ 352\ 00 \\ \underline{246\ 90} \\ 105\ 100 \\ \underline{98\ 760} \end{array}$$

5. 0.428 to the nearest hundredth is 0.43.

Ans. 0.43

Note: The procedure we shall follow in this text will be to carry the operation one place beyond that which is required. If that last place obtained is 5 or more, it is dropped but 1 is added to the digit in the required place. If the last place is less than 5, we just drop it. In this problem the result was to be carried to the nearest hundredth. We carried the operation to thousandths, the digit 8. Since this was greater than 5 we dropped it and added 1 to the digit 2 in the hundredths place. The result to the nearest hundredth is 0.43.

Summary

To divide decimal numerals:

1. Arrange the numerals as you would when dividing whole numbers, putting the product in the box and the factor outside.

EXAMPLE

Divide 2.0763 by 1.84. Carry the result to the nearest hundredth.

1. $1.84 \overline{\smash{)}2.0763}$

2. Move the decimal point in the factor to the extreme right of this numeral, here two places. Move the decimal point in the product to the right the same number of places.

3. Divide as for whole numbers. Make sure the decimal point in the result is immediately above the decimal point in the product. To obtain the result to the nearest hundredth (second decimal place), carry the division one place beyond. Annex zeros if necessary.

4. If the digit in the last place is 5 or more, drop it but add one to the digit in the preceding place. If the digit in the last place is less than 5, drop it and make no change in the preceding digit.

```
                  1.128
2, 3.    184 ) 207.630
               184
               236
               184
               523
               368
              1550
              1472
```

4. 1.128 to the nearest hundredths is 1.13.

Ans. 1.13

EXERCISES

Divide Problems 1-5 completely; divide 6-10 to the nearest tenth and 11-15 to the nearest hundredth.

1. 7.42 by 0.8
2. 9.24 by 0.004
3. 5.6 by 2.5
4. 196 by 3.125
5. 87.4225 by 5.5

6. 6.29 by 4.625
7. 15.2 by 2.023
8. 44.7 by 10.108
9. 2.61 by 5.7
10. 0.426 by 2.08

11. 5.27 by 2.14
12. 8.116 by 5.07
13. 9.25 by 3.7
14. 49.28 by 5.602
15. 15.48 by 6.45

PROBLEMS

23. A shipment of No. 00 pure, solid, bare copper wire weighs 1,407 lbs. If wire of this size weighs 0.402 lb. per linear ft., how many feet were in the shipment?

24. How many pins, having the dimensions indicated in Fig. 3-28, can be made from a 12′ bar of steel? The allowance for cutting each pin is 0.125″.

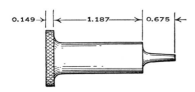

Fig. 3-28. Pin.

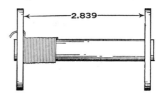

Fig. 3-29. Electromagnet core.

25. How many turns of No. 22 double-cotton-covered magnet wire can be wound on the magnet core in one layer (Fig. 3-29)? The diameter of the wire including insulation is 0.0334".

26. How many cubic feet of steel are in a steel shaft 2 1/4" in diameter and 9 ft. long if steel of this diameter weighs 13.59 lbs. per linear ft.? Steel weighs 489.6 lb. per cu. ft.

27. A radio service man finds it necessary to make a Bakelite terminal strip having the dimensions specified in Fig. 3-30. (a) What will be the distance between the centers of the holes? (b) How long is the terminal strip? (c) If it is necessary to make another Bakelite terminal strip having 2 1/2 times the dimensions specified in Fig. 3-30, what would be the distance between centers of the holes? (d) What would be the total length of the strip?

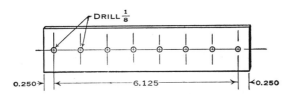

Fig. 3-30. Terminal strip.

28. The tapered section of a cylindrical taper gage (Fig. 3-31) is 9 in. long, and the large diameter is 1.6875". Find the taper per foot if the small diameter is 1.500". (*Hint:* Divide the difference of the diameters by the length, then multiply the result by 12.)

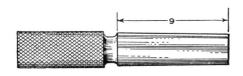

Fig. 3-31. Cylindrical taper gage.

Expressing decimal numerals as common fractions

Any decimal numeral may be written as a common fraction.

Example 1. Express 2.75 as an equivalent common fraction.

THINK	DO THIS
1. 2.75 is read two and seventy-five hundredths. As a common fraction this may be written 2 75/100.	1. $2.75 = 2\frac{75}{100}$
2. Reduce the fraction to simplest form.	2. $2\frac{75}{100} = 2\frac{3}{4}$

Ans. 2.75 = 2 3/4

Note that the decimal part of the given number, the 75, was divided by 1 followed by as many zeros as there are places in the decimal, here two.*

Example 2. Express 0.3125 as a common fraction.

THINK	DO THIS
1. Since there are 4 decimal places in 0.3125, we write 3125 divided by 1 followed by as many zeros as there are decimal places, in this case four.*	1. $0.3125 = \frac{3125}{10000}$
2. Reduce the fraction. We can divide both numerator and denominator by 25.	2. $\frac{3125}{10000} = \frac{125}{400}$
3. We can divide both by 25 again.	3. $\frac{125}{400} = \frac{5}{16}$

Ans. 0.3125 = 5/16

*This follows from the principle of 1. In Example 1, if we multiply 2.75 by 100 we obtain 275 with no decimal point. But by the principle of 1 we must multiply by 100/100, or 1, in order not to change the number. Then 2.75 x 100/100 = 275/100 = 2 75/100.

Summary

To express a decimal numeral as a common fraction:

1. Write the whole-number part of the numeral; then write the decimal part divided by 1 followed by as many zeros as there are places in the decimal part of the numeral; if the decimal part is hundredths, divide by 100; if thousandths, divide by 1000, etc.*

2. Reduce the resulting fraction to simplest form.

EXAMPLE

Express 26.865 as a common fraction.

1. $26.865 = 26\frac{865}{1000}$

2. $26\frac{865}{1000} = 26\frac{173}{200}$

Ans. $26.865 = 26\ 173/200$

EXERCISES

Express each of the following decimals as common fractions:

1. 0.5	6. 0.1875	11. 823.45	16. 25.8435
2. 0.375	7. 14.0625	12. 96.4675	17. 43.96875
3. 2.125	8. 10.384	13. 1.00875	18. 32.43372
4. 9.65	9. 6.5625	14. 10.4925	19. 17.62549
5. 1.875	10. 2.15625	15. 419.8125	20. 362.7854

Expressing a decimal as a fraction with a predetermined denominator

It is often convenient or necessary to express a decimal like 0.234 as a fraction with a stated denominator such as 8. For example, we have an object that is 0.234″ long and wish to know its length in 8ths of an inch. Frequently this cannot be done exactly. In any case, it is useful to know the number of 8ths that is closest to 0.234.

Example 1. Change 0.234 to a fraction with denominator 8. That is, find the number of eighths that is nearest to 0.234.

THINK

1. Express the given decimal as a common fraction.

DO THIS

1. $0.234 = \frac{234}{1000}$ (See Note 1.)

---

*This follows from the principle of 1. In Example 1, if we multiply 2.75 by 100 we obtain 275 with no decimal point. But by the principle of 1 we must multiply by 100/100, or 1, in order not to change the number. Then 2.75 x 100/100 = 275/100 = 2 75/100.

2. The fraction now has the denominator 1000, but we require that the denominator be 8. What number multiplied by 1000 will equal 8? The answer, 8/1000. By the principle of 1, we may multiply numerator and denominator of 234/1000 by 8/8 and then divide this new fraction by 1000/1000.

2. $\dfrac{234}{1000} \times \dfrac{8}{8} = \dfrac{1872}{8000}$

$\dfrac{1872}{8000} \div \dfrac{1000}{1000} = \dfrac{1.872}{8}$

(See Note 2.)

3. By choosing the whole number nearest to this numerator, here 2, we will have the number of 8ths nearest to 0.234.

3. 2/8

Ans. Number of eights nearest to 0.234 is 2/8.

Note 1: We can justify this equivalent form of 0.234 by use of the principle of 1. To express 0.234 without the decimal we must multiply it by 1,000. However, if we do this alone we change the value of 0.234. To retain the value we multiply by 1 in the form of 1,000/1,000.

Note 2: From Step 2, we could have obtained the value of 0.234 in terms of 8ths merely by multiplying the given number, 0.234, by 1 in the form of 8/8.

Example 2. Change 0.234 to the nearest number of 25ths.

THINK | DO THIS

1. Since we require 25ths, we multiply 0.234 by 1 in the form of 25/25.

1. $0.234 \times \dfrac{25}{25} = \dfrac{5.850}{25}$

2. Select the whole number nearest to the resulting numerator, here 6.

2. 6/25
Ans. Number of 25ths nearest to 0.234 is 6/25.

Example 3. Change 0.186 to the nearest number of 16ths.

THINK | DO THIS

1. Since we require 16ths we multiply 0.186 by 1 in the form of 16/16.

1. $0.186 \times \dfrac{16}{16} = \dfrac{2.976}{16}$

2. Select the whole number nearest to the resulting numerator, here 3.

2. 3/16
Ans. Number of 16ths nearest to 0.186 is 3/16.

Summary

To obtain a fraction with a specfied denominator that is closest to a given decimal number, take the following steps:

1. Multiply the decimal by the specified denominator divided by itself. (This equals 1.)

2. Replace the numerator of the result of the first step by the nearest whole number. This result is the required fraction.

EXAMPLE

Obtain the number of 32nds nearest to 0.563.

1. $0.563 \times \dfrac{32}{32} = \dfrac{18.016}{32}$

2. The whole number nearest to 18.016 is 18. Then 18/32 is the 32nd closest to 0.563.

Ans. Number of 32nds nearest to 0.563 is 18/32.

EXERCISES

Express each of these decimals as fractions with the denominators indicated.

1. 0.0256 as 16ths
2. 0.438 as 8ths
3. 0.1296 as 32nds
4. 0.1560 as 25ths
5. 0.7455 as 25ths

6. 0.734 as 40ths
7. 0.666 as 64ths
8. 0.385 as 40ths
9. 0.986 as 64ths
10. 0.546 as 8ths

PROBLEMS

29. Make a sketch of the jaw shown in Fig. 3-32. Show all dimensions to the nearest 1/32".

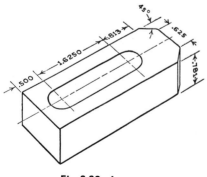

Fig. 3-32. Jaw.

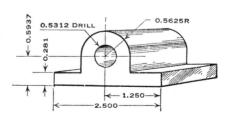

Fig. 3-33. Socket.

30. A pin made to fit a hole 21/64" in diameter was turned to a finished diameter of 0.343". Is the pin too large or too small? Express the difference to the nearest 1/64".

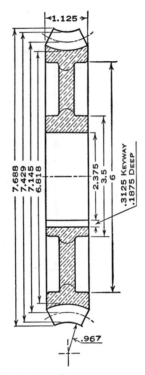

Fig. 3-34. Section of a bronze worm gear.

31. Express each of the dimensions of the socket shown in Fig. 3-33 to the nearest 1/64".
32. Change all decimal dimensions in Fig. 3-34 to the nearest 1/64".
33. The diameter of a tap drill for a 7/8-9 American National coarse (NC) thread is 0.7656". What size drill is this, to the nearest 1/64"?
34. Change the decimal dimensions in Fig. 3-35 to equivalent fractional forms.
35. A drawing calls for a piece of stock to be turned to a diameter of 0.3593". The lathe operator reduces the stock to a diameter of 23/64". Does the finished diameter conform to the specifications?
36. What is the diameter of a steel pin expressed in 64ths of an inch if the pin measures 0.5156"?

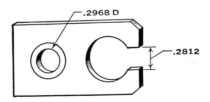

Fig. 3-35. Special ring and snap gage.

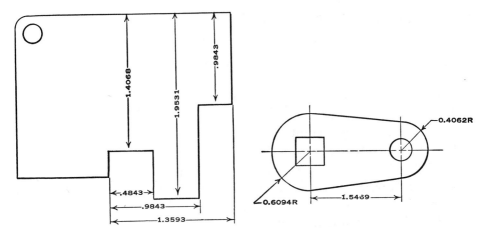

Fig. 3-36. Template.   Fig. 3-37. Lever.

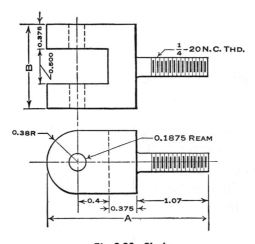

Fig. 3-38. Clevis.

37. Redraw the template in Fig. 3-36 and express all dimensions to the nearest 1/64″.

38. Find the over-all length of the lever illustrated in Fig. 3-37 to the nearest 1/64″.

39. (a) Find the value of dimensions A and B in Fig. 3-38 to the nearest 32nd.

    (b) If a No. 13 drill having a diameter of 0.1850″ is used to drill the hole, how much metal (thickness) will have to be reamed out?

Expressing fractions in decimal form

Example. Express the fraction 7/16 in decimal form.

THINK	DO THIS
1. Since 7/16 means 7 divided by 16, to express it in decimal form we place a decimal point after the numerator, 7, and affix zeros to this.	1. $\dfrac{7}{16} = \dfrac{7.0000}{16}$
2. Divide this result by the denominator of the fraction. The factor obtained is the required decimal.	2. $\begin{array}{r} 0.4375 \\ 16\,)\overline{7.0000} \\ \underline{6\,4} \\ 60 \\ \underline{48} \\ 120 \\ \underline{112} \\ 80 \\ \underline{80} \end{array}$

Ans. 7/16 = 0.4375

Summary	EXAMPLE
To express a fraction in decimal form:	Express 9/32 in decimal form.
1. Place a decimal point after the numerator and affix zeros to it.	1. $\dfrac{9}{32} = \dfrac{9.0000}{32}$
2. Divide by the denominator. The result of this division is the required decimal form.	2. $\begin{array}{r} 0.28125 \\ 32\,)\overline{9.00000} \\ \underline{6\,4} \\ 2\,60 \\ \underline{2\,56} \\ 40 \\ \underline{32} \\ 80 \\ \underline{64} \\ 160 \\ \underline{160} \end{array}$

Ans. 9/32 = 0.28125

Note: The number of zeros affixed depends upon the situation in which the problem may occur.

EXERCISES

Express each of the following as decimals:

1. 1/4	6. 5/8	11. 4 1/64
2. 7/8	7. 5/12	12. 3 1/2
3. 3/10	8. 3/32	13. 9 11/12
4. 17/64	9. 11/32	14. 75 3/16
5. 9/16	10. 9/64	15. 15 19/64

PROBLEMS

40. A bolt has a 1 3/8-6 NC thread. If the major and minor diameters are 1 3/8" and 1.1585", respectively, what is the depth of the thread?

41. A dimension on a drawing reads 7/32". The finished part, when checked by the inspector, measures 0.221". What will be the actual difference in these dimensions in thousandths of an inch?

42. Change all the fractional dimensions in Fig. 3-39 to decimals.

43. The tapered end of a No. 9 *B & S* taper plug gage is 4 3/8" long. The diameter of the small end is 0.937". What is the diameter of the large end if the taper per inch is 0.0467"?

44. A hole is to be tapped for a 1" thread and a 55/64" tap drill is to be used. How much material will be left for the tap to remove? (Express the answer in decimal form.)

45. How many laminations, each 0.035" thick, are required to make a pole 4 3/8" long (Fig. 3-40) for a direct-current generator?

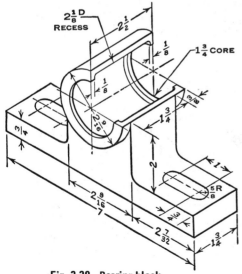

Fig. 3-39. Bearing block.

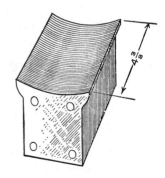

Fig. 3-40. Laminated pole.

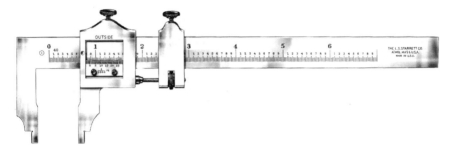

Fig. 3-41. Vernier caliper. *(Courtesy L.S. Starrett Company.)*

The vernier caliper

Another instrument often used to measure to the nearest thousandth of an inch is the vernier caliper (Fig. 3-41). This instrument may be used for accurate measurements both "outside" and "inside" (see Figs. 3-42 and 3-43).

Notice that there is a scale, B, on the tool itself, and another scale, A, on the slide (Fig. 3-44). This scale A is called a *vernier.* What follows suggests how the vernier is used.

The B scale is in inches. From 1 to 2 is one inch (here enlarged). The inch is first divided into *ten* equal parts. Hence from 1 to 2, or from 2 to 3, is a tenth of an inch (1/10″). But each of these spaces is again divided into *four* equal parts. Then each of these smallest divisions is 1/4 of 1/10 in., or 1/40 in., or 0.025 in. (This is the same as the scale on the sleeve of the micrometer. See page 65.)

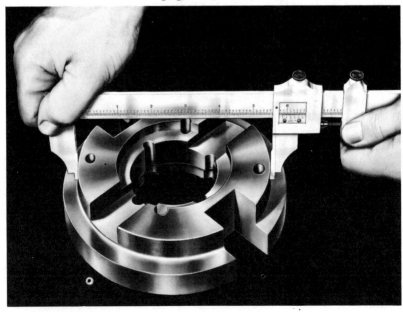

Fig. 3-42. Making an outside-diameter measurement. *(Courtesy L.S. Starrett Company.)*

Fig. 3-43. Making an inside-diameter measurement. *(Courtesy L.S. Starrett Company.)*

Now look at scale A. It is divided into *twenty-five* equal parts, and it covers six of the larger spaces of scale B. Hence the entire scale A is 6/10 or 0.6 in. long. Since scale A is divided into twenty-five parts, each smallest division of this scale equals 1/25 of 6/10, or 6/250, or 24/1000 inch. This may also be written 0.024".

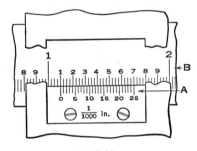

Fig. 3-44.

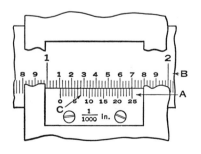

Fig. 3-45.

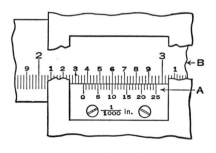

Fig. 3-46.

Hence each smallest division on scale B is 0.025″ and each smallest division on the vernier scale A is 0.024″.

By studying the examples on the pages that follow, you will learn how to use the vernier correctly.

Example 1. Read the vernier caliper shown in Fig. 3-45.

THINK	DO THIS
Here the 0 of the vernier (A) scale is a little to the right of the small 1 of the B scale. We want the distance from 1 of the B scale to 0 of the A scale. We obtain this distance by the following procedure:	
1. Look along the A scale until you come to the first division line of this scale that is directly in line with one of the division lines of the B scale. This is the line marked C.	1. The division lines are in line at C.
2. Count the number of spaces on the A scale from 0 to this line marked C.	2. There are 7 spaces 0 to C on the A scale.
3. Each one of these 7 spaces indicates 0.001″ on the B scale from 1 of the B scale to 0 of the A scale.	3. The distance from 1 of the B scale to 0 of the A scale is 0.001″ x 7, or 0.007″.
4. The entire measurement then is that indicated by the 1 in. and the 0.1 in. on the B scale plus this reading on the vernier scale.	4. 1 + 0.1 + 0.007 = 1.107 *Ans.* Reading is 1.107″

Example 2. State the reading on the vernier in Fig. 3-46.

THINK	DO THIS
1. Read the number of larger spaces on the B scale.	1. 2 larger spaces = 2″
2. From 2 to 3 is 3 x 0.1″.	2. 3 smaller spaces = 3 x 0.1 = 0.3″
3. Then from 3 there are two complete smaller spaces of the B scale to 0 of the vernier.	3. 2 x 0.025 = 0.050″

4. Look along the vernier until you see a line of the vernier that is directly in line with one of the B scale; the 9th line is the one.

4. 9 x 0.001″ = 0.009″

5. The measurement indicated is the sum of these four lengths.

5. 2 + 0.3 + 0.050 + 0.009 = 2.359
Ans. Reading is 2.359″

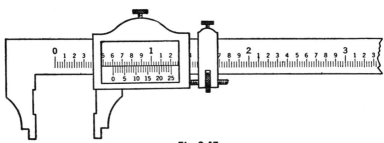

Fig. 3-47.

Summary

To make a reading with the vernier caliper:

1. Read the number of largest spaces on the scale of the caliper. Each of these is 1″.

2. Read the number of complete smaller spaces from the last largest space, here 0, to the 0 of the vernier, here 6. Each of these spaces is 0.1″. Then 6 of them measures 0.6″.

3. Read the number of complete smallest spaces on the tool from the last numbered space, 6, to the 0 on the vernier, here 0. Each of these spaces is 0.025″.

4. Count the number of spaces on the vernier from 0 to the line which is exactly in line with a line of the scale on the tool, here 4. Each of these represents 0.001″.

5. Add these partial measurements to obtain the required reading.

EXAMPLE

Read the measurement shown on the vernier caliper above.

1. 0 x 1″ = 0″

2. 6 x 0.1″ = 0.6″

3. 0 x 0.025″ = 0.000″

4. 4 x 0.001″ = 0.004″

5. 0 + 0.6 + 0.000 + 0.004 = 0.604
Ans. Reading is 0.604″

EXERCISES

Make the readings on the vernier calipers shown. (These figures are all slightly enlarged for ease in reading.)

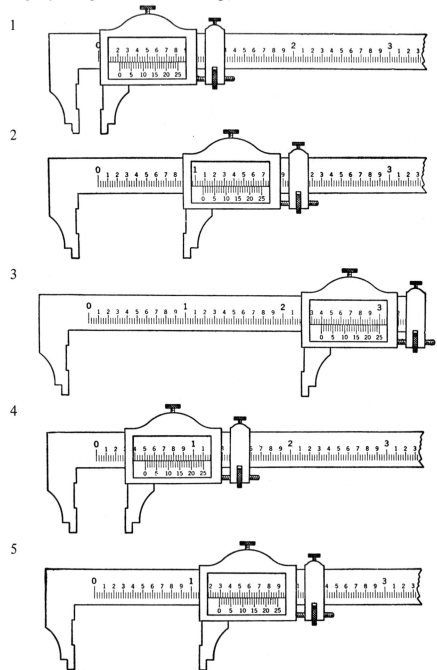

6

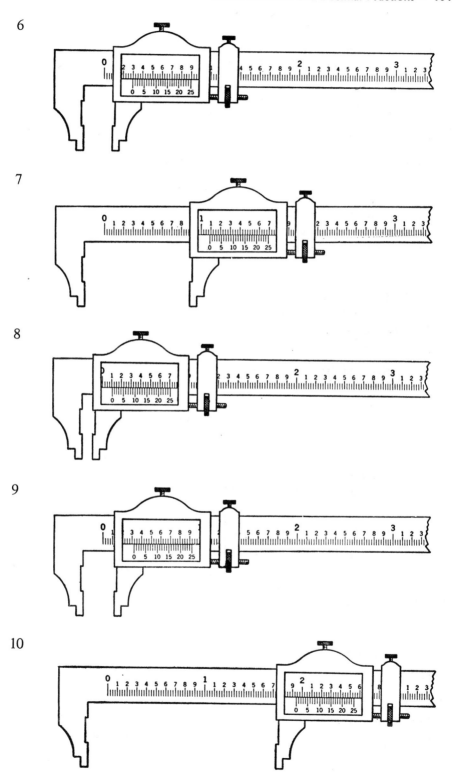

7

8

9

10

Percent, a form of decimal

Certain decimals have been given special names. Hundredths are often called "percent." For example, 0.09, or nine hundredths, is often written as 9% and called "nine percent," and when we say 9% of a number we mean 0.09 times the number.

Expressing percents as decimal numerals

Example 1. Express 52% as a decimal numeral.

THINK	DO THIS
By the definition of percent, 52% means 0.52. Then to write a percent as a decimal numeral:	
1. Drop the "%" symbol.	1. 52
2. Move the decimal point two places to the left. Ordinarily these two things are done in one step.	2. 0.52
	Ans. 52% = 0.52

When no decimal point is written it is understood to be at the right end of the numeral.

Example 2. Express 2 1/2% as a decimal numeral.

THINK	DO THIS
1. If a fraction occurs in the percent, change the fraction to decimal form.	1. 2 1/2% = 2.5%
2. Drop the "%" symbol and move the decimal point two places to the left.	2. 2.5% = 0.025
	Ans. 2 1/2% = 0.025

Example 3. Express 1/8% as a decimal numeral.

THINK	DO THIS
1. Express the fraction in decimal form.	1. 1/8% = 0.125%
2. Drop the "%" symbol and move the decimal point two places to the left.	2. 0.125% = 0.00125
	Ans. 1/8% = 0.000125

Summary

To express a percent as a decimal numeral:

1. If the percent contains a fraction, express that fraction as a decimal numeral.
2. Drop the percent symbol and move the decimal point *two places to the left.*

EXAMPLE

81 1/4% is what decimal numeral?

1. 81 1/4% = 81.25%

2. 81.25% = 0.8125

Ans. 81 1/4% = 0.8125

EXERCISES

Express each of these percents as a decimal numeral.

1. 75%
2. 23%
3. 84%
4. 6%
5. 3 1/2%
6. 1 1/2%
7. 2 1/4%
8. 5 3/4%
9. 2%
10. 12 1/2%
11. 1.5%
12. 150%
13. 3/4%
14. 1/2%
15. 16 2/3%
16. 1/3%

Expressing decimal numerals as percents

Example 1. Express 0.25 as a percent.

THINK

1. Using the principle of 1 we see that 0.25 x 100/100 = 25.00/100 = 25/100.
2. Then by definition of percent 25/100 = 25%.

DO THIS

1. $0.25 \times \frac{100}{100} = \frac{25}{100}$

2. $\frac{25}{100} = 25\%$

Ans. 0.25 = 25%

If we compare this with 0.25 we note that a short cut to the answer would be to move the decimal point *two places to the right* and annex the percent symbol, %.

Example 2. Express 0.0225 as a percent.

THINK

1. Using the short cut, move the decimal point two places to the right.
2. Annex the percent sign.

DO THIS

1. 0.0225

2. 0.0225 = 2.25%

Ans. 0.0225 = 2.25%

Summary	EXAMPLE
To express a decimal numeral as a percent:	1.625 is what percent?
1. Move the decimal point two places to the right.	1. 1.6̲2̲5
2. Annex the percent symbol.	2. 1.625 = 162.5%
	Ans. 1.625 = 162.5%

EXERCISES

Express each of these decimals as a percent.

1.	0.03	5.	0.015	9.	0.6
2.	0.75	6.	0.333	10.	0.02
3.	1.25	7.	0.005	11.	0.0725
4.	0.08	8.	0.10	12.	0.00625

Working with percents

On page 24 we saw that 1/2 of 3/4 equals 3/8. 1/2 of 3/4 means $1/2 \times 3/4 = 3/8$. In the same way when we say 20% of $15 we mean $0.20 \times \$15$, or $20/100 \times \$15 = \$3$. Also by 10% of 320 tons we mean 0.10×320 tons, or $1/10 \times 320$ tons, or 32 tons, etc.

Note carefully here that 0.20 and 15 are *factors* and the result, 3, is the *product* of those factors, also that 0.10 and 320 are factors, and 32 is their product.

Example 1. What is 18% of 250?

THINK	DO THIS
1. Express the percent as a decimal numeral or as a fraction with denominator 100.	1. 18% = 0.18
2. Find the product of this number and the other factor.	2. $0.18 \times 250 = 45$
	Ans. 18% of 250 = 45

Example 2. A company's total costs are $75,000 per week. What is the insurance expense if it is 4.4% of the total cost?

THINK	DO THIS
Here 4.4% of the total cost means 4.4% of $75,000.	
1. Express 4.4% as a decimal numeral.	1. 4.4% = 0.044.
2. Find the product of this number and the other factor, 75,000.	2. $0.044 \times \$75,000 = \$3,300$
	Ans. Insurance expense is $3,300

Example 3. What will be the price of 2 doz. sets of drawing instruments if the list price is $60 per doz. and trade discounts of 20% and 10% are allowed.

THINK

Trade discounts are allowed on the list price. This means that 20% and 10% are allowed on $60. One way to find this is:

1. Find 20% of $60 and deduct it from $60.

2. Find 10% *of this result* and deduct it from $48. This result is the price of one dozen.

3. Find the selling price for 2 doz.

DO THIS

1. 20% of $60 =
   0.20 x $60 = $12.00
   $60 − $12 = $48

2. 10% of $48 =
   0.10 x $48 = $4.80
   $48 − $4.80 = $43.20

3. $43.20 x 2 = $86.40
   Ans. 2 doz. cost $86.40

An alternate solution:

1. Deduct 20% from 100% or 1.
2. Find 10% of this result, and deduct it.

3. Find the resulting percent of the list price. This gives the price of 1 doz.
4. Find the selling price of 2 doz.

1. 100% − 20% = 80%
2. 10% of 80% =
   1/10 x 80% = 8%
   80% − 8% = 72%
3. 0.72 x $60 = $43.20

4. $43.20 x 2 = $86.40
   Ans. 2 doz. cost $86.40

Summary

To find a percent of a number:

1. Express the percent as a decimal numeral or as a fraction.
2. Multiply the number by this result.
3. This product is the required number.

EXAMPLE

What is 4 1/4% of $450?

1. 4 1/4% = 4.25% = 0.0425

2. $450 x 0.0425 = $19.125 or $19.13

Ans. 4 1/4% of $450 = $19.13

EXERCISES

Find the values of the following:

1. 15% of $25
2. 30% of 482 sq. ft.
3. 25% of 7,428 ft.
4. 2 1/2% of 1,275 lbs.
5. 18 1/2% of 2,862 tons

6. 7.3% of 4,277 lbs.
7. 0.5% of 2.175
8. 1.2% of 81,996
9. 0.08% of 2.7304
10. 142% of $29.64

PROBLEMS

46. It is estimated that 12,200 bricks are required for a job. If 15% is allowed for waste, how many additional bricks must be ordered?
47. A building costs $78,500 to construct. What is the excavation cost if it is 2.4% of the total construction cost?
48. An engine costs $1,765 less a 2% discount for cash in ten days. What is the actual cost of the engine if the bill is paid within the ten day period?
49. An electrical contractor buys a supply of wire for a total price of $725 before a price increase of 16.5%. What would he have had to pay for the wire after the price increase?
50. A certain bearing bronze is composed of 82% copper, 16% tin, and 2% zinc. How many pounds of each of these metals are there in 12 cored bronze bars that weigh 96.89 lbs. apiece?
51. The list price of a certain make of 6″ side-cutting pliers (Fig. 3-48) is $35.00 per dozen less discounts of 20%, 10%, and 5%. What is the net cost of a dozen pairs of pliers? *Hint:* Take the first discount, 20%, off the $35.00. Then take the second discount, 10%, off this result. Then take the 5% off the last result.

Fig. 3-48. Side-cutting pliers.

52. A popular make of portable TV lists for $135.75. A dealer offers this set at 7 1/4% off for cash. How much money is saved by paying cash?
53. The efficiency of a generator is 82%. If the input is 75,000 watts, what is the output?
54. What will be the price for 5 dozen 4″ hermaphrodite calipers if the list price is $26.50 per dozen and the trade discounts are 24% and 5% with 2% off for cash?

55. A firm purchases machinery and contracts to pay an interest rate of 6.25% per year on a loan. If purchases of $17,800, $27,000, and $41,850 were made and the loan paid in full nine months later, how much total interest must the firm pay on the loan?

Example 4. A salesman sold $378.50 worth of merchandise on Monday, which was 12% of his average weekly sales. What were his average weekly sales?

THINK

Read this problem carefully. It tells us that 12% of the salesman's average weekly sales amounted to $378.50.

1. Write the statement in mathematical form.

2. Since we know the *product* ($378.50) and *one of the factors* (0.12), we can find the *other factor* by dividing the product 378.50 by the known factor.

DO THIS

1. 12% of [?] = $378.50
   or 0.12 x [?] = $378.50

2. 378.50 ÷ 0.12 =

$$.12\overline{)378.50.00}$$
$$\quad\quad 31\ 54.17$$

Ans. Avg. weekly sales = $3,154.17

Note: One can check the correctness of this solution by showing that 12% of $3,154.17 = $378.50.

Example 5. A contractor received $276.50 for a masonry job. If $82.20 of this was his profit, what was his percent profit?

THINK

Read the problem carefully. Note that $82.20 = [?]% of $276.50.

1. Write the statement in mathematical form.

2. The *product* is 82.20. *One of the factors* is 276.50. *Divide the product by the known factor* to obtain the other factor.

3. Express the result in percent form.

DO THIS

1. 82.20 = [?] x 276.50

2. $\frac{82.20}{276.50}$ = 0.297 approx.

3. .297 = 29.7%
   Ans. $82.20 = 29.7% of $276.50

Summary

To solve problems that involve percents:

1. Read the problem carefully to determine the number of which the percent is taken.

2. Write the statement of the problem in mathematical form.

3. If the product is not known, multiply the factors.

3'. If the product and one factor are known, divide the product by the known factor.

4. Express the result in percent form if required.

EXAMPLE A

Find 6% of 893.

1. 6% of 893

2. 0.06 x 893 = ?

3. 0.06 x 893 = 53.58

          Ans. 6% of 893 = 53.58

EXAMPLE B

84 is 15% of what number?

1. 15% of the required number

2. 84 = 0.15 x [?]

3'. $\dfrac{84}{.15}$ = 560

        Ans. 15% of 560 = 84

EXAMPLE C

125 is what percent of 1,000?

1. The percent is of 1,000.

2. 125 = [?] x 1,000

3'. 125/1000 = 1/8 = 0.125

4. 0.125 = 12.5%

      Ans. 125 is 12.5% of 1,000.

EXERCISES

1. 48 lbs. is 3% of how many pounds?
2. $.85 is 2 1/2% of what amount of money?
3. 375 board feet is 33% of how many board feet?
4. 96 is 2% of what number?
5. 1.237 is 1/2% of what number?
6. What percent of 84 is 42?
7. What percent of $1.25 is $.25?
8. What percent of 18 in. is 4 in.?
9. What percent of a quart is a pint?
10. What percent of 0.185 is 0.0005?
11. What percent of 0.726 is 0.0003?

Often you will have occasion to express in percent certain fractions that occur frequently. Sometimes they are given as decimal numbers and must be expressed in percent or as common fractions. The most common of these equivalents are given in the following table. Complete this table and memorize the equivalents you might use in your work.

Common Fraction	Decimal Numeral	Percent	Common Fraction	Decimal Numeral	Percent
1/2	0.5	50%	?	0.01	?
1/4	0.25	25%	?	0.02	?
1/8	0.125	12 1/2%	?	0.75	?
1/16	?	?	?	?	37 1/2%
1/3	?	?	?	?	62 1/2%
1/6	?	?	?	?	87 1/2%
1/40	?	?	?	0.833	?
1/20	?	?	?	0.075	?

PROBLEMS

56. In the manufacture of semifinished hexagonal steel nuts, 72 nuts were found defective. This was 2% of the total quantity produced. How many hexagonal steel nuts were made?

57. A set of tools was purchased by a mechanic at a savings of $4.50. This savings was a 5% discount of the marked price. What did he pay for the tools?

58. A lathe hand engaged in turning elevator pulleys produced 280 finished pieces on a certain day. This was 80% of his normal output. How many pulleys does he usually produce per day?

59. Because of an overload, the speed of an electric motor is decreased by 75 r.p.m. (revolutions per minute). This represents a loss in speed of 5% of the full load speed. What is the full load speed of the motor?

60. A contractor receives a bill for building materials which is $125 or 3% in excess of the original estimate. Find the amount of the original estimate and also the total amount of the bill.

61. A building contractor is paid $81,800 for a building which his firm constructed. His expenses are $37,300 for labor, $32,900 for materials and $4,100 for miscellaneous expenses. What percent profit does he make on this job?

62. In order to make a tank, $42.78 was spent on materials and $63.80 on labor. What percent of the total cost was spent on materials? What percent was spent on labor?

63. During the month of September the number of accidents in a certain factory employing 675 men was 50. At the close of the following month after an intensive safety drive the number of accidents was reduced to 5. Express in percent the number of accidents each month. By what percent were the accidents reduced?

64. A lumber and building materials dealer receives a shipment of lumber that is billed to him at $3,450. If he sells this lumber for $4,750, what percent of the cost price does he receive to cover expenses and profit?

65. An engine is rebored and increased in compression ratio. Prior to the engine rebuild, the engine developed 275 horsepower. The rebuild resulted in an increase of 12 horsepower. By what percent is the original horsepower increased?

66. If 1,250 tons of iron ore yield 125 tons of cast iron, the weight of the iron is what percent of the weight of the ore?

67. Ninety lbs. of tin, 7 lbs. of antimony, and 3 lbs. of copper are required to make 100 lbs. of Babbitt metal. Find the percent of each of the metals in Babbitt metal.

68. The input to an electrical motor is 95 h.p. (horsepower). The actual output of the motor at full load is 87 h.p. What percent of the input (% efficiency) is actually available for use?

REVIEW PROBLEMS

1. A shop receives an order to make 65 copper oxyacetylene cutting tips (Fig. 3-49). What is the decimal equivalent of each of the dimensions given in fractional form? How many feet of copper will be required for this job if 0.125″ is to be allowed for cutting off and facing? What diameter stock should be ordered if a total cut of 0.035″ is required to reduce the stock to the 0.745″ diameter?

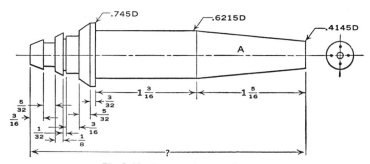

Fig. 3-49. Oxyacetylene cutting tip.

2. Supply the missing dimension in Fig. 3-50.

3. Redraw Fig. 3-51 and give all dimensions in fractional form. (Supply the missing dimensions in decimal form first.)

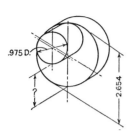

Fig. 3-50. Eccentric pin.

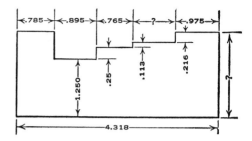

Fig. 3-51. Gage.

4. A shipyard employing 3,450 men finds it necessary, because of increased business, to increase the number of its employees by 18%. How many additional employees are required?

5. A shop order calls for 540 steel pins each 1.385″ long. If 1/8″ is to be allowed for cutting off and facing, how many pins can be obtained from a 12′ length of material? How many 12′ lengths are needed for the order?

6. Johansson gage blocks, similar to those shown in Fig. 3-52, were used to check a dovetail: 1.000″, 0.750″, 0.1003″, and 0.119″. What was the distance measured by these blocks?

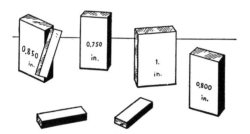

Fig. 3-52. Johansson gage blocks.

7. The actual internal diameter of 5″ standard wrought pipe (Fig. 3-53) is 5.047″. The external diameter is 5.563″. What is the thickness of the wall?

8. Compute the depth of the tooth of the gear in Fig. 3-54.

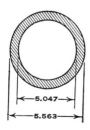

Fig. 3-53. Cross section of a 5″ wrought pipe.

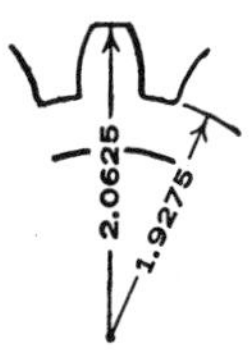

Fig. 3-54. Gear tooth.

Fig. 3-55. Taper.

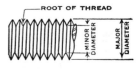

Fig. 3-56.

9. Find the taper (difference in diameters) of the piece of stock in Fig. 3-55. If the stock is 8″ long, what is the taper per inch?

10. The distance across the flats of a finished hexagonal-head bolt is 0.6875″. The stock used to make these bolts is 15/16″ in diameter. Find the depth of the cut required to mill each flat.

11. Find the diameter at the root of the thread (the minor diameter) shown in Fig. 3-56 if the major diameter is 0.0723″ and the depth of the thread is 0.0096″.

12. A high-speed steel rod 1 1/4″ in diameter weighs 0.393 lb. per lineal inch. How many feet are there in a shipment weighing 250 lbs.?

13. In Fig. 3-57 find (a) the length of the threaded section and (b) the length of the knurl.

14. The copper cable terminal in Fig. 3-58 has the following dimensions: C = 0.263″, K = 0.200″, G = 0.9375″.

    (a) Find the thickness of the wall of the tubing from which it is made.

    (b) How many terminals can be made from a 12 ft. length of tubing? Allow 1/16″ for cutting off each terminal.

    (c) What is the length of the piece left over?

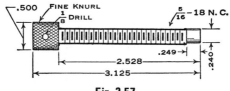

Fig. 3-57.

15. Linotype metal is composed of 12% antimony, 5% tin, and 83% lead. How many pounds of each metal are required to make 355 lbs. of this alloy?

16. A patternmaker allows 1/8″ per lineal foot for shrinkage. Find the percent allowance.

17. If the clearance between the punch and die for perforating hard rolled steel (Fig. 3-59) is equal to 1/14 of the thickness of the stock to be perforated, find the clearance for stock 0.80″ thick.

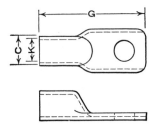

Fig. 3-58. Copper cable terminal.

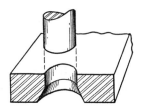

Fig. 3-59. Punch and die.

18. Supply the dimensions that are missing from the gage shown in Fig. 3-60.

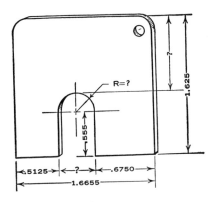

Fig. 3-60. Gage.

19. Figure 3-61 shows an American National Form Thread with a pitch of 1/7″. Find the flat at the top and bottom of the thread if the flat is 0.125 times the pitch. Express your answer to the nearest ten-thousandth of an inch.

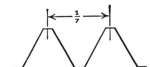

Fig. 3-61. American National Form thread.

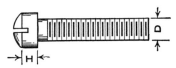

Fig. 3-62. Flat fillister-head machine screw.

20. A storage tank (capacity 66 cu. ft.) is filled with gasoline costing $.749 per gal. What does it cost to fill the tank? (Use 1 cu. ft. = 7 1/2 gal.)

21. Of 3,673 pins produced by an automatic screw machine, 169 were rejected by the inspectors as not conforming to specifications. What percent of the total output was rejected?

22. An electric motor delivers 7.5 h.p., which is 82% of the input. What is the h.p. input to the nearest hundredth?

23. Flat steel 1/8″ thick and 5″ wide is listed as weighing 2.13 lb. per lineal foot. What is the weight of 28 1/2″?

24. The thickness of the head of a flat fillister head machine screw (Fig. 3-62) is 0.660 times the diameter of the body minus 0.002″. What is the thickness of the head of a screw having a body diameter of 0.32″? ($H = 0.660D - 0.002$.)

4 Measuring Surfaces and Finding Volumes and Capacities

In the preceding units we measured the lengths of lines and saw how, in order to do that job well, we must know how to work with fractions and decimals. Frequently in industry we need to find the amount of surface contained within a figure. Sometimes we must find the capacity of a box or can or tank. We shall learn how to do these things in the present unit.

Measuring the surface in a rectangular area

We measure a surface by finding the number of surface units contained in it. One unit of surface is the *square yard*. This is a plane surface equivalent to that enclosed by a square which is one yard on each side.

The *square foot* is a plane surface equivalent to that enclosed by a square which is one foot on each side.

The *square inch* is a plane surface equivalent to that enclosed by a square which is one inch on each side. Figure 4-1 shows a square that is one inch on each side. The area contained within this square is a *square inch.*

Fig. 4-1. One square inch.

Units of Area

			Abbreviated		
1 square yard	= 1,296 square inches		1 sq. yd.	= 1,296 sq. in.
1 square yard	= 9 square feet		1 sq. yd.	= 9 sq. ft.
1 square foot	= 144 square inches		1 sq. ft.	= 144 sq. in.

EXERCISES

1. How many square inches are there in 2 sq. ft.? In 3 1/2 sq. ft.?
2. How many square feet are there in 5 sq. yd.? In 73 sq. yd.?
3. How many square feet are there in 720 sq. in.?
4. How many square yards are there in 216 sq. ft.?
5. How many square yards are there in 514 sq. ft.? Express the result in decimal form.

Another common, but much larger, unit of area is the *acre*, which contains 43,560 sq. ft.

Finding the area of a rectangle

A *rectangle* is a four-sided plane figure with four right angles (Fig. 4-2).

Fig. 4-2. Rectangle.

A *square* is a special kind of rectangle in which the sides are all equal. A square that is 5 in. on each side is sometimes referred to as a 5-in. square; a square that is 4 in. on each side may be called a 4-in. square.

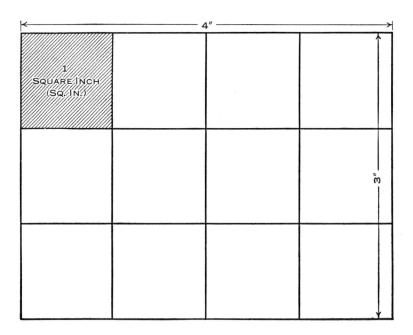

Fig. 4-3.

To find the area of a rectangle we must find the number of surface units contained in the rectangle.

In Fig. 4-3 we note that the length is 4 in. and the width 3 in. There are 4 sq. in. in each row and three such rows. The rectangle contains 3 x 4 or 12 sq. in. We multiply the number of units in the length by the number of units in the width and express the result in square units. This is often stated simply as "multiply the length by the width."

Example. Find the area of a rectangular room that is 8 1/2 ft. wide and 12 ft. long.

THINK	DO THIS
Since the area of a rectangle is the product of the length and width, multiply 8 1/2 by 12 and express the result in square feet.	8 1/2 x 12 = 102 *Ans.* 102 sq. ft.

Summary	EXAMPLE
The area (A) or the number of units of surface in a rectangle equals the length (l) times the width (w). The length and width must both be expressed in terms of the same linear unit. The resulting area must be expressed in square units. ($A = lw$ square units.)	Find the area of a rectangle that is 8 1/2 in. wide and 11 1/3 in. long. Solution: $11\frac{1}{3} \times 8\frac{1}{2} = \frac{34}{3} \times \frac{17}{2}$ $\frac{34}{3} \times \frac{17}{2} = \frac{578}{6}$ or $96\frac{1}{3}$ *Ans.* Area = 96 1/3 sq. in.

EXERCISES

Find the areas of rectangles with the following dimensions:

1. Length 6″, width 12″
2. Length 15″, width 18″
3. Length 8 1/2″, width 15″
4. Length 7.3′, width 8.7′
5. Length 9.7′, width 12.8′
6. Length 4.25″, width 11.2″
7. Length 9 1/3′, width 8 1/2′
8. Length 12.6′, width 12.6′
9. Length 18′-4″, width 19′-3″*
10. Length 5′-9″, width 7′-6″

11. What is the difference in square inches between the area of a 5-inch square and 5 square inches?

Find the widths of rectangles with the following dimensions:

12. Area 127.5 sq. in., length 15 in.
13. Area 63.51 sq. ft., length 8.7 ft.
14. Area 7,639.7 sq. ft., length 120.5 ft.

*Change these dimensions to feet before finding the area.

PROBLEMS

1. Each of the full-size drawings in Fig. 4-4 represents a cross section of a hard-drawn copper rectangular bus bar similar to that used in manufacturing electrical equipment for power stations. Measure each to the nearest 1/16″, then find each cross-sectional area.

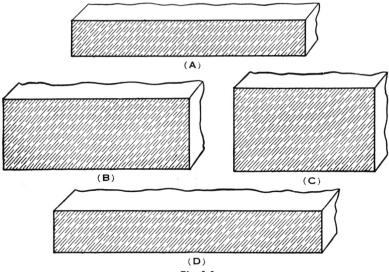

(A)

(B)　(C)

(D)

Fig. 4-4.

2. The driveway leading to the garage at the rear of a house is 95 ft. long and 13 ft. wide. How many square yards of the lot are used for this driveway?

3. (a) How many square yards of linoleum are required to cover the floor of the kitchen shown in Fig. 4-5? Allow 5% for matching.
   (b) Matched flooring is to be used in the living room. If 33 1/3% is allowed for waste and matching, how many square feet of flooring are required?

Fig. 4-5. Floor plan of a one-family house.

(c) If the basement measures 21'-6" x 29'-6", how many tiles 9" square will be required to cover the concrete floor?

4. A one-family house has seventeen double-hung window frames. Each sash contains a light of glass 28" x 34". How many square feet of glass were used for the windows of this house?

5. Find the cross-sectional area of the step block shown in Fig. 4-6. *Hint:* Assume that surface A of the step block is made up of two rectangles.

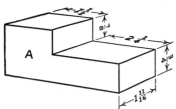

Fig. 4-6. Step block.

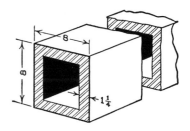

Fig. 4-7. Section of a hollow steel beam.

6. What is the cross-sectional area of the steel in the hollow beam shown in Fig. 4-7?

7. A partition wall 9' high and 23' long is to be plastered on both sides. If an allowance for a door opening 4' x 7' must be made, how many square yards of surface will be plastered?

8. A 30" x 60" sheet of flat plain copper weighs 3 1/2 lb. per square foot. Find the area of one sheet in square feet. How much would 17 sheets of copper weigh?

9. How many bricks are required to build the wall shown in Fig. 4-8? Assume that there are 15 bricks per square foot for a wall 8" thick.

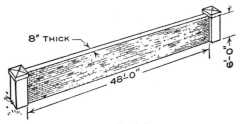

Fig. 4-8.

10. Find the number of sheets of plasterboard required to cover a ceiling 19 1/2' wide and 39' long if each piece is 4' x 8'. What will be the cost of this material if plasterboard of the type used lists at $3.92 per sheet?

Finding the area of a parallelogram

Parallel lines are lines that are in the same plane but never meet even if extended indefinitely. The line segments in Fig. 4-9 are parallel.

Fig. 4-9. Parallel line segments.

A *parallelogram* is a four-sided plane figure that has two pairs of parallel sides. Figure 4-10 is a parallelogram. The perpendicular distance *a* between two of the parallel sides is the altitude of the parallelogram. The side AB is the base.

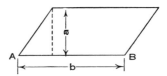

Fig. 4-10. Parallelogram.

The *area of a parallelogram* is the produce of the number of units in the base and the number of units in the altitude (A_p = *ba*) expressed in square units.

Example. Find the area of a parallelogram that has an altitude of 7 3" and a base of 8.4".

THINK	DO THIS
The area of the parallelogram is the product of the base and altitude, both expressed in terms of the same linear unit. We multiply the base 8.4" by the altitude 7.3".	8.4 7.3 252 588 61.32 or 61.3
We expressed the area in tenths since the length and width were expressed to tenths. A product is expressed to the same number of decimal places as the dimension with the smaller number of places.	*Ans.* Area = 61.3 sq. in.

Summary

The area of a parallelogram (A_p) equals the base (b) times the altitude (a). The result must be expressed in square units.

1. Express the dimensions in the same linear unit.
2. Find their product and express this result in square units.

EXAMPLE

A parallelogram is 2'-6" wide and 5'-3" long. What is its area?

1. width 2'-6" = 2 1/2' = alt.
   length 5'-3" = 5 1/4' = base

2. $5\frac{1}{4} \times 2\frac{1}{2} = \frac{21}{4} \times \frac{5}{2} = \frac{105}{8}$

$\frac{105}{8} = 13\frac{1}{8}$ or $13\frac{1}{8}$ sq. ft.

Ans. Area = 13 1/8 sq. ft.

EXERCISES

Find the areas of the parallelograms given in Exercises 1-10.

1. Altitude 6", base 20"
2. Altitude 15", base 54"
3. Altitude 11", base 24"
4. Altitude 9 3/4", base 10 1/2"
5. Altitude 5 1/4", base 8"

6. Altitude 6.4", base 8.3"
7. Altitude 8', base 3'-6"
8. Altitude 7.2', base 8.9'
9. Altitude 6'-4", base 10"-9"
10. Base 8.7', altitude 4.5'

Find the altitudes of the parallelograms in Exercises 11-14. *Hint:* Since $A_p = ba$, $a = A_p \div b$.

11. Area 810 sq. in., base 54 in.
12. Area 264 sq. in., base 24 in.
13. Area 28 sq. ft., base 3 1/2 ft.
14. Area 4,677.12 sq. ft., base 128 ft.

PROBLEMS

11. Find the cross-sectional area of the piece of tool steel shown in Fig. 4-11.

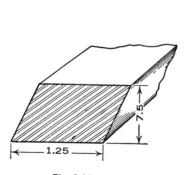

Fig. 4-11.

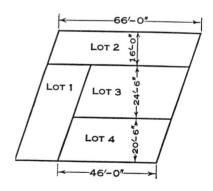

Fig. 4-12.

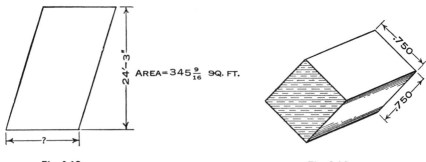

Fig. 4-13.   Fig. 4-14.

12. A large plot of ground is to be divided into four smaller lots, each having the shape of a parallelogram as in Fig. 4-12. What will be the area of each smaller lot? What is the area of the large lot?

13. The area of a parallelogram is 125 sq. in. The altitude is 27.8 in. Find the base.

14. Find the base of the parallelogram in Fig. 4-13.

15. Find the cross-sectional area of the square brass rod in Fig. 4-14.

Finding the area of a triangle

A *triangle* is a figure of the sort shown in Fig. 4-15. It is a plane figure having three sides and three angles. Each point where two sides meet is called a vertex of the triangle.

Look carefully at Fig. 4-16. We started with the triangle ABC. Then we drew line BD parallel to AC and line CD parallel to AB. We thus made a parallelogram. The altitude of the parallelogram is a, and its base is AC.

The altitude of the triangle is a and its base is AC.

The area of the triangle is just half the area of the parallelogram.

The area of the parallelogram is the product of the base and altitude $(A_p = ba)$.

Since the triangle is just half the parallelogram, its area is one-half the product of the base and altitude $(A_t = 1/2ba)$ expressed in square units.

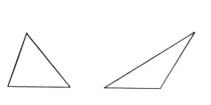

Fig. 4-15. Triangles.

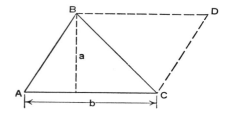

Fig. 4-16.

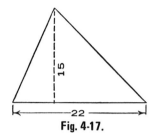

Fig. 4-17.

Example. Find the area of the triangle represented in Fig. 4-17.

THINK	DO THIS
The area of a triangle is half the product of its base and altitude.	1/2 x 22 x 15 = 165 *Ans.* Area = 165 sq. in.

Summary	EXAMPLE
The altitude of a triangle is the length of the perpendicular from one vertex to the opposite side. The opposite side we call the base of the triangle.	A triangle has a base of 8'-6" and the altitude to this base is 6'-6". What is its area?
The area of the triangle is half the product of the base and altitude ($A_t = 1/2ba$) expressed in square units.	
1. Express each dimension in the same linear unit.	1. base = 8'-6" or 8 1/2' alt. = 6'-6" or 6 1/2'
2. Find the product of 1/2 the base and the altitude, and express this result in square units.	2. Area = 1/2 x 8 1/2 x 6 1/2 $\frac{1}{2}$ x $\frac{17}{2}$ x $\frac{13}{2}$ = $27\frac{5}{8}$ *Ans.* Area = 27 5/8 sq. ft.

EXERCISES

Find the areas of the triangles whose dimensions are given in Exercises 1-10.

1. Altitude 8", base 5"
2. Altitude 12 1/2", base 6"
3. Altitude 18", base 9.3"
4. Altitude 15 ft., base 12 ft.
5. Altitude 10 ft. 6 in., base 14 ft. 6 in.

6. Altitude 18.3 ft., base 24.6 ft.
7. Altitude 4.51 ft., base 8.62 ft.
8. Altitude 22.8 ft., base 18.4 ft.
9. Altitude 10.62 ft., base 3.12 ft.
10. Altitude 8 3/8", base 4 5/16"

PROBLEMS

16. Find the area in square yards of the strip of land shown in Fig. 4-18.
17. What is the cross-sectional area of the phosphor bronze pivot shown in Fig. 4-19? The altitude is 0.190".
18. Find the area of section *A* of the roof of the one-car garage shown in Fig. 4-20. (Altitude of section *A* is approximately 10 ft. 1 in.) Also find the area of the floor for this garage in square yards.

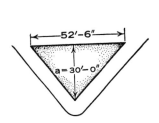

Fig. 4-18. Triangular strip of land.

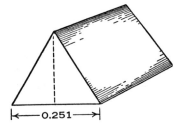

Fig. 4-19. Phosphor bronze pivot greatly enlarged.

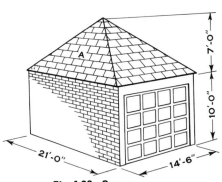

Fig. 4-20. Garage.

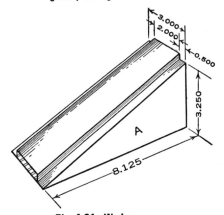

Fig. 4-21. Wedge.

19. Find the area of surface *A* of the wedge shown in Fig. 4-21.
20. (a) Find the area of the gable shown in Fig. 4-22. Make the necessary deduction for the window.
    (b) Find the total area of this end of the building to be covered by shingles. Deduct for all openings.
21. Find the area of the triangular plot of ground in Fig. 4-23.

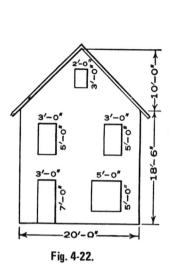

Fig. 4-22.

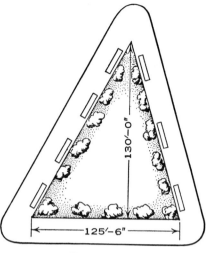

Fig. 4-23.

Finding the area of a trapezoid

A *trapezoid* is a four-sided plane figure with one pair of parallel sides. *AECD* in Fig. 4-24 is a trapezoid. The parallel sides are *AE* and *DC.* The other two sides are not parallel.

The altitude of a trapezoid is the perpendicular distance between the two parallel sides. In trapezoid *AECD*, the altitude is *a.* The parallel sides *AE* and *CD* are the bases of the trapezoid.

The area of a trapezoid is one-half the product of the altitude and the sum of the bases:

$$A_{tz} = 1/2a\,(b + B)$$

Example. Find the area of the trapezoid represented in Fig. 4-25.

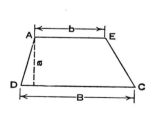

Fig. 4-24. Trapezoid.

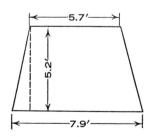

Fig. 4-25.

THINK	DO THIS

The area of a trapezoid is one-half the product of the altitude and the sum of the bases.

1. Find the sum of the bases.

2. Find one-half the product of the altitude and this sum.

1. $5.7 + 7.9 = 13.6$

2. $1/2 \times 5.2 \times 13.6 =$

$$\begin{array}{r} 13.6 \\ 5.2 \\ \hline 27\,2 \\ 680 \\ \hline 70.72 \end{array}$$

$\frac{1}{2} \times 70.72 = \frac{70.72}{2} = 35.36$

3. Since the dimensions are given to tenths, the product is expressed to the nearest tenth.

3. 35.36 to nearest tenth = 35.4

Ans. Area = 35.4 sq. ft.

Summary

EXAMPLE

A trapezoid is a four-sided plane figure with one pair of parallel sides. The area of a trapezoid is one-half the product of the altitude and sum of the bases, $A_{tz} = 1/2a\,(b + B)$, expressed in square units.

A trapezoid has altitude 28.6″ and bases 14.3″ and 19.4″. Find its area.

1. Express all dimensions in the same linear unit.

1. $b = 14.3; B = 19.4$

2. Find the sum of the bases.

2. $b + B = 14.3 + 19.4 = 33.7$

3. Find one-half the product of the altitude and the sum of the bases, and express this result in square units.

3. Area = $1/2 \times 28.6 \times 33.7 = 481.91$
Area expressed to tenths = 481.9

Ans. Area = 481.9 sq. in.

EXERCISES

Find the areas of the trapezoids whose dimensions are given in Exercises 1-8.

1. Bases 8″ and 12″, altitude 4″
2. Bases 15″ and 18″, altitude 8 1/2″
3. Bases 72′ and 84′, altitude 53 1/2′
4. Altitude 32 yd., bases 75 yd. and 125 yd.

5. Altitude 51 yd., bases 100 yd. and 150 yd.
6. Bases 152 ft. 6 in. and 275 ft. 6 in., altitude 87 ft. 4 in.
7. Bases 92 ft. 9 in. and 146 ft. 8 in., altitude 68 ft. 6 in.
8. Altitude 43.6″, bases 61.5″ and 58.1″

PROBLEMS

22. Find the cross-sectional area of the metal that was removed from the rectangular bar of steel to make the dovetail slide shown in Fig. 4-26.

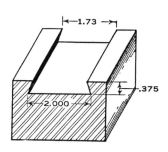

Fig. 4-26. Dovetail slide.

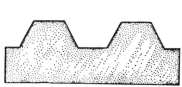

Fig. 4-27. Stairway.

23. How many square yards are in the section of the wall beneath the stairway in Fig. 4-27?

24. The hip roof illustrated in Fig. 4-28 is to be covered with slate-surfaced asphalt hexagonal shingles (Fig. 4-29). If the shingles to be used come in bundles covering 33 1/3 sq. ft. and weighing 73 lb. per bundle, find (a) the number of bundles required, and (b) the amount of load on the roof created by the shingles.

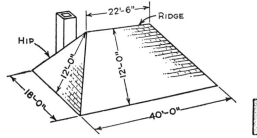

Fig. 4-28. Hip roof.

Fig. 4-29. Hexagonal shingle.

25. Find the cross-sectional area of the concrete dam shown in Fig. 4-30.
26. Find the cross-sectional area of the steel slide bar shown in Fig. 4-31.
27. Find the area of the face A of the key shown in Fig. 4-32.
28. Figure 4-33 shows a portion of a plan of city lots. Find the area of each plot. Express your answer in square feet and square yards. What is the area of the entire plot in acres? (1 acre = 43,560 sq. ft.)

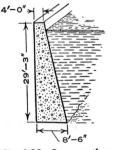

Fig. 4-30. Cross section of a
concrete dam.

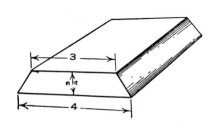

Fig. 4-31. Steel slide bar.

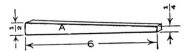

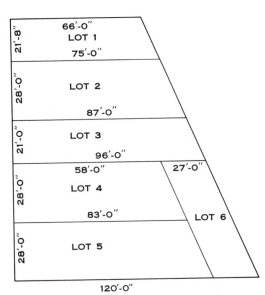

Fig. 4-33. Plan of city lots.

29. How many rolls of slate-surfaced asphalt roofing will be required to cover the construction shed illustrated in Fig. 4-34 if each roll actually covers 100 sq. ft., including allowance for lapping? The door and roof are to be covered. The shed has only one window.

30. Find the area of the template shown in Fig. 4-35.

31. How many bundles of four-in-one shingles will be required to cover the end of a house with a gambrel-type roof (Fig. 4-36) if there are six window openings, 2'-10" x 5'-10", and one door opening, 3'0" x 6'-8"? Each bundle will cover 50 sq. ft. and provide a 2" headlap. Exposed foundation is 4' high.

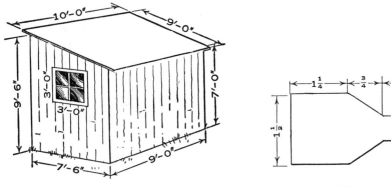

Fig. 4-34. Construction shed.   Fig. 4-35. Template.

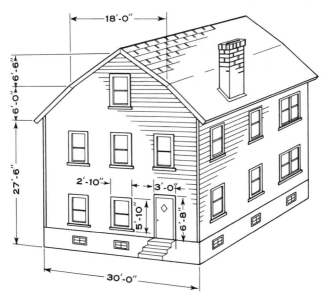

Fig. 4-36. House with gambrel roof.

Finding the area of a circle

A *circle* is a curve, every point of which is at a fixed distance from a point called the *center*. In Fig. 4-37, *ABCRDE* is a circle whose center is *O*.

Fig. 4-37. Circle.

A *radius* (plural, radii) is a line drawn from the center O to the circle. OB is a radius.

A *diameter* is the line drawn from one point on the circle to another point on the circle and passing through the center of the circle. AR is a diameter. Note that the length of the diameter equals the length of two radii of the same circle ($d = 2r$).

The area of a circle is equal to the square of the radius times π. π is a Greek letter pronounced "pi," and its value is the ratio of the *circumference*, or length, of the circle to its diameter, approximately 3.1416.

$$A_c = \pi r^2$$

Note that r^2 (read "*r squared*"), means r x r.

The area of a circle is also equal to 0.7854 times the square of the diameter. $A_c = 0.785d^2$. $(0.7854 = 3.1416 \div 4)$

We often use the word radius to mean the length of the radius. Likewise, we often use the term diameter to mean the length of the diameter and circumference to mean the length of the circle.

Example. Find the area of a circle which has a radius of 7.5".

THINK	DO THIS
The area of a circle equals π (3.1416) times the square of the radius, or 3.1416 times r x r.	
1. Square the radius; that is r x r.	1. r x r = 7.5 x 7.5 = 56.25
2. Multiply this result by π (3.1416). Since the radius of the circle in this example is given in tenths, it is sufficiently accurate to use 3.14 instead of 3.1416 here.	2. 56.25 x 3.14 56.25 3.14 22500 5625 16875 176.6250
3. Express this result in square units, to the nearest tenth.	3. 176.625 to the nearest tenth equals 176.6. *Ans.* Area = 176.6 sq. in

Summary

The area of a circle is the square of the radius expressed in square units, multiplied by π.

$A_c = \pi r^2$. An approximate value for π is 3.1416.

It is approximate to express π to one decimal place more than is present in the value of the radius.

EXAMPLE

What is the area of a circle of radius 6.4'?

1. $r = 6.4'$

2. $r^2 = 6.4 \times 6.4 = 40.96$

$$
\begin{array}{r}
6.4 \\
6.4 \\
\hline
256 \\
384 \\
\hline
40.96
\end{array}
\qquad
\begin{array}{r}
40.96 \\
3.14 \\
\hline
16384 \\
4096 \\
12288 \\
\hline
128.6144 \text{ or } 128.6
\end{array}
$$

Ans. Area = 128.6 sq. ft.

EXERCISES

Find the area of each of the circles indicated in Exercises 1-10.

1. Radius 12"
2. Radius 15"
3. Radius 4.5"
4. Radius 5.8"
5. Radius 7'

6. Radius 7'-6"
7. Radius 6.8'
8. Diameter 15.4'
9. Diameter 9 1/2'
10. Diameter 10.9'

Measure to the nearest 1/16" the diameter of each of the following circles and find the area:

11.

12.

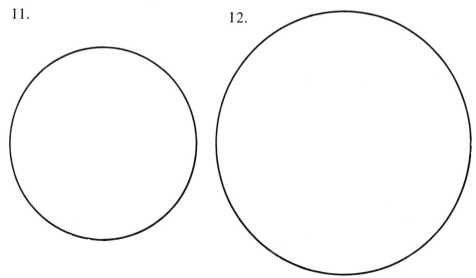

13. 14.

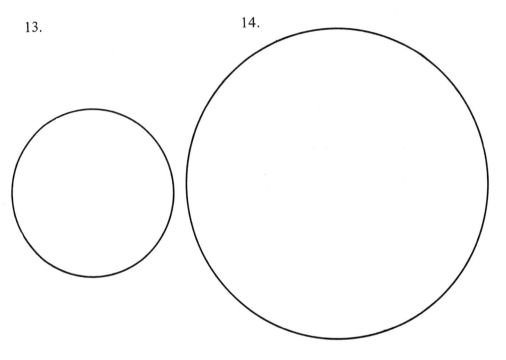

PROBLEMS

32. The bore of each cylinder of a six-cylinder automobile engine requires a piston having a diameter of 3 7/16″ (Fig. 4-38). What is the area of each piston head? Express your answer correct to three decimal places.

Fig. 4-38. Piston.

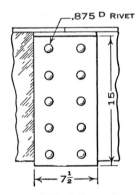

Fig. 4-39. Splice plate.

33. Find the cross-sectional area of each of the rivets used in the splice plate shown in Fig. 4-39. What is the total area occupied by rivets? What is the area of the splice plate, not including the area of the rivet holes?

34. Find the cross-sectional area of each size of the hot galvanized, black enameled electrical conduit (Fig. 4-40) shown in the following table:

| Size in inches | Diameters in inches | | Area of steel in sq. in. |
	External	Internal	
1/2	0.840	0.622	?
3/4	1.050	0.824	?
1	1.315	1.049	?
1 1/2	1.900	1.610	?

Express your answers correct to three decimal places. (*Hint:* Find area of outside circle, subtract area of inside circle.)

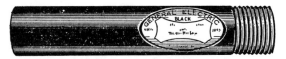

Fig. 4-40. Hot galvanized, black enameled conduit.

35. The diameter of the small end of a taper plug gage (Fig. 4-41) is 3.2″. The diameter of the large end of the taper is 4.5″. Find the area at each end of the tapered section, to the nearest 1/10 sq. in. If the taper is 5″ long, what is the decrease per inch?

Fig. 4-41. Taper plug gage.

36. Find the area of each of the openings in the rod support illustrated in Fig. 4-42.

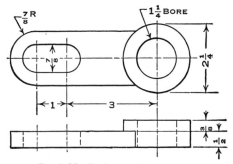

Fig. 4-42. Rod support.

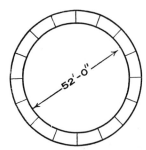

Fig. 4-43. Circular cement walk.

37. How much will it cost to construct a circular concrete walk 5 1/2′ wide (Fig. 4-43) if the cost of the labor and the materials is to be estimated at $.55 per square foot?

38. A gasoline storage tank 7' in diameter and 18' long (Fig. 4-44) is to be painted with two coats of paint, each gallon of which has a maximum coverage of 400 sq. ft. for two coats. If paint of this quality sells for $5.79 per gallon, what will be the cost of the paint required to do the job? (Compute to the nearest quarter gallon.) The area of the curved surface of the cylinder equals circumference times length. Find circumference (length of the circle) by multiplying π times diameter. Remember that the total area equals the area of the curved surface plus the area of the two bases.

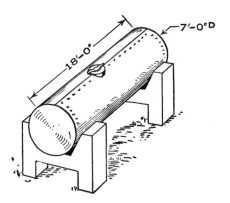

Fig. 4-44. Gasoline storage tank.

39. Find the actual area of the end of the jig shown in Fig. 4-45. The center hole of the jig is 1″ in diameter, whereas the remaining four holes are each 3/8″ in diameter.

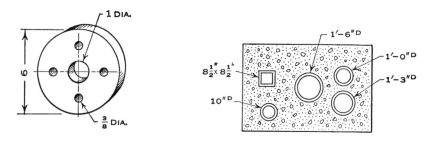

Fig. 4-45. Jig.   **Fig. 4-46. Cross section of a chimney.**

40. A 6'-6″ x 4'-3″ chimney made of stone and rubble has four circular openings or flues and one square flue, as shown in Fig. 4-46. Find the area of each opening and the cross-sectional area of the stone and rubble used to make the chimney. The thickness of the wall of each flue is 1 1/2″.

41. Find the area of the template shown in Fig. 4-47.

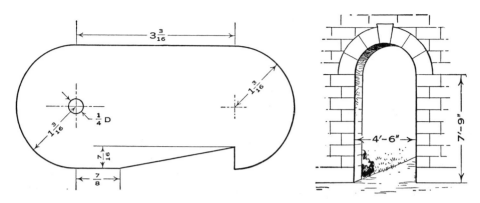

Fig. 4-47. Template.   Fig. 4-48. Opening in brick wall.

42. What is the area of the opening in the brick wall illustrated in Fig. 4-48?

Finding the volume or cubical contents of a box

A *cube* is a rectangular box whose sides or faces are all squares (see Fig. 4-49). If you count carefully, you will find that a cube has six faces, twelve edges and eight vertices.

A *cubic yard* is the volume equivalent to that enclosed by a cube one yard on each edge.

A *cubic foot* is the volume or amount of space equivalent to that enclosed by a cube one foot on each edge.

The volume or amount of space equivalent to that enclosed by a cube one inch on each edge is a *cubic inch*. Figure 4-49 represents a cube each face of which is a one-inch square. Each edge of the cube is one inch long. The space enclosed by the cube pictured in Fig. 4-49 is a cubic inch.

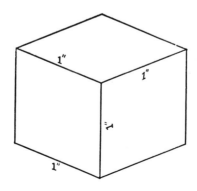

Fig. 4-49. 1-inch cube, or 1 cubic inch.

Units of Volume

		Abbreviated
1 cubic yard	= 46,656 cubic inches	1 cu. yd. = 46,656 cu. in.
1 cubic yard	= 27 cubic feet	1 cu. yd. = 27 cu. ft.
1 cubic foot	= 1,728 cubic inches	1 cu. ft. = 1,728 cu. in.

A box-like figure whose sides are all rectangles is a *rectangular prism*, or simply a prism.

To find the volume of a box or prism, study Fig. 4-50. Notice that on the bottom layer there are 4 cubic inches in a row and 2 such rows, or 4 x 2 = 8 cubic inches. There are three such layers. Contained in this box are 4 x 2 x 3 or 24 cubic inches. The volume of the box is 24 cubic inches.

We could have obtained this result by multiplying the length, width, and thickness or depth of the box and expressing the result in cubic units.

The volume of any rectangular object or box is the length times the width times the thickness or height $(V_b = lwh)$. The result must be expressed in cubic units such as the cubic inch, cubic foot or cubic yard.

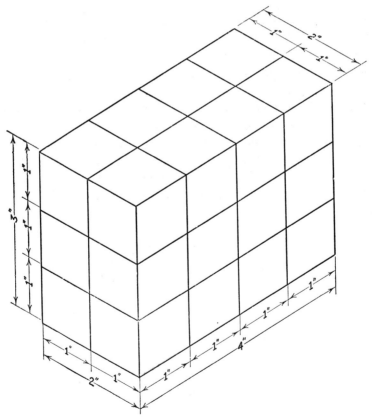

Fig. 4-50. Rectangular prism.

Example. Find the volume of a rectangular bin that is 8 ft. long, 6.5 ft. wide, and 4.5 ft. deep.

THINK

The rectangular bin is a rectangular prism. Its volume is the product of its length, width and depth.

1. Express all dimensions in the same linear units.

2. Find the product of the length, width and depth.

3. Express the product in cubic units.

Note: Since the length is given in whole feet, the result is expressed the same way, in whole cubic feet.

DO THIS

1. length = 8'
   width = 6.5'
   depth = 4.5'

2. Volume = 8 x 6.5 x 4.5

   $$\begin{array}{rr} 6.5 & 52.0 \\ \underline{8} & \underline{4.5} \\ 52.0 & 2600 \\ & \underline{2080} \\ & 234.00 \end{array}$$

3. Volume = 234 cu. ft.

Ans. 234 cu. ft.

Summary

To find the volume of an object in the shape of a rectangular prism:

1. Express all dimensions in the same linear unit.

2. Find the product of the length, width and thickness, or depth:
   $V_{rp} = lwh$.

3. Express the product in cubic units.

EXAMPLE

Find the volume of a rectangular packing case that is 4.6' long, 2.8' wide and 2.6' deep.

1. length = 4.6'
   width = 2.8'
   depth = 2.6'

2. Volume = 4.6 x 2.8 x 2.6

   $$\begin{array}{rr} 4.6 & 12.88 \\ \underline{2.8} & \underline{2.6} \\ 368 & 7\,728 \\ \underline{92} & \underline{25\,76} \\ 12.88 & 33.488 \text{ or } 33.5 \end{array}$$

3. Volume = 33.5 cu. ft.

Ans. 33.5 cu. ft.

EXERCISES

1. How many cubic feet are there in 18 cu. yd.? In 40 cu. yd? In 52 cu. yd.?
2. If a truck carries 2 cu. yd., how many cubic feet does it contain?
3. How many cubic yards are there in 81 cu. ft.? In 243 cu. ft.? In 275 cu. ft.?
4. How many cubic inches are there in 4 cu. ft.? In 10.5 cu. ft.?
5. How many cubic feet are there in 864 cu. in.? In 20,763 cu. in?

Find the volume of each of the rectangular boxes with dimensions as follows:

6. Length 8″, width 15″, depth 10″
7. Length 15″, width 24″, depth 12″
8. Length 4′, width 6′, height 8′
9. Length 8.2′, width 4.5′, depth 10′
10. Length 6 ft.-4 in., breadth 6 ft.-4 in., depth 12 ft.-6 in.
11. Length 3 yd., breadth 18 yd., depth 5 yd.

PROBLEMS

43. How many cubic yards of earth were removed from an excavation 70 ft. long, 24 ft. wide, and 15 ft. deep? How many truck loads of earth were carried away if each truck had a capacity of 2 1/2 cubic yards?
44. How many cubic yards of concrete were used in the construction of the foundation of a brick building 50′ x 90′? A section of the foundation is shown in Fig. 4-51. If concrete costs $28.50 per cubic yard, how much does the concrete in this foundation cost?

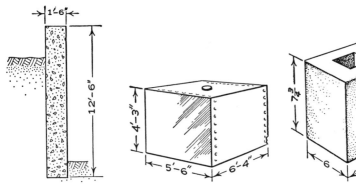

Fig. 4-51. Section of a concrete foundation.

Fig. 4-52. Reserve tank.

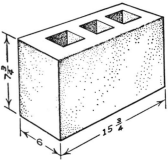

Fig. 4-53. Concrete block.

45. A rectangular tank (Fig. 4-52) is 5'-6" x 4'-3" x 6'-4". How many cubic feet of kerosene oil will it hold? What is the capacity of this tank in gallons?

46. Estimate the cost of the following bill of materials if steel sells at $175.00 per ton. Assume that steel weighs 0.28 lb. per cubic inch.

5 bars	7/8" x 7/8"	12' long	S.A.E. 1020
2 bars	5/8" x 2"	12' long	S.A.E. 1112
3 bars	1/4" x 1 1/2"	12' long	S.A.E. 1112X

47. How many cubic yards of concrete are required to make 265 three-core partition blocks (Fig. 4-53)? Assume that each block has 40% air space.

48. How many cubic inches of steel are in the groove strip shown in Fig. 4-54?

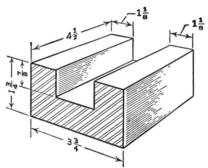

Fig. 4-54. Groove strip.

49. If a cubic inch of cast iron weighs 0.26 lb., what will be the weight of 75 corner plate castings (Fig. 4-55)?

50. What is the weight of 24 oak table tops that measure 1 1/2" in thickness, 28" in width, and 48" in length (Fig. 4-56)? Assume oak to weigh 0.029 lb. per cubic inch.

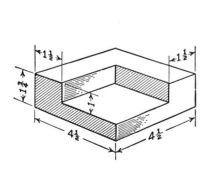

Fig. 4-55. Cast iron corner plate.

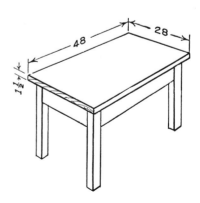

Fig. 4-56. Oak table.

51. An excavation for a cellar (Fig. 4-57) is 5 ft. deep, 23 ft. wide, and 45 ft. long. How many tons of earth must be removed? Assume that 1 cu. ft. of packed earth weighs 100 lb.

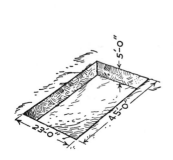

Fig. 4-57. Cellar foundation.

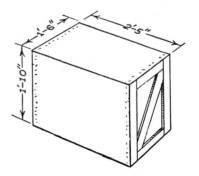

Fig. 4-58. Shipping case.

52. How many cubic feet of space are occupied by the plywood shipping case illustrated in Fig. 4-58?

Finding the volume of a cylinder

A *circular cylinder* is a figure bounded by a lateral curved surface and two flat circular ends or bases. Figure 4-59 represents a circular cylinder. The bases are the circles *ABCD* and *EHFG*, and the lateral curved surface is the portion *M N*.

The height or length of the cylinder is the perpendicular distance between the bases, such as *AE*.

The volume of a cylinder* is the area of the circular base times the height. The result must be expressed in cubic units.

Example. A cylindrical tank is 4'-0" in diameter and 4'-6" long (Fig. 4-60). Find its volume. Express the volume first in cubic feet, then in cubic inches.

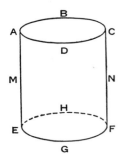

Fig. 4-59. Cylinder.

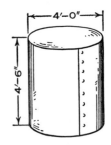

Fig. 4-60. Cylindrical tank.

*In this text a cylinder will mean circular cylinder.

THINK	DO THIS
1. See that all dimensions are expressed in the same linear unit.	1. Diameter of base = 4' Radius = 2' Length = 4'6" = 4 1/2' = 4.5'
2. Find the area of the circular base: find π x square of the radius (p. 130).	2. Area of base = 2^2 x π 2^2 x π = 4 x π = 4 x 3.14 = 12.56
3. Then the volume is the product of the area of the base and the length (or height).	3. Volume = 12.56 x 4.5 = 56.520
4. Express the result in cubic units.	4. 56.52 or 56.5 cu. ft. The result may also be written 4π x 4.5 or 18π cu. ft. *Ans.* Volume = 56.5 cu. ft.

Summary	EXAMPLE
To find the volume of a cylinder:	What is the volume of a cylindrical tank that is 10'6" high and 15'4" in diameter?
1. See that all dimensions are expressed in the same linear unit. Here the dimensions are expressed in feet.	1. Height = 10 1/2' = 10.5' Diameter of base = 15 1/3' = 15.33'
2. Find the product of the area of the base and the length or height ($V = \pi r^2 l$).	2. Volume = area of base x height = $$\pi \ x \ \left(\frac{15.33}{2}\right)^2 \ x \ 10.5$$
3. Express the result in cubic units.	3. 3.14 x 7.67^2 x 10.5 = 1,939.7 cu. ft. *Ans.* Volume = 1,939.7 cu. ft.

EXERCISES

Find the volumes of the cylinders with dimensions as follows:

1. Radius of base 6", height 9"
2. Radius of base 10", height 5"
3. Radius of base 7.5", length 48"
4. Radius of base 2'-6", length 4'-9"
5. Diameter of base 4', length 8.6'
6. Diameter of base 6'-3", length 14'-7"

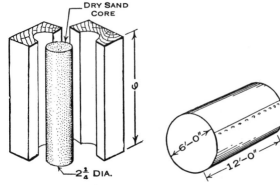

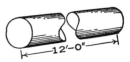

Fig. 4-61. Whole core box.   Fig. 4-62. Fuel oil storage tank.   Fig. 4-63. Steel shaft.

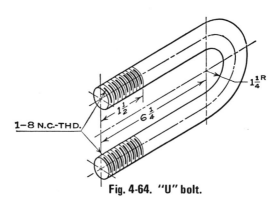

Fig. 4-64. "U" bolt.

PROBLEMS

53. Find the volume of the cylindrical core that can be made in the whole core box shown in Fig. 4-61.

54. A cylindrical tank (Fig. 4-62) whose diameter is 6 ft. and whose length is 12 ft. is filled with fuel oil. How many cubic feet of oil does it contain? If a cubic foot contains 7 1/2 gallons, how many gallons of oil are there in the tank? At $.445 per gallon, what is the value of a tank full of oil?

55. Find the volume of steel in a shaft 1 3/4" in diameter and 12' long (Fig. 4-63). What is the weight of this shaft if steel weighs 489.6 lb. per cubic foot?

56. How many cubic inches of steel are required to make the "U" bolt illustrated (Fig. 4-64)? How much does it weigh? The length is to be measured along the center line.

57. Find the number of cubic inches of brass in the bushing in Fig. 4-65.

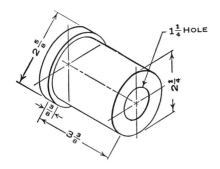

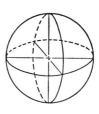

Fig. 4-65. Bushing.   Fig. 4-66. Sphere.

Finding the volume of a sphere

A *sphere* (Fig. 4-66) is a figure bounded by a curved surface, every point of which is at a fixed distance from an internal point known as the center. A baseball is a familiar example of a sphere.

A line from the center to the surface is a *radius*, and a line from any one point on the surface to another through the center is a *diameter*. Compare these terms with those applied to the circle, page 130.

The volume of a sphere is 4/3 times π times the cube of the length of the radius. The result must be expressed in cubic units. The cube of a number such as 8, written 8^3, means 8 x 8 x 8; the cube of 5, written 5^3, means 5 x 5 x 5; the cube of r, written r^3, means r x r x r. The cube of the radius is the result of multiplying three numbers. All three represent the radius.

The term "cube" used in this connection must not be confused with Fig. 4-49 on p. 135, called "a cube."

Example.   Find the volume of a spherical drum that has a diameter of 24".

THINK	DO THIS
The volume of a sphere is equal to 4/3 times π times the cube of the radius of the sphere.	
1. Find the radius.	1. 24/2″ = 12″ radius
2. Find the cube of the radius.	2. 12 x 12 x 12 = 1,728
3. Find the product of 4/3 times π times the cube of the radius.	3. 4/3 x 3.14 x 1,728 = 7,234.56
4. We express this product in cubic units.	4. Volume = 7,234.6 cu. in. *Ans.* 7,234.6 cu. in.

Summary

To find the volume (V) of a sphere with radius (r) known, multiply the cube of the radius by 4/3 x π, written

$$V = 4/3\pi r^3 *$$

Express the product in cubic units.

EXAMPLE

A spherical water tank is 10'6" in diameter. What is its volume?

1. Radius = 5'3" or 5 1/4'

2. Radius cubed = $5\frac{1}{4}$ x $5\frac{1}{4}$ x $5\frac{1}{4}$

   or $\frac{21}{4}$ x $\frac{21}{4}$ x $\frac{21}{4}$ = 144.7

3. V = 4/3 x 3.14 x 144.7 = 605.8
   Ans. Volume = 605.8 cu. ft.

EXERCISES

Find the volumes of the balls (spheres) with dimensions as follows:

1. Radius 6"
2. Radius 8"
3. Radius 3.5"
4. Diameter 5.2"
5. Diameter 4.8"

6. Diameter 4'
7. Diameter 2'4"
8. Diameter 2' 10"
9. Radius 0.28"
10. Radius 0.625"

PROBLEMS

58. A copper float (Fig. 4-67) has an inside diameter of 3.8 in. How many cubic inches of air does it contain?

59. A thrust collar bearing (Fig. 4-68) contains 20 steel balls 7/8" in diameter. What is the total weight of the balls in this bearing? Assume that steel weighs 0.281 lb. per cubic inch.

Fig. 4-67. Copper float.

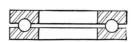

Fig. 4-68. Thrust collar bearing.

60. What will be the weight of 375 round head steel rivets made according to the diagram in Fig. 4-69? Assume that rivets weigh 0.274 lb. per cubic inch.

61. A spherical tank used to store gas (Fig. 4-70) is 75 ft. in diameter. How many cubic feet of gas does this tank contain?

62. A steel ball used to check a dovetail slide (Fig. 4-71) has a diameter of 0.375". What is its volume in cubic inches?

*Remember that quantities are to be multiplied when no plus, +, or minus, -, sign stands between them.

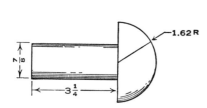

Fig. 4-69. Round head steel rivet.

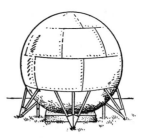

Fig. 4-70. Spherical tank.

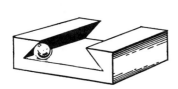

Fig. 4-71. Steel ball for checking a dovetail slide.

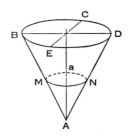

Fig. 4-72. Cone.

Finding the volume of a cone

A *circular cone* is an object bounded by a curved surface, one end of which is a point, the vertex, the other a circle, the base. Fig. 4-72 represents a cone. The circular base is *BCDE* and the vertex is *A*. The curved surface is *MN*. There are many kinds of cones, but in this text we shall consider only circular cones.

The altitude of the cone is the perpendicular distance from the vertex *A* to the base *BCDE*. The altitude of the cone in Fig. 4-72 is *a*.

The volume of a cone is one-third of the area of the base times the altitude. The result must be expressed in cubic units.

Example. Find the volume of a conical tank that has an altitude of 5′ and diameter of the base 4′-6″.

THINK	DO THIS
The volume of a circular cone is 1/3 times the area of the base times the altitude.	
1. See that dimensions are expressed in terms of the same linear unit.	1. Altitude = 5′ Diameter of base = 4 1/2′ Radius of the base = 2 1/4′
2. Find the area of the base.	2. Area of base = 3.14 x 9/4 x 9/4 = 15.9

3. Multiply 1/3 and the area of the base and the altitude.

4. Express the result in cubic units.

3. $\frac{1}{3}$ x 15.9 x 5 = $\frac{79.5}{3}$ = 26.5

4. Volume = 26.5 cu. ft.

Ans. 26.5 cu. ft.

Summary

To find the volume of a circular cone, first express the dimensions in the same linear unit. Then the volume is 1/3 times the area of the base times the altitude:

$$V = 1/3\pi r^2 a$$

Express the product in cubic units.

EXAMPLE

Find the volume of a circular cone that has altitude 36″ and diameter of base 30″.

1. Altitude = 36″
   Radius of base = 15″

2. Area of base = 3.14 x 15² = 706.5

3. V = 1/3 x 706.5 x 36 = 8,478.0

Ans. Volume = 8,478.0 cu. in.

EXERCISES

Find the volumes of the cones with dimensions as follows:

1. Radius of base 3″, height 5″
2. Radius of base 6″, height 18″
3. Radius of base 7.5″, height 15″
4. Diameter of base 33″, altitude 40″
5. Diameter of base 42″, altitude 60″
6. Diameter of base 6′-6″, altitude 8′-4″

PROBLEMS

63. How many cubic feet of air are there in the tent shown in Fig. 4-73? The diameter of the base is 12′ and the altitude is 10.5′.

64. A No. 3 taper standard lathe center has a tungsten carbide tip with the dimensions indicated in Fig. 4-74. How many cubic inches of carbide are there in the exposed part of the tip?

Fig. 4-73. Tent.

Fig. 4-74. Lathe center with tungsten carbide tip.

Fig. 4-75. Inclined conveyor and crane. Note shape of pile of sand.

65. The sand pile in Fig. 4-75 is in the shape of a cone 35 ft. in diameter and 20 ft. high. How many cubic yards of sand are in the pile?

Capacity

The *capacity* of a container such as a tank is the number of units of material it can contain.

The unit of capacity most frequently used is the gallon.

A gallon is 231 cubic inches.

A quart is one-quarter of a gallon. (Do you see why it is called a quart?)

Units of Capacity

		Abbreviated	
1 gallon	= 231 cubic inches	1 gal.	= 231 cu. in.
1 gallon	= 4 quarts	1 gal.	= 4 qt.
1 cubic foot	= 7.5 gallons	1 cu. ft.	= 7.5 gal.

To find the capacity of a container in terms of the number of gallons it can contain, first express the volume in cubic inches, and then divide the result by 231; or express the volume in cubic feet, and multiply the result by 7.5.

Example. What is the capacity of a cylindrical tank that is 4' in diameter and 8' long?

THINK	DO THIS
1. Find the volume of the tank. Since this tank is a cylinder use $V = \pi r^2 l$.	1. Volume of a cylinder $= \pi r^2 l$. $V = 3.14 \times 2^2 \times 8 = 100.5$ cu. ft.
2. Since the volume is expressed in cubic feet, multiply this result by 7.5 (the number of gallons in one cu. ft.).	2. $100.5 \times 7.5 = 753.75$ or 754 *Ans.* Capacity = 754 gallons

Summary	EXAMPLE
To find the capacity in gallons of a container, find the volume of the container in cubic feet or cubic inches.	What is the capacity of a conical tank of height 2'-6" and diameter of base 3'?
1. If the volume is expressed in cubic feet, multiply the volume by 7.5. $C = V$ in cu. ft. $\times 7.5$	1. We express the volume in cu. ft. $V = \frac{1}{3} \times 3.14 \times \left(\frac{3}{2}\right)^2 \times \frac{5}{2} = \frac{47.1}{8}$ cu. ft. $C = \frac{47.1}{8} \times 7.5 = 44$ gallons
2. If the volume is expressed in cubic inches, divide the volume by 231. $C = \frac{V \text{ in cu. in.}}{231}$	2. We express the volume in cu. in. $V = 1/3 \times 3.14 \times (18)^2 \times 30 = 10{,}173.6$ cu. in. $C = \frac{10{,}173.6}{231} = 44$ gallons *Ans.* Capacity = 44 gallons

EXERCISES

Find the capacity in gallons of each of the following:
1. A rectangular tank 4' by 6' by 3'.
2. A square tank 3.5' on a side and 56" high.
3. A cylindrical tank 3'-4" in diameter and 8'-6" long.
4. A cylindrical can 6" in diameter and 12" high.
5. A conical tank with a base 5' in diameter and 7'-6" altitude.
6. A rectangular container that is 5' x 7' x 3'.
7. A rectangular swimming pool that is 25' x 50' and averages 7' deep.
8. A fruit juice can that is 4" in diameter and 7 1/2" high.
9. A milk can that is 14" in diameter and 28" high.

PROBLEMS

66. How many gallons of lubricating oil will a cylindrical tank hold if the diameter of the tank is 3'-8" and the length 5'-0"?

67. Find the number of cubic inches of cutting oil in a 50-gallon drum.

68. A sheet metal container 9" x 9" x 14 1/4" is filled with turpentine. How many gallons were required to fill this container?

69. The tank for a hardening bath measures 1'-6" x 2'-4" and is 2'-0" deep. Find the number of gallons of water that will fill the tank to within 4" of the top.

70. A fuel oil tank holding 275 gallons of oil measures 28" in diameter. What is its length?

71. A kitchen floor measuring 8'-6" x 12'-6" (Fig. 4-76) is to be covered with plain linoleum costing $4.05 per square yard. One gallon of linoleum cement is required for each 10 sq. yds. of linoleum and sells for $2.25 per quart. What is the cost of the linoleum and cement required for this floor? Your estimate must be based on full square yards and quarts.

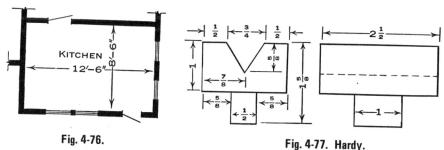

Fig. 4-76.

Fig. 4-77. Hardy.

72. How many pounds of steel are required to make 7 hardies of the type illustrated in figure 4-77? (Weight of steel is 489.6 lb. per cu. ft.)

Metric measure

The *metric system* of measurement is used by all but a few nations of the world. For many years the metric system has been used in this country in the manufacture of munitions, optics, and medicine. Metrics are being more widely used in order to compete with other countries in the sale of manufactured goods. Producers of machinery, tools, and automobiles are increasingly converting to metric measurement. It appears that the trend toward replacing the English system with the metric system will continue. In the near future, the metric system may replace the English system as the <u>official</u> system of measurement in the United States.

The metric system is superior to the English system for computational purposes. Metrics are easier to learn since the system is based on units of ten. For example, it is easier to remember that 1 000 meters equals one kilometer than to remember that 1,760 yards equals one mile.

Metrics are used for all units of measure such as length, area, volume, and weight. The standard unit of length is the meter, which equals about 39.37 inches. The standard unit for area is the square meter; for volume it is the cubic meter; and for weight it is the kilogram.

Prefixes (based on powers of ten) used in relation to the meter follow.

milli- means one-thousandth (1/1 000)   deka- means ten (10)
centi- means one-hundredth (1/100)   hecto- means hundred (100)
deci- means one-tenth (1/10)   kilo- means thousand (1 000)

Linear Measure

Table 4-1 shows the units of length with their abbreviations. These units are based on the meter. Notice that each unit shown in the table is 10 times greater than the unit directly above it.

Metric Units of Linear Measure		
1 millimeter (mm)	=	0.001 meter (m)
1 centimeter (cm)	=	0.01 meter (m)
1 decimeter (dm)	=	0.1 meter (m)
1 meter (m)	=	1 meter (m)
1 dekameter (dam)	=	10 meters (m)
1 hectometer (hm)	=	100 meters (m)
1 kilometer (km)	=	1 000 meters (m)
1 000 millimeters (mm)	=	1 meter (m)
100 centimeters (cm)	=	1 meter (m)
10 decimeters (dm)	=	1 meter (m)
1 meter (m)	=	1 meter (m)
0.1 dekameter (dam)	=	1 meter (m)
0.01 hectometer (hm)	=	1 meter (m)
0.001 kilometer (km)	=	1 meter (m)

Table 4-1.

Expressing linear measurements in different units

To express a unit of length in another unit (either larger or smaller), multiply by one of the values given in Table 4-1. Moving a decimal point a certain number of places either to the left or right accomplishes the same purpose.

Example 1. Express 50 decimeters (dm) in meters (m)

Procedure: Since 1 dm = 0.1 m, 50 dm = 50 x 0.1 m = 5 m. *Ans.* Multiplying by 0.1 is the same as moving the decimal point one place to the left: 5 0.; 50 dm = 5 m

Example 2. Express 0.237 meters (m) in millimeters (mm).

Procedure: Since 1 000 mm = 1 m; 0.237 m = 0.237 x 1 000 mm = 237 mm. *Ans.* Multiplying by 1000 is the same as moving the decimal point 3 places to the right: 0.237; 0.237 m = 237 mm.

Example 3. Express 0.12 decimeters (dm) in centimeters (cm).

Procedure: Since centimeter is the next smaller unit to decimeters, 1 dm = 10 cm. Therefore, 0.12 dm = 0.12 x 10 cm = 1.2 cm. *Ans.* Multiplying by 10 is the same as moving the decimal point one place to the right; 0.12; 0.12 dm = 1.2 cm.

Example 4. Add 0.3 meters (m), 4.6 decimeters (dm), 23 centimeters (cm), and 124 millimeters (mm). Give answer in millimeters.

There is more than one way of solving this example. One procedure is given as follows:

Step 1. Express each value as the equivalent value in meters. 0.3 m = 0.3 m; 4.6 dm = 4.6 x 0.1 m = 0.46 m; 23 cm = 23 x 0.01 m = 0.23 m; 124 mm = 124 x 0.001 m = 0.124 m.

Step 2. Add   0.3   m
              0.46 m
              0.23 m
            + 0.124 m
              1.114 m

Step 3. Express 1.114 meters in millimeters. Since 1 m = 1 000 mm, 1.114 m = 1.114 x 1 000 mm = 1 114 mm. *Ans.*

EXERCISES

Express the following lengths in meters.

1. 25 decimeters
2. 0.62 decimeters
3. 300 centimeters
4. 9 250 millimeters
5. 3.7 dekameters

6. 32.5 hectometers
7. 0.08 kilometers
8. 58 centimeters
9. 2 885 millimeters

Express each of the following values in the indicated units.

10. 5 decimeters to centimeters
11. 30 millimeters to centimeters
12. 573 dekameters to kilometers

13. 0.8 hectometers to decimeters
14. 500 centimeters to dekameters
15. 0.07 kilometers to centimeters

PROBLEMS

73. Determine, in decimeters, dimensions A and B in Fig. 4-78.
74. Determine, in millimeters, dimensions C and D in Fig. 4-79.

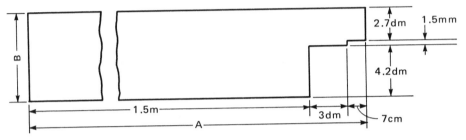

Fig. 4-78.

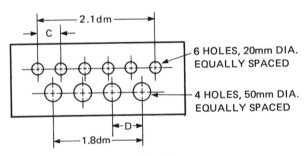

Fig. 4-79.

75. Two pieces of stock measuring 4.5 decimeters and 10.8 centimeters are cut from a board 2.7 meters long. How many meters long is the remaining piece?

76. Find the total length in meters of the wall section shown in Fig. 4-80.

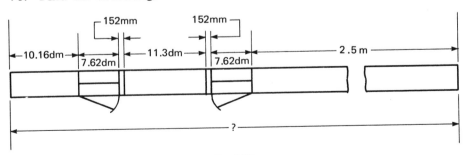

Fig. 4-80.

Converting linear units from one system to the other

It is sometimes necessary to convert values between the English system and the metric system. Some basic linear conversion factors are given in Table 4-2.

Linear Measure

Metric to English Units		
1 kilometer (km)	=	0.6214 mile
1 meter (m)	=	39.37 inches
1 meter (m)	=	3.2808 feet
1 centimeter (cm)	=	0.3937 inch
1 millimeter (mm)	=	0.03937 inch

English to Metric Units		
1 mile	=	1.609 kilometers (km)
1 yard	=	0.914 4 meter (m)
1 foot	=	0.304 8 meter (m)
1 inch	=	2.54 centimeters (cm)
1 inch	=	25.4 millimeters (mm)

Table 4-2

Figure 4-81 shows the relationship of English and metric units by comparing decimal-inch and metric scales.

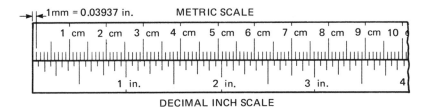

Fig. 4-81.

To convert units from one system to the other, multiply by the appropriate conversion factor.

Examples of converting values between systems (refer to Table 4-2):

Example 1. Convert 0.25 meters to inches.

Procedure: Since 1 m = 39.37 in.; 0.25 m = 0.25 x 39.37 = 9.8425 in. *Ans.*

Example 2. Convert 3.800 inches to centimeters.

Procedure: Since 1 in. = 2.54 cm; 3.800 in. = 3.800 x 2.54 cm
= 9.652 cm. *Ans.*

Example 3. Convert the dimensions given in Fig. 4-82 to millimeters.

Procedure: a. Since 1 in. = 25.4 mm and 1' 8" = 20 in.; 1' 8" = 20
x 25.4 mm = 508 mm. *Ans.*

b. Since 1 in. = 25.4 mm; 6.100" = 6.100 x 25.4 mm
= 154.94 mm. *Ans.*

c. Since 1 in. = 25.4 mm and 2 3/16" = 2.1875"; 2 3/16 in.
= 2.1875 x 25.4 mm = 55.5625 mm. *Ans.*

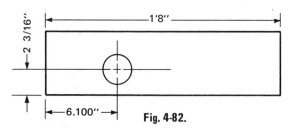

Fig. 4-82.

Example 4. A pipe is 8 ft. 3 1/2 in. long. How many meters long is the
pipe?

Procedure: Express 8 ft. as inches and add: 8 ft. = 8 x 12 in. = 96 in.;
96 in. + 3 1/2 in. = 99 1/2 in. = 99.5 in. Since 1 in.
= 2.54 cm, 99.5 in. = 99.5 x 2.54 cm = 252.73 cm.
Since 1 cm = 0.01 m; 252.73 cm = 252.73 x 0.01 m
= 2.527 3 m. *Ans.*

EXERCISES

Convert the following metric lengths to the indicated English units.
Where necessary, round off answers to 3 decimal places.

1. 10 mm to in.	6. 7 km to mi.	11. 7.2 cm to in.
2. 0.1 m to in.	7. 0.6 m to ft.	12. 68 cm to ft.
3. 20 cm to in.	8. 0.08 m to in.	13. 108 km to mi.
4. 3 m to ft.	9. 55.2 mm to in.	14. 2 mm to in.
5. 5 m to yd.	10. 172 mm to ft.	15. 4.2 m to yd.

Convert the following English lengths to the indicated metric units.
Where necessary, round off answers to 3 decimal places.

16. 3 in. to mm	21. 3 mi. to km	26. 18 ft.-3 in. to m
17. 12 in. to cm	22. 0.8 mi. to km	27. 3 1/2 yd. to m
18. 2 ft. to m	23. 7 yd. to m	28. 100 mi. to km
19. 0.5 ft. to m	24. 8.3 in. to cm	29. 2 ft. 1 1/2 in. to m
20. 0.08 in. to mm	25. 3 ft. to cm	30. 3/16 in. to mm

PROBLEMS

77. The part shown in Fig. 4-83 is dimensioned in millimeters. Convert each dimension to inches and redimension the drawing. Round off dimensions to the closest thousandth of an inch.

78. The floor plan shown in Fig. 4-84 is dimensioned in feet and inches. Convert each dimension to meters. Round off dimensions to the closest thousandth of a meter.

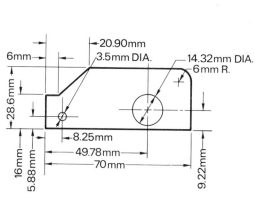

Fig. 4-83.

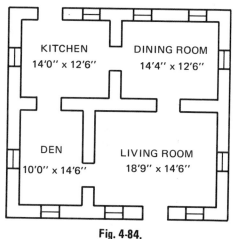

Fig. 4-84.

Square measure

The method of computing surface measure problems is the same in the metric system as in the English system. The product of two linear measures produces *square measures*. For example, in the English system 3 inches x 5 inches = 15 square inches. In the metric system, 3 centimeters x 5 centimeters = 15 square centimeters. The only difference is in the use of metric rather than English units.

The basic unit of area is the square meter. One square meter is equal to the area of a square having sides one meter in length as shown in Fig. 4-85. Other metric surface measure units are based on the square meter. For example, in Fig. 4-86, the relationship between a square meter and a

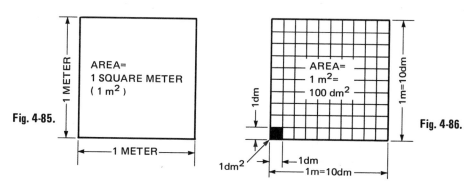

Fig. 4-85.

Fig. 4-86.

square decimeter is shown. Observe that one meter equals 10 decimeters, but one square meter equals 10 decimeters x 10 decimeters or 100 square decimeters.

Table 4-3 shows the units of surface measure with their abbreviations. These units are based on the square meter. Notice that each unit in the table is 100 times greater than the unit directly above it.

Metric Units of Square Measure
1 square millimeter (mm^2) = 0.000 00 1 square meter (m^2)
1 square centimeter (cm^2) = 0.000 1 square meter (m^2)
1 square decimeter (dm^2) = 0.01 square meter (m^2)
1 square meter (m^2) = 1 square meter (m^2)
1 square dekameter (dam^2) = 100 square meters (m^2)
1 square hectometer (hm^2) = 10 000 square meters (m^2)
1 square kilometer (km^2) = 1 000 000 square meters (m^2)
1 000 000 square millimeters (mm^2) = 1 square meter (m^2)
10 000 square centimeters (cm^2) = 1 square meter (m^2)
100 square decimeters (dm^2) = 1 square meter (m^2)
1 square meter (m^2) = 1 square meter (m^2)
0.01 square dekameter (dam^2) = 1 square meter (m^2)
0.000 1 square hectometer (hm^2) = 1 square meter (m^2)
0.000 001 square kilometer (km^2) = 1 square meter (m^2)

Table 4-3.

Expressing area measurements in different units

To express a unit of area in another unit (either larger or smaller), multiply by one of the values given in Table 4-3. Moving a decimal point a certain number of places either to the left or right accomplishes the same purpose.

Example 1. Express 730.8 square decimeters (dm^2) in square meters (m^2).

Procedure: Since 1 dm^2 = 0.01 m^2; 730 dm^2 = 730.8 x 0.01 m^2 = 7.308 m^2. *Ans.* Multiplying by 0.01 is the same as moving the decimal point two places to the left: 7.30.8; 730.8 dm^2 = 7.308 m^2.

Example 2. Express 0.023 square centimeters (cm^2) to square millimeters (mm^2).

Procedure: Since square millimeters is the next smaller unit to square centimeters, 1 cm^2 = 100 mm^2. Therefore, 0.023 cm^2 = 0.023 x 100 mm^2 = 2.3 mm^2. *Ans.* Multiplying by 100 is the same as moving the decimal point two places to the right: 0.02 3; 0.023 cm^2 = 2.3 mm^2.

Example 3. Find the area in square decimeters of the part shown in Fig. 4-87. There is more than one way of solving this example. One procedure is given as follows:

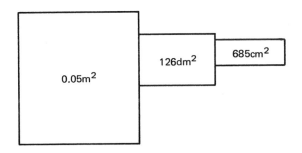

Fig. 4-87.

Step 1. Express each value as the equivalent value in square meters: 0.5 m^2 = 0.5 m^2; 12.6 dm^2 = 12.6 x 0.01 m^2 = 0.126 m^2. 685 cm^2 = 685 x 0.000 1 m^2 = 0.068 5 m^2.

Step 2. Add: 0.5 m^2
             0.126 m^2
          + 0.068 5 m^2
             0.694 5 m^2

Step 3. Express 0.694 5 square meters as square decimeters. Since 1 m^2 = 100 dm^2; 0.694 5 m^2 = 0.694 5 x 100 dm = 69.45 dm^2. *Ans.*

EXERCISES

Express each of the following values in square meters.

1. 107.3 dm^2
2. 29 850 cm^2
3. 12 dam^2
4. 0.008 km^2
5. 0.78 hm^2
6. 250 000 mm^2
7. 3.2 dm^2
8. 783 cm^2

Express each of the following values in the indicated units.

9. 17.8 cm² to mm²
10. 3 mm² to cm²
11. 393 m² to dm²
12. 0.87 dam² to hm²
13. 38.5 km² to hm²
14. 0.078 m² to dm²

15. 5 600 mm² to dm²
16. 3 km² to dam²
17. 785 cm² to hm²
18. 0.009 m² to mm²
19. 56 dam² to dm²
20. 0.07 hm² to km²

PROBLEMS

Solve the following problems. Note: Some problems require unlike units to be changed to like units before proceeding with arithmetic operations. Give answers to 3 decimal places.

79. Determine the area of a building 16 meters wide and 35 meters long.
80. A piece of steel 25 centimeters wide has an area of 90 square centimeters. How long is the sheet?
81. A bathroom 2.8 meters long and 2.2 meters wide is to have four walls covered with tile to a height of 1.3 meters. An allowance of 1.5 square meters is made for door and window openings. How many square meters of tile are required?
82. A rectangular strip of aluminum is 12 millimeters wide and 18 centimeters long. What is the area in square centimeters?
83. Find the number of square meters in a triangle with a base of 0.75 meters and an altitude of 25.5 centimeters.

Find the area of the following figures. It may be necessary to divide a figure into two or more simple figures. Give answers to 3 decimal places.

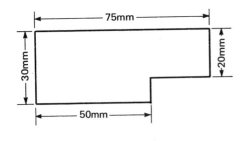

Fig. 4-88.

84. Area = ? mm²

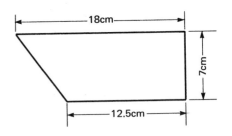

Fig. 4-89.

85. Area = ? cm²

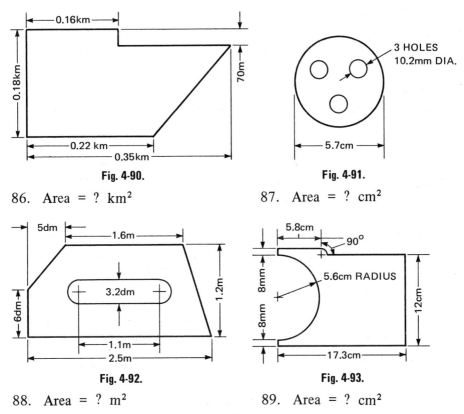

Fig. 4-90.

86. Area = ? km²

Fig. 4-91.

87. Area = ? cm²

Fig. 4-92.

88. Area = ? m²

Fig. 4-93.

89. Area = ? cm²

Converting area units from one system to the other

Table 4-4 gives the basic conversion factors for changing metric area units to English units and English units to metric units.

Area Measure	
1 square kilometer (km²)	= 0.3861 sq. mile (sq. mi.)
1 square meter (m²)	= 1.196 sq. yards (sq. yd.)
1 square meter (m²)	= 10.764 sq. feet (sq. ft.)
1 sq. centimeter (cm²)	= 0.155 sq. inch (sq. in.)
1 sq. millimeter (mm²)	= 0.00155 sq. inch (sq. in.)
1 sq. mile (sq. mi.)	= 2.589 9 square kilometers (km²)
1 sq. yard (sq. yd.)	= 0.836 square meter (m²)
1 sq. foot (sq. ft.)	= 0.092 9 square meter (m²)
1 sq. inch (sq. in.)	= 6.452 square centimeters (cm²)
1 sq. inch (sq. in.)	= 645.2 square millimeters (mm²)

Table 4-4.

To convert units from one system to the other, multiply by the appropriate conversion factor.

Examples of converting values between systems (refer to Table 4-4):

Example 1. Convert 30 square feet to square meters.

Procedure: Since 1 sq. ft. = 0.092 9 m^2 ; 30 sq. ft. = 30 x 0.092 9 m^2
= 2.787 m^2. *Ans.*

Example 2. Convert 50.6 square centimeters to square inches.

Procedure: Since 1 cm^2 = 0.155 sq. in.; 50.6 cm^2 = 50.6 x 0.155
sq. in. = 7.843 sq. in. *Ans.*

The values of problems are sometimes given in units in one system, but the answer is required in units of another system. To reduce the amount of conversion do the operations using the units as given. Convert only the answers to the other system.

Example 1. Find the area in square meters of the rectangular slab shown in Fig. 4-94.

Step 1.   Multiply using the given units. 3.4 yd. x 5.2 yd. = 17.68 sq. yd.

Step 2.   Convert 17.68 sq. yd. to m^2. Since 1 sq. yd. = 0.836 m^2; 17.68 sq. yd. = 17.68 x 0.836 m^2 = 14.780 m^2. *Ans.*

Example 2. Determine the area in square inches of the triangle shown in Fig. 4-95.

Step 1.   Find the area using the given units. Area = 1/2 ba; Area = 1/2 x 16 cm x 12 cm = 96 cm^2.

Step 2.   Convert 96 cm^2 to sq. in. Since 1 cm^2 = 0.155 sq. in.; 96 cm^2 = 96 x 0.155 sq. in. = 14.88 sq. in. *Ans.*

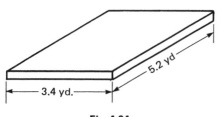

Fig. 4-94.

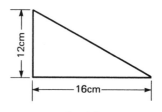

Fig. 4-95.

EXERCISES

Convert the following metric areas to the indicated English areas. Where necessary, round off answers to 3 decimal places.

1. 100 mm² to sq. in.
2. 75 cm² to sq. in.
3. 20 m² to sq. yd.
4. 12 km² to sq. mi.
5. 0.5 m² to sq. ft.

6. 7.6 cm² to sq. in.
7. 5.3 m² to sq. yd.
8. 376 mm² to sq. in.
9. 2.8 km² to sq. mi.

Convert the following English areas to the indicated metric areas. Where necessary, round off answers to 3 decimal places.

10. 16 sq. yd. to m²
11. 0.8 sq. in. to mm²
12. 7.3 sq. mi. to km²
13. 130 sq. ft. to m²
14. 4.9 sq. in. to cm²

15. 215 sq. yd. to m²
16. 6 sq. mi. to km²
17. 810 sq. in. to mm²
18. 40.7 sq. ft. to m²

PROBLEMS

Solve the following problems. Give answers to 2 decimal places.

90. Determine the area in square meters of a sheet of plywood 8 feet long and 4 feet wide.
91. What is the area in square yards of the walls of a building 20 meters long, 14.8 meters wide, and 7.4 meters high?
92. Six 1/2 inch diameter holes are punched in a triangular piece of stock with a 6 1/4 inch base and a 2 1/8 inch altitude. Determine the number of square centimeters of material remaining.
93. What is the area in square meters of the gambrel roof shown in Fig. 4-96?
94. Determine the area in square inches of material required for the sheet metal duct shown in Fig. 4-97.

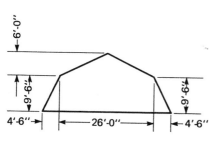

Fig. 4-96.

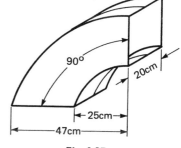

Fig. 4-97.

Volume measure

The methods of computing volume measure problems are the same in the metric system as in the English system. The product of three linear measures produces *cubic measure*. For example, in the English system 2 inches x 4 inches x 5 inches = 40 cubic inches. In the metric system, 2 centimeters x 4 centimeters x 5 centimeters = 40 cubic centimeters. The only difference is in the use of metric rather than English units.

The basic unit of volume is the cubic meter. One cubic meter is equal to the volume of a cube having sides one meter in length as shown in Fig. 4-98. Other metric volume measure units are based on the cubic meter. For example, in Fig. 4-99, the relationship between a cubic meter and a cubic decimeter is shown. Observe that one meter equals 10 decimeters, but one cubic meter equals 10 decimeters x 10 decimeters x 10 decimeters or 1 000 cubic decimeters.

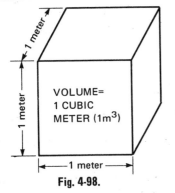

Fig. 4-98.

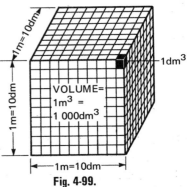

Fig. 4-99.

Table 4-5 shows the conversion units of volume measure with their abbreviations. These units are based on the cubic meter. Notice that each unit in the table is 1,000 times greater than the unit directly above it.

Metric Units of Volume Measure
1 cubic millimeter (mm³) = 0.000 000 001 cubic meter (m³)
1 cubic centimeter (cm³) = 0.000 001 cubic meter (m³)
1 cubic decimeter (dm³) = 0.001 cubic meter (m³)
1 cubic meter (m³) = 1 cubic meter (m³)
1 000 000 000 cubic millimeters (mm³) = 1 cubic meter (m³)
1 000 000 cubic centimeters (cm³) = 1 cubic meter (m³)
1 000 cubic decimeters (dm³) = 1 cubic meter (m³)
1 cubic meter (m³) = 1 cubic meter (m³)

Table 4-5.

Expressing volume in different units of measure

To express a unit of volume in another unit (either larger or smaller) multiply by one of the values given in Table 4-5. Moving a decimal point a certain number of places to the left or right accomplishes the same purpose.

Example 1. Express 2.3 cubic meters (m^3) in cubic decimeters (dm^3).

Procedure: Since 1 m^3 = 1 000 dm^3; 2.3 dm^3 = 2.3 x 1 000 dm^3 = 2 300 dm^3. *Ans.*   Multiplying by 1 000 is the same as moving the decimal point three places to the right: 2.300; 2.3 m^3 = 2 300 dm^3.

Example 2. In machining a piece of steel, 250 cubic millimeters of material are removed.   How many cubic centimeters are removed?

Procedure: Since cubic centimeters is the next larger unit to cubic millimeters, 1 cm^3 = 1 000 mm^3 or 0.001 cm^3 = 1 mm^3. Therefore, 250 mm^3 = 250 x 0.001 cm^3 = 0.250 cm^3. *Ans.*   Multiplying by 0.001 is the same as moving the decimal point three places to the left: 250. ; 250 mm^3 = 0.250 cm^3.

Example 3. Find the volume, in cubic meters, of the concrete block shown in Fig. 4-100.

Step 1.   Determine volume in cubic decimeters. *Volume = l x w x h, V* = 18 dm x 4 dm x 3.2 dm = 230.4 dm^3.

Step 2.   Express cubic decimeters in cubic meters. Since 1 dm^3 = 0.001 m^3; 230.4 cm^3 = 230.4 x 0.001 m^3 = 0.230 4 m^3. *Ans.*

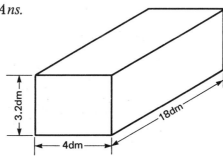

Fig. 4-100.

EXERCISES

Express each of the following values in cubic meters.

1. 107 dm³
2. 270 000 cm³

3. 560 000 mm³
4. 0.03 dm³

Express each of the following values in the indicated units.

5. 287 cm³ to mm³
6. 370 mm³ to cm³
7. 3.85 m³ to dm³
8. 585 000 mm³ to cm³

9. 0.005 m³ to dm³
10. 3 dm³ to cm³
11. 3.85 cm³ to mm³
12. 815 mm³ to cm³

PROBLEMS

Solve the following problems. Give answers to 3 decimal places.

95. An automobile gasoline tank has a volume of 58 cubic decimeters. Determine the volume in cubic meters.

96. An automobile engine with a displacement of 4 200 cubic centimeters is re-bored which increases its displacement by 0.3 cubic decimeters. After re-boring, what is the engine displacement in cubic meters?

97. A pump is capable of removing 85 cubic decimeters of water per minute. How many cubic meters of water can be removed in 2 hours?

98. A rectangular tank 30 decimeters long and 21.5 decimeters wide contains 9 cubic meters of fuel. Determine the depth of fuel in the tank.

99. Compute the length of a piston which has a volume of 410 cubic centimeters and a diameter of 7.6 centimeters.

Refer to the following illustrations and solve for the unknown value in each one. It may be necessary to divide a figure into two or more simpler figures. Compute answers to two decimal places.

100. Volume of angle iron
= ? cm³

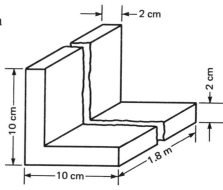

Fig. 4-101.

101. Volume of Concrete retaining wall = ? m³

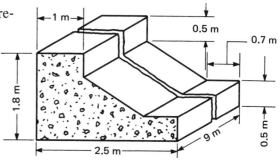

Fig. 4-102.

102. Volume of material in
steel base plate = ? cm³

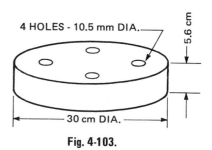

Fig. 4-103.

103. Volume of building
column and footing = ?
m³

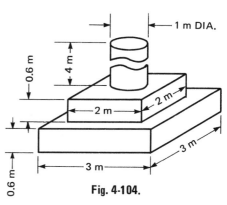

Fig. 4-104.

104. Volume of space in
tubing = 80 cm³

Inside diameter of tub-
ing = ? cm

Fig. 4-105.

Changing volume units from one system to the other

Table 4-6 gives the basic conversion factors for changing metric area units to English units and English units to metric units.

Volume Measure
Metric to English Units
1 cubic meter (m^3)   = 1.308 cubic yards (cu. yd.) 1 cubic meter (m^3)   = 35.314 cubic feet (cu. ft.) 1 cubic centimeter (cm^3)= 0.061 cubic inch (cu. in.)
English to Metric Units
1 cubic yard (cu. yd.) = 0.764 5 cubic meter (m^3) 1 cubic foot (cu. ft.) = 0.028 3 cubic meter (m^3) 1 cubic inch (cu. in.) = 16.387 2 cubic centimeters (cm^3)

Table 4-6.

To convert units from one system to the other, multiply by the appropriate conversion factor.

Examples of converting values between systems (refer to Table 4-6):

Example 1. Convert 15 cubic yards (cu. yd.) to cubic meters (m^3).

Procedure: Since 1 cu. yd. = 0.764 5 m^3, 15 cu. yd. = 15 x 0.764 5 m^3 = 11.468 m^3. *Ans.*

Example 2. Convert 120 cubic centimeters (cm^3) to cubic inches (cu. in.).

Procedure: Since 1 cm^3 = 0.061 cu. in.; 120 cm^3 = 120 x 0.061 cu. in. = 7.320 cu. in. *Ans.*

As with linear and square measures, values of volume problems are sometimes given in one system, but the answer is required in units of another system. The amount of conversion is reduced if the operations are performed using the units as given. Convert only the answer to the other system.

Example. Determine the volume in cubic centimeters (cm^3) of the corner plate shown in fig. 4-106.

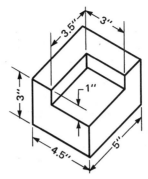

Fig. 4-106.

Step 1. Determine the volume of the plate disregarding the cut-out using the given units: V = l x w x h; V = 5 in. x 4.5 in. x 3 in. = 67.5 cu. in.

Step 2. Determine the volume of the cut-out using the given units: V = l x w x h; V = 3.5 in. x 3 in. x 1 in. = 10.5 cu. in.

Step 3. Determine the volume of the plate: Volume = 67.5 cu. in. – 10.5 cu. in. = 57 cu. in.

Step 4. Convert 57 cu. in. to cm³: Since 1 cu. in. = 16.387 2 cm³; 57 cu. in. = 57 x 16.387 2 cm³ = 934.070 cm³. *Ans.*

EXERCISES

Convert the following metric volumes to the indicated English volumes. Where necessary, round off answers to 3 decimal places.

1. 750 cm³ to cu. in.
2. 3.4 m³ to cu. ft.
3. 0.8 m³ to cu. yd.
4. 97.3 cm³ to cu. in.
5. 0.05 m³ to cu. ft.
6. 132 m³ to cu. yd.

Convert the following English volumes to the indicated metric volumes. Where necessary, round off answers to 3 decimal places.

7. 8.2 cu. yd. to m³
8. 10 cu. in. to cm³
9. 380 cu. ft. to m³
10. 0.75 cu. yd. to m³
11. 0.25 cu. in. to cm³
12. 23 cu. yd. to m³

PROBLEMS

Solve the following problems. Give answers to 2 decimal places unless otherwise specified.

105. A bin contains 12.3 cubic feet of sand. How many cubic meters are contained in it?

106. A structural draftsman estimates the volume of a building as 1 650 cubic meters. Determine the volume in cubic feet.

107. A certain stainless steel alloy contains 12 percent chromium and 2.5 percent nickel by volume. How many cubic centimeters of chromium and nickel are contained in a steel block having a volume of 0.02 cubic feet?

108. A bricklayer is required to estimate the number of bricks needed to construct a wall which has a volume of 12 cubic meters. Each brick contains 64 cubic inches. Determine the number of bricks required to the nearest whole brick.

109. A plumber installs a cylindrical tank 8 meters long and 1.4 meters in diameter in a pneumatic water system. How many cubic feet of water does the tank contain?

110. Determine the number of cubic centimeters of material contained in the construction column form shown in Fig. 4-107.

111. Determine the number of cubic inches of material contained in the flanged collar shown in Fig. 4-108.

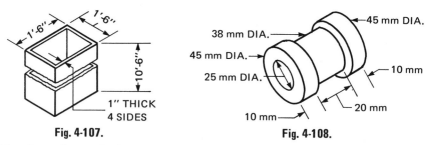

1'·6" 1'·6" 1'·6" 10'-6" 1" THICK 4 SIDES

Fig. 4-107.

38 mm DIA. 45 mm DIA. 45 mm DIA. 25 mm DIA. 10 mm 20 mm 10 mm

Fig. 4-108.

Liquid volume (capacity) measure

The *liter* is the standard unit of liquid volume measure in the metric system. In the English system, gasoline is purchased in gallons; in the metric system it is purchased in liters. A liter is equal to 1 000 cubic centimeters or 1 cubic decimeter. Since 1 000 cubic decimeters (dm^3) equals 1 cubic meter (m^3), 1 000 liter (ℓ) also equals 1 cubic meter (m^3).

Table 4-7 shows the units of liquid volume measure. These units are based on the liter. Notice that each unit is 10 times greater than the unit directly above it.

Metric Units of Liquid Volume Measure		
1 milliliter (mℓ)	=	0.001 liter (ℓ)
1 centiliter (cℓ)	=	0.01 liter (ℓ)
1 deciliter (dℓ)	=	0.1 liter (ℓ)
1 liter (ℓ)	=	1 liter (ℓ)
1 dekaliter (daℓ)	=	10 liter (ℓ)
1 hectoliter (hℓ)	=	100 liter (ℓ)
1 kiloliter (kℓ)	=	1 000 liter (ℓ)

Table 4-7.

1 000 milliliter (ml)	=	1 liter (l)
100 centiliters (cl)	=	1 liter (l)
10 deciliters (dl)	=	1 liter (l)
1 liter (l)	=	1 liter (l)
0.1 dekaliter (dal)	=	1 liter (l)
0.01 hectoliter (hl)	=	1 liter (l)
0.001 kiloliter (kl)	=	1 liter (l)

Table 4-7. (Con't)

Expressing capacity measure in different units

To change a unit of liquid volume to another unit (either larger or smaller), multiply by one of the values given in Table 4-7. Moving a decimal point a certain number of places to the left or right accomplishes the same purpose.

Example 1. Change 15.5 deciliters (dl) to liters (l).

Procedure: Since 1 dl = 0.1 l; 15.5 dl = 15.5 x 0.1 l = 1.55 l. *Ans.* Multiplying by 0.1 is the same as moving the decimal point one place to the left: 1.5.5; 15.5 dl = 1.55 l.

Example 2. Three containers of 0.56 liters (l), 12 centiliters (cl), and 40 milliliters (ml) are filled with water. Determine the total number of deciliters of water in the containers. There is more than one way of solving this example: One procedure is given as follows:

Step 1. Express each value as the equivalent value in liters. 0.56 l = 0.56 l; 12 cl = 12 x 0.01 l = 0.12 l; 40 ml = 40 x 0.001 l = 0.04 l.

Step 2. Add: 0.56 l
              0.12 l
            + 0.04 l
              0.72 l

Step 3. Express 0.72 liters as deciliters. Since 1 l = 10 dl, 0.72 l = 0.72 x 10 dl = 7.2 dl. *Ans.*

Example 3. A tank has a volume of 500 cubic meters (m^3). How many liters of gasoline can be stored in the tank?

Procedure: Since 1 000 liters (l) = 1 cubic meter (m^3), 500 m^3 = 500 x 1 000 l = 500 000 l. *Ans.*

EXERCISES

Express the following values in liters.

1. 12 centiliters (cℓ)
2. 7.5 hectoliters (hℓ)
3. 7 850 milliliters (mℓ)

4. 0.08 kiloliters (kℓ)
5. 385 deciliters (dℓ)
6. 29 dekaliters (daℓ)

Express each of the values in the indicated values.

7. 6 ℓ to mℓ
8. 125 cℓ to dℓ
9. 53 kℓ to hℓ
10. 0.07 ℓ to cℓ
11. 5 700 mℓ to cℓ

12. 3.9 daℓ to hℓ
13. 28.7 ℓ to daℓ
14. 6.3 mℓ to cℓ
15. 0.85 cℓ to dℓ

PROBLEMS

Solve the following problems. Give answers to 3 decimal places where necessary.

112. What is the total volume of twelve tanks, each containing 10.75 liters?
113. A gasoline storage tank which has a capacity of 50 kiloliters is 22 percent full. How many liters of gasoline are in the tank?
114. What is the sum in liters of the following volumes: 0.35 dekaliter, 3.3 deciliters, and 86 centiliters?
115. How many liters are contained in 7.85 cubic meters?
116. The cooling system of an automobile contains 8.5 liters of water. If 38 deciliters of water are drained from the system, how many liters remain?
117. A pump discharges 150 liters of water per minute. If the pump operates for 3 hours and 15 minutes, how many kiloliters are discharged?
118. Fuel oil is pumped through 3 pipes into a tank which has a capacity of 75 kiloliters. How many hours will it take to fill the tank if the fuel is pumped through each pipe at the rate of 20 deciliters per second?

Converting liquid volume (capacity) from one system to the other

Table 4-8 gives the basic conversion factors for changing metric capacity units to English units and English units to metric units.

Capacity Measure
Metric to English Units
1 liter (ℓ) = 0.2642 gallon (gal.)   1 liter (ℓ) = 1.0567 quarts (qt.)
English to Metric Units
1 gallon (gal.) = 3.785 liters (ℓ)   1 quart (qt.) = 0.946 liter (ℓ)

Table 4-8.

To convert units from one system to the other, multiply by the appropriate conversion factor.

Examples of converting values between systems (refer to Table 4-8):

Example 1.   Convert 8.6 quarts (qt.) to liters (ℓ).
  Procedure: Since 1 qt. = 0.946 ℓ; 8.6 qt. = 8.6 x 0.946 ℓ = 8.135 6 ℓ.
    Ans.

Example 2.   Convert 37.4 liters (ℓ) to gallons (gal.).
  Procedure: Since 1 ℓ = 0.2642 gal.; 37.4 ℓ = 37.4 x 0.2642 gal. =
    9.8811 gal. *Ans.*

Values of problems of capacity measure are sometimes given in one system, but the answer is required in units of another system. The amount of conversion is reduced if the operations are performed using the units as given. Convert only the answer to the other system. Example: How many liters are there in the sum of 8 quarts, 0.7 quart, and 12.8 quarts?

  Step 1.   Add the given units: 8 qt. + 0.7 qt. + 12.8 qt. = 21.5 qt.
  Step 2.   Convert 21.5 qt. to ℓ: Since 1 qt. = 0.946 ℓ, 21.5 qt. =
    21.5 x 0.946 ℓ = 20.339 ℓ. *Ans.*

EXERCISES
Convert each value to the indicated value in Table 4-9.

	Gallon	Quart	Liter		Gallon	Quart	Liter
1.	5	–	?	5.	520 1/2	–	?
2.	–	16.8	?	6.	–	3/4	?
3.	?	–	23.75	7.	?	–	0.09
4.	–	?	104.3	8.	–	?	74.3

Table 4-9.

9. 7 qt. + 23 qt. + 0.7 qt. = ? ℓ   11.   3 1/2 gal. + 1/4 gal. + 5 gal. = ? ℓ
10. 33 ℓ + 18 ℓ + 0.3 ℓ = ? gal.   12.   1.6 ℓ + 35 ℓ + 0.2 ℓ = ? qt.

PROBLEMS

120. A container has a volume of 346.5 cubic inches. How many liters of water can the container hold? (1 gal. = 231 cu. in.)
121. How many liters of fuel does a rectangular tank 3 ft. x 5.5 ft. x 6 ft. contain when half full? (1 cu. ft. = 7.5 gal.) Give answer to 3 decimal places.
122. Determine the number of liters of water in the cylinder shown in Fig. 4-109 which is 3/4 full. (1 cu. ft. = 7.5 gal.) Give answer to 3 decimal places.

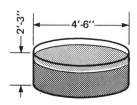

Fig. 4-109.

Units of weight (mass)

The *kilogram* is the standard unit of weight in the metric system. Things that are weighed in pounds in the English system are weighed in kilograms in the metric system.

Table 4-10 shows the units of weight. These units are based on the gram. Notice that each unit is 10 times greater than the unit directly above it.

Metric Units of Weight (Mass) Measure		
1 milligram (mg)	=	0.001 gram (g)
1 centigram (cg)	=	0.01 gram (g)
1 decigram (dg)	=	0.1 gram (g)
1 gram (g)	=	1 gram (g)
1 dekagram (dag)	=	10 grams (g)
1 hectogram (hg)	=	100 grams (g)
1 kilogram (kg)	=	1 000 grams (g)
1 000 milligrams (mg)	=	1 gram (g)
100 centigrams (cg)	=	1 gram (g)
10 decigrams (dg)	=	1 gram (g)
1 gram (g)	=	1 gram (g)
0.1 dekagram (dag)	=	1 gram (g)
0.01 hectogram (hg)	=	1 gram (g)
0.001 kilogram (kg)	=	1 gram (g)

Table 4-10.

For large measures of weight, the metric ton is commonly used. One metric ton equals 1 000 kilograms.

Expressing weight in different units

To change a unit of weight to another unit (either larger or smaller), multiply by one of the values given in Table 4-10. Moving a decimal point a certain number of places to the left or right accomplishes the same purpose.

Example 1. Express 525 centigrams (cg) in grams (g).

Procedure: Since 1 cg = 0.01 g; 525 cg = 525 x 0.01 g = 5.25 g *Ans.* Multiplying by 0.01 is the same as moving the decimal point two places to the left: 5 25.; 525 cg = 5.25 cg.

Example 2. Determine the total weight in dekagrams (dag) of the following weights: 22 grams (g), 0.67 hectograms (hg), and 534 centigrams.

There is more than one way of solving this example. One procedure is given as follows:

Step 1. Express each value as the equivalent value in grams. 22 g = 22 g; 0.67 hg = 0.67 x 100 g = 67 g; 534 cg = 534 x 0.01 g = 5.34 g.

Step 2. Add: 22   g
               67   g
             + 5.34 g
             94.34 g

Step 3. Change 94.34 grams to dekagrams. Since 1 g = 0.1 dag, 94.34 g = 94.34 x 0.1 dg = 9.434 dg. *Ans.*

EXERCISES

Express the following values in grams.

1. 68 decigrams (dg)
2. 890 centigrams (cg)
3. 0.8 dekagrams (dag)
4. 4.7 kilograms (kg)
5. 932 milligrams (mg)
6. 55 hectograms (hg)
7. 87.9 centigrams (cg)
8. 0.075 kilograms (kg)
9. 9 853 milligrams (mg)

Express each of the values in the indicated value.

10. 10.3 mg to cg
11. 9 hg to kg
12. 0.07 kg to hg
13. 15 g to dg
14. 0.03 g to mg

15. 873 cg to dag
16. 1.8 hg to cg
17. 2 000 dg to hg
18. 612 g to kg

19. Add the following weights: 13 cg; 0.7 cg; 3.6 cg; 0.03 cg.
20. Add the following weights: 3 g; 790 cg; 0.62 hg; 48 dg. Give the answer in grams.
21. Add the following weights: 285 mg; 0.7 g; 34 cg; 5 dg. Give the answer in centigrams.

PROBLEMS

123. What is the total weight in kilograms of 175 stampings, each weighing 27 dekagrams?
124. Two hundred identical strips are sheared from a steel sheet weighing 16 kilograms. What is the weight, in grams, of one strip? Give answer to 2 decimal places.
125. What is the weight, in metric tons, of four pieces of scrap metal weighing 45 hectograms, 18.6 hectograms, 520 dekagrams, and 98 dekagrams?
126. Determine the weight, in kilograms, of a sheet of 8 gage (3.264 mm thick) aluminum, 1 meter wide and 2.75 meters long. Aluminum weighs 2 707 kilograms per cubic meter. Give answer to 3 decimal places.
127. Figure 4-110 shows a tank which is filled with water. One liter (1 000 cm³) of water weighs 0.99 kilograms. Find the total number of kilograms of water contained in the tank to 2 decimal places.
128. Determine the total weight in metric tons of the concrete wall illustrated in Fig. 4-111. The concrete used weighs 2 325 kilograms per cubic meter.

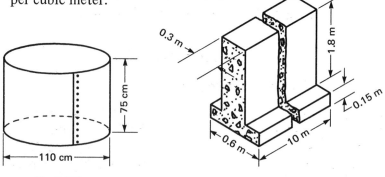

Fig. 4-110. **Fig. 4-111.**

Converting weight units from one system to the other

Table 4-11 gives the basic conversion factors for changing metric weight units to English units and English units to metric units.

Weight Measure
Metric to English Units
1 metric ton   = 0.9842 long ton (2240 pounds) 1 kilogram (kg) = 2.2046 pounds (lb.) 1 gram (g)   = 0.0353 ounce (oz.)
English to Metric Units
1 long ton   = 1.016 1 metric tons 1 pound (lb.) = 0.453 6 kilograms (kg) 1 ounce (oz.) = 28.348 grams (g)

Table 4-11.

To convert units from one system to the other, multiply by the appropriate conversion factor.

Examples of converting values between systems (refer to Table 4-11):

Example 1. Convert 15.2 pounds to kilograms.

Procedure: Since 1 lb. = 0.453 6 kg; 15.2 lb. = 15.2 x 0.453 6 kg = 6.895 kg. *Ans.*

Example 2. Convert 7.8 metric tons to long tons.

Procedure: Since 1 metric ton = 0.9842 long ton; 7.8 metric tons = 7.8 x 0.9842 long tons = 7.677 long tons. *Ans.*

Values for problems in weight are sometimes given in one system, but the answer is required in units of another system. The amount of conversion is reduced if the operations are performed using the units as given. Convert only the answer to the other system.

Example: How many grams are there in the sum of 3 ounces, 0.7 ounces and 1.4 ounces?

Step 1. Add the given units: 3 oz. + 0.7 oz. + 1.4 oz. = 5.1 oz.

Step 2. Convert 5.1 oz. to g: Since 1 oz. = 28.348 g; 5.1 oz. = 5.1 x 28.348 g = 144.575 g. *Ans.*

EXERCISES

Convert each value to the indicated value in Table 4-12. Give answers to 2 decimal places.

	Gram	Kilogram	Ounce	Pound
1.	26.5	–	?	–
2.	–	3.7	–	?
3.	?	–	0.33	–
4.	–	?	–	34.4

	Kilogram	Metric Ton	Pound	Long Ton
5.	22.2	–	?	–
6.	?	–	500	–
7.	–	175	–	?
8.	–	?	–	355

Table 4-12.

9. 13 lb. + 0.6 lb. + 9 lb. = ? kg
10. 27 kg + 13 kg + 0.4 kg = ? lb.
11. 3 oz. + 0.9 oz. + 1.8 oz. = ? g
12. 42 g + 3.1 g + 0.6 g = ? oz.

PROBLEMS

129. A steel beam weighs 875 pounds. What is the weight of 8 beams, in kilograms?
130. A truck is loaded with 4.8 long tons of lumber. The truck is found to be 15 percent over the maximum legal weight limit. How many metric tons of lumber must be removed?
131. A machined part weighs 9 ounces before a drilling operation. Ten holes are drilled with 0.55 ounce of stock removed with each hole. What is the weight in grams of the finished part?

REVIEW PROBLEMS

1. Determine the weight of 4,500 1 1/8" spherical-head rivets (Fig. 4-112). The steel used to make these rivets weighs 475.5 lb. per cubic foot.

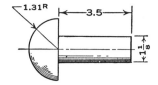

Fig. 4-112. Rivet.

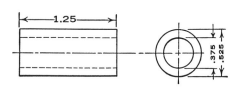

Fig. 4-113. Brass bushing.

2. The brass used to make the bushing shown in Fig. 4-113 weighs 546.2 lb. per cubic foot. Find the weight of 2,200 bushings.
3. The sand removed from a sand pit (Fig. 4-114) is in a pile shaped like a cone. The pile is 25 ft. high and 45 ft. in diameter. How many cubic yards of sand are in the pile?
4. Determine the area of the stamping in Fig. 4-115.
5. Find the weight of 75 steel balls (Fig. 4-116) each of which is 1 13/16″ in diameter. One cubic foot of the steel used weighs 476 lb.

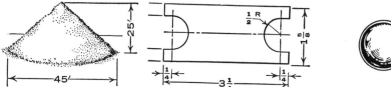

| Fig. 4-114. | Fig. 4-115. | Fig. 4-116. Steel ball for bearing. |

6. The diameter of a cast-iron pulley is 49 in. Find the circumference in feet.
7. A water tank has an inside diameter of 12′-6″ and is 15′-6″ high. How many gallons of water will it hold?
8. A concrete sidewalk is 55′-0″ long and 7′-6″ wide. Estimate the cost of the sidewalk if the labor and material will be supplied for $0.54 per square foot.
9. Determine the cross-sectional area of the dovetail in Fig. 4-117.
10. How many barrels of oil will a tank hold that is 4′-5″ in diameter and 6′-0″ long? There are 31 1/2 gallons to a barrel.
11. A carpenter orders a roll of 12-mesh copper wire cloth (Fig. 4-118) for window screens. The roll is 30 in. wide and 60 ft. long. Material of this type lists for $0.85 per square foot less 25% when purchased in lots of 100 or more square feet. What does he pay for the cloth?
12. A roll of sheet copper 0.042 in. thick, 18 in. wide, and 3 1/2 yds. long weighs 40 oz. per square foot. Find the weight in pounds of five rolls of copper this size. (16 oz. = 1 lb.)

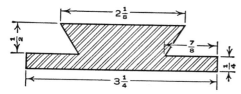

Fig. 4-117. Cross section at dovetail of the cross slide of a lathe.

Fig. 4-118. Wire cloth.

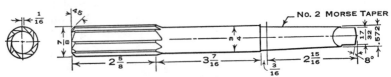

Fig. 4-119. Taper shank reamer.

13. Compute the weight of 75 cast-iron blocks 1 1/8" x 2 1/2" x 6". Cast iron weighs 490 lb. per cubic foot.

14. Tool steel 1" in diameter is used to make the reamer in Fig. 4-119. If this steel weighs 1.84 lb. per lineal foot, find:

   (a) The number of cubic inches of steel in one lineal foot.
   (b) The weight of a cubic inch of tool steel.
   (c) The number of pounds of steel needed to make 48 reamers of the size specified in the drawing if 3/16" is allowed for cutting off and facing.

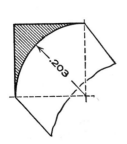

Fig. 4-120. Spandrel or fillet.

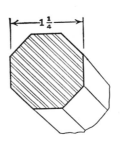

Fig. 4-121.

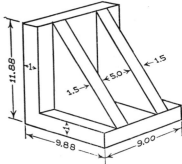

Fig. 4-122. Angle iron.

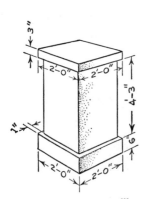

Fig. 4-123. Concrete pillar.

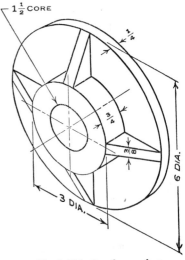

Fig. 4-124. Bearing washer.

15. What is the area of the cross section of the spandrel or fillet in Fig. 4-120?

16. How many cubic inches are there in one lineal foot of octagonal tool steel (Fig. 4-121), the distance across the flats of which is 1 1/4"? How many feet are there in a shipment of 785 lb.? The weight of a cubic inch of this steel is 0.278 lb. One side of the octagon is 0.517".

17. Find the weight of the cast-iron angle iron in Fig. 4-122. Cast iron weighs 0.284 lb. per cubic inch.

18. Find the number of cubic yards of concrete required to make six pillars such as the one shown in Fig. 4-123.

19. Estimate the weight of 6 cast-iron bearing washers having the dimensions shown in Fig. 4-124. (Cast iron weighs 0.284 lb. per cubic inch.)

20. Find the length in meters of the wall section shown in Fig. 4-125.

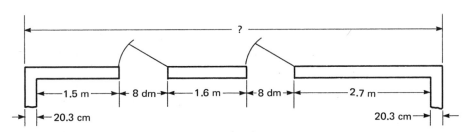

Fig. 4-125.

21. Three pieces, 3.8 decimeters, 0.9 decimeters, and 12.4 centimeters, are cut from a pipe which is 5.6 meters long. How many meters long is the remaining pipe?

22. Determine the number of square decimeters to 2 decimal places of material in the sheet metal piece shown in Fig. 4-126.

23. A room measures 4.4 meters wide, 6.8 meters long, and 2.6 meters high. Allowing 0.9 square meters for each of three windows and 1.7 square meters for a doorway, determine the actual number of square meters of wall surface in the room. Give answer to 2 decimal places.

24. Find the number of square centimeters of material contained in the flange plate illustrated in Fig. 4-127. Give answer to three decimal places.

25. A cylindrical tank containing fuel oil is 65 percent full. The tank is 5.6 meters high and 3.9 meters in diameter. How many cubic meters of fuel oil are required to fill the tank. Give answer to 2 decimal places.

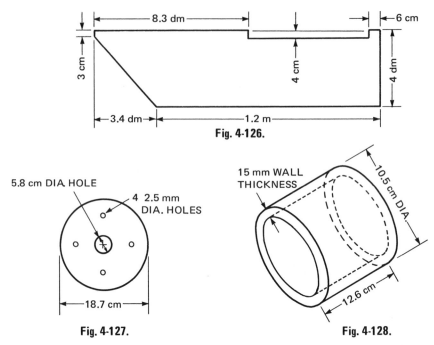

Fig. 4-126.

5.8 cm DIA HOLE

4 2.5 mm DIA. HOLES

18.7 cm

Fig. 4-127.

15 mm WALL THICKNESS

10.5 cm DIA.

12.6 cm

Fig. 4-128.

26. Determine the number of cubic centimeters of zinc contained in the casting shown in Fig. 4-128. Give answer to 3 decimal places.

27. A system of drain pipes is capable of discharging 200 liters of water per minute. How many kiloliters of water are discharged in 35 minutes?

28. Determine the total number deciliters of liquid in three vessels which contain 0.25 liter, 14.8 centiliters, and 84 milliliters, respectively.

29. The steel beam shown in Fig. 4-129 weighs 7 800 kilograms per cubic meter. Determine the weight of 50 beams in metric tons to three decimal places.

30. A triangular plate is formed of copper which is 1.2 centimeters thick. The base of the plate is 14.6 centimeters and the altitude is 11.5 centimeters. Copper weighs 8.8 grams per cubic centimeter. Determine the weight of the plate in kilograms to 3 decimal places.

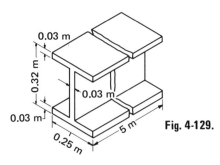

0.03 m

0.32 m

0.03 m

0.03 m

0.25 m

5 m

Fig. 4-129.

5 Formulas and Important Ideas from Algebra

Algebra is a branch of mathematics in which letters are used to represent numbers. In Unit 4 problems are solved using expressions such as $A = lw$ and $A_{tz} = 1/2\, a\,(B + b)$. By the use of letters, general rules called *formulas* can be stated mathematically. Many operations in shop and industrial work are expressed as formulas. A knowledge of algebra fundamentals is essential in the use of trade handbooks. Algebra is often required in solving geometry and trigonometry problems.

The basic principles of algebra presented in this text are intended to provide a practical background for trade applications. Algebra is an extension of arithmetic; therefore, the rules and procedures of arithmetic also apply to algebra. Many problems which are difficult or impossible to solve by arithmetic are easily solved by algebra.

Symbolism

Symbols are the language of algebra. Both arithmetic numbers and literal numbers are used in algebra.

Arithmetic numbers are numbers which have definite numerical values, such as 5, 7.2, and 3/4.

Literal numbers are letters which represent arithmetic numbers, such as *a*, *x*, *V*, and *P*. Depending on how it is used, a literal number can represent one particular arithmetic number, a wide range of numerical values, or all numerical values.

Customarily the multiplication sign (x) is not used in algebra because it can be misinterpreted as the letter *x*. When a literal number (letter) is multiplied by a numerical value, or when two or more literal numbers are multiplied, no sign of operation is required. For example, 6 x *b* is written 6*b*; *a* x *b* is written *ab*, 5 x *V* x *T* is written 5*VT*.

Parentheses ( ) or raised dots (·) are sometimes used in place of the multiplication sign (x) when numerical values are multiplied. For example, 2 x 5 can be written as 2(5) or 2 · 5, and 1/2 x 2.3 x 9 can be written as 1/2 (2.3)(9) or 1/2 · 2.3 · 9.

An *algebraic expression* is a word statement put into mathematical form by using literal numbers, arithmetic numbers, and signs of operation.

Examples of algebraic expressions

Example 1. The statement "subtract x from 12" is expressed algebraically as 12-x.

Example 2. The statement "the product of 5 and x divided by y" is expressed algebraically as $5x \div y$ or $5x/y$.

Example 3. Figure 5-1 shows a rectangle whose length is increased by 3/4 inch. If ℓ is the original length, the increased length is $\ell + 3/4$. The word statement is expressed algebraically as $\ell + 3/4$.

Example 4. If one machine produces d pieces per hour, five machines produce five times the number of pieces per hour which is stated algebraically as $5d$ pieces.

Example 5. A concrete slab is shown in Figure 5-2. Distance A is x feet, distance B is 1/2 of distance A or $1/2x$, and distance C is twice distance A or $2x$. The total length is expressed algebraically as $x + 1/2x + 2x$ or $3\,1/2x$.

Note: If no arithmetic number appears before a literal number, it is assumed that the value is the same as if a one (1) appeared before the letter, $x = 1x$.

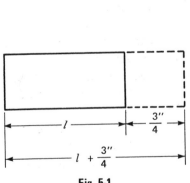

Fig. 5-1.

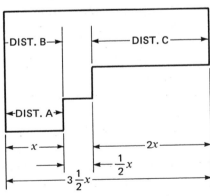

Fig. 5-2.

Example 6. Figure 5-3 shows a plate with four drilled holes. The distance from the left edge of the plate to hole 1 is b inches. The distance between holes 1 and 2, holes 2 and 3, and holes 3 and 4 are each equal to c inches. The distance from hole 4 to the right edge of the plate is d inches. The total length of the plate is $b + c + c + c + d$, or $b + 3c + d$.

Note: Only like literal numbers may be arithmetically added.

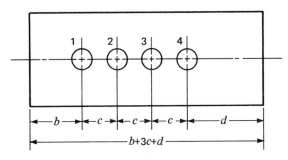

Fig. 5-3.

EXERCISES

Express each of the following exercises as algebraic expressions.

1. Add 5 to x.
2. The sum of a and b.
3. Subtract 3 from c.
4. Subtract c from 3.
5. The product of 9 and y.
6. Multiply V times P.
7. Divide r by S.

8. Twice b minus 2.
9. The product of 3 and x, plus y.
10. The sum of d and e, divided by f.
11. Divide 15 by x and subtract 2.
12. Add b to the product of one-half c and d.

PROBLEMS

Express each of the following problems as algebraic expressions:

1. a. What is the total length of this part?
   b. What is the length from point A to point B?

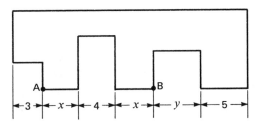

2. What is the distance between the indicated points?
   a. point A to point B
   b. point F to point C
   c. point B to point C
   d. point D to point E

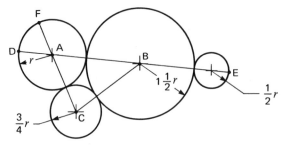

3. What is the length of each of the following dimensions?
   a. dimension A
   b. dimension B
   c. dimension C

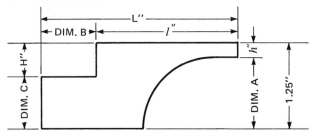

4. Stock is removed in two operations from a block n inches thick. A milling operation removes p inches and a grinding operation removes t inches. What is the final thickness of the block?

5. Given: s as the length of a side of a regular hexagon, r as the radius of the inside circle, and R as the radius of the outside circle.
   a. What is the length of r if r equals the product of 0.866 and the length of a side of the hexagon?
   b. What is the length of R if R equals the product of 1.155 and the radius of the inside circle?

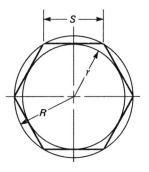

Expressing formulas as rules

A *formula* is a brief way of stating a general rule or a relation among numbers. When one knows the name of the quantity that each letter (literal number) represents, one should be able to state the rule expressed by the formula. Examples of expressing formulas as rules:

Example 1. If A represents the area of a rectangle, ℓ its length, and w its width, state the rule expressed by the formula $A = \ell w$.

THINK	DO THIS
To state the rule we say "area" where this formula has A, "length" where it has ℓ and "width" where it has w.	$A = \ell w$ means: "The area equals ℓ times w," or "to find the area of a rectangle multiply the length by the width."

Example 2. If P represents the perimeter of a rectangle, ℓ its length, and w its width, state the rule expressed by the formula $P = 2\ell + 2w$.

THINK	DO THIS
To state the rule we say "perimeter" where this formula has P, "length" where the formula has ℓ, and "width" where it has w.	$P = 2\ell + 2w$ means: "The perimeter equals 2 times ℓ plus 2 times w"; or "To find the perimeter of a rectangle, add twice the length and twice the width."

Example 3. If A represents the average of two numbers and b and c represent the two numbers, state the rule that is expressed by the formula $A = \dfrac{b+c}{2}$.

THINK	DO THIS
The horizontal line under $b + c$ means divide, and $b + c$ means "add b and c."	$A = \dfrac{b+c}{2}$ means that the average of two numbers is the sum of the numbers divided by 2; or "To find the average A, of two numbers b and c, add the two numbers and divide the sum by "2".

Summary

To state in words the rule given by a formula:

1. Determine what quantity each of the letters of the formula represents.
2. Using the words in place of the letters, state the rule by expressing the operation given by the signs of the formula.

EXERCISES

State the rule given by each of the following formulas:

1. $P = 4s$, where P is the perimeter of a square and s is the length of a side.

2. $D = 30t$, where D represents the distance in miles and t the number of hours.

3. $C = \pi d$, where C is the circumference of a circle and d its diameter.

4. $P_c = (\pi D)/N$, where P_c, the circular pitch, is the distance between centers of adjoining teeth on a gear, D is the diameter of the pitch circle of the gear, and N the number of teeth in the gear.

5. $S = (a + b + c)/2$, where S is the semiperimeter of a triangle and a, b, and c are the three sides of the triangle.

6. $S = C \times rpm$, where S represents the cutting speed of a drill, C its circumference, rpm the number of revolutions per minute the drill makes.

7. $G = S - C$, where G represents the gain on an article, S the selling price, and C the cost of the article.

8. $W = fs$, where W represents the work in foot-pounds, s the distance in feet, and f the force exerted in pounds.

9. $H = Qt$, where H represents the amount of heat in BTU (British thermal units), Q the quantity of water in pounds, and t the change in temperature in degrees Fahrenheit.

10. $R = (K\ell)/C.M.$, where R represents the resistance in ohms, K the resistance of one mil-foot of wire, $C.M.$ the cross section of the wire in circular mils, and ℓ the length of the wire in feet.

11. $T = (D - d)/L$, where T equals the taper per inch, D the diameter of the large end, d the diameter of the small end, and L the length of the taper.

12. $F = 9/5\ C + 32$, where F represents degrees Fahrenheit and C degrees Celsius.

13. $I = (E_a - E_g)/R$, where I represents armature current flowing through a motor armature, E_a applied voltage, E_g counter-voltage, and R resistance.

14. $R = 1/(\mu A)$, where R represents the reluctance of a magnetic circuit, μ (read "mu") the permeability, and A the cross-sectional area of the magnetic circuit.

Solving formulas not only requires the individual operations of addition, subtraction, multiplication, and division. It is also important that the student learn how to perform the operations of powers and roots and is able to solve expressions consisting of combinations of several operations.

Powers

Two or more numbers multiplied together to produce a given number are *factors* of the given number. Factors of 10 are 2 and 5, 2 x 5 = 10; factors of 24 are 4 and 6, 4 x 6 = 24; factors of 6y are 6 and y; factors of abc are a, b, and c.

A *power* is the product of two or more equal factors.

3 x 3 is the second power of 3; 5 x 5 x 5 is the third power of 5; and d x d x d x d is the fourth power of d.

An *exponent* shows the number of times a number is to be taken as a factor. It is written above and to the right of the number.

Example 1. 4^2 means 4 x 4. The exponent 2 shows that 4 is taken as a factor twice. It is read "4 to the second power" or "4 squared"; $4^2 = 16$.

Example 2. 2^5 means 2 x 2 x 2 x 2 x 2. It is read as "two to the fifth power"; $2^5 = 32$.

Example 3. 0.8^3 means 0.8 x 0.8 x 0.8. It is read as "eight-tenths cubed"; $0.8^3 = 0.512$.

Example 4. $(7/8)^2$ means 7/8 x 7/8. It is read as "seven-eighths squared"; $(7/8)^2 = 49/64$.

Note: Both the numerator and denominator are squared.

Example 5. y^4 means y x y x y x y. It is read as "y to the fourth power."

When no exponent is written, the exponent is understood to be 1; $3^1 = 3$, $x^1 = x$.

EXERCISES

Raise each of the following problems to the indicated power.

1. 5^2	6. 0.1^4	11. 0.3^5	16. $(3/4)^2$
2. 1.6^2	7. 0.02^2	12. 2.1^5	17. $(1/2)^3$
3. 6^3	8. 0.02^3	13. 1^5	18. $(1/4)^1$
4. 2.1^3	9. 15^1	14. 1^6	19. $(4\ 1/2)^2$
5. 10^4	10. 3^5	15. 0.2^6	20. $(8\ 1/10)^3$

Roots

The *root* of a number is a quantity which is taken two or more times as an equal factor of the number. Determining a root is the opposite operation of determining a power.

The symbol $\sqrt{}$, called a *radical sign*, indicates a root of a number.

A number called the *index* is written to the left and above the radical sign. The index indicates the number of times that a root is to be taken as an equal factor to produce the given number. For example, the square root of 9 is written $\sqrt[2]{9}$. It asks the question, "What number multiplied by itself equals 9?" Since 3 x 3 = 9, 3 is the square root of 9. Therefore $\sqrt[2]{9}$ = 3.

Note: The index ² is usually omitted in problems requiring square roots; $\sqrt[2]{9}$ is written $\sqrt{9}$.

Example 1. Compute $\sqrt{36}$

THINK	DO THIS
What is the square root of 36? What number multiplied by itself equals 36?	6 x 6 = 36; therefore, $\sqrt{36}$ = 6 *Ans.*

Example 2. Compute $\sqrt[3]{125}$.

THINK	DO THIS
What is the cube root of 125? What number used as a factor 3 times equals 125?	5 x 5 x 5 = 125; therefore, $\sqrt[3]{125}$ = 5. *Ans.*

Example 3. Compute $\sqrt[4]{81}$.

THINK	DO THIS
What is the fourth root of 81? What number used as a factor 4 times equals 81?	3 x 3 x 3 x 3 = 81; therefore, $\sqrt[4]{81}$ = 3. *Ans.*

If a root of a fraction is to be computed, both the numerator and denominator must be enclosed within the radical sign. For example, $\sqrt{16/25}$ shows that the square root of the complete fraction is to be taken. Compute the square root of both the numerator and denominator: $\sqrt{16}$ = 4; $\sqrt{25}$ = 5; 4/5 x 4/5 = 16/25. Therefore, $\sqrt{16/25}$ = 4/5.

When working with roots of fractions, be careful to distinguish whether the root of the entire fraction or only part of the fraction is to be taken.

Example 1. Only the root of the numerator is taken.

$$\frac{\sqrt{16}}{25} = \frac{\sqrt{4 \times 4}}{25} = \frac{4}{25} \; Ans.$$

Example 2. Only the root of the denominator is taken.

$$\frac{16}{\sqrt{25}} = \frac{16}{\sqrt{5 \times 5}} = \frac{16}{5} = 3 \; 1/5 \; Ans.$$

Expressions sometimes consist of two or more operations within the radical sign. For example, the expression $\sqrt{a + b}$ is read as "the square root of a plus b." It is solved by first adding a and b, then computing the square root of the sum.

To solve problems which involve operations within the radical sign, apply the following procedure

1. Perform the operations within the radical sign first.
2. Then, compute the root.

Example 1. Compute $\sqrt{3 \times 12}$.

THINK	DO THIS
1. First, multiply 3 x 12	1. $3 \times 12 = 36$
2. Then, compute the root. What number multiplied by itself equals 36?	2. $6 \times 6 = 36$, therefore, $\sqrt{36} = 6$ *Ans.*

Example 2. Compute $\sqrt{5 + 59}$.

Procedure: $\sqrt{5 + 59} = \sqrt{64} = \sqrt{8 \times 8}$; therefore $\sqrt{5 + 59} = 8$ *Ans.*

Example 3. Compute $\sqrt{128 - 7}$.

Procedure: $\sqrt{128 - 7} = \sqrt{121} = \sqrt{11 \times 11}$; therefore $\sqrt{128 - 7} = 11.$ *Ans.*

Example 4. Compute $\sqrt{32/2}$.

Procedure: $\sqrt{32/2} = \sqrt{16} = \sqrt{4 \times 4}$; therefore $\sqrt{32/2} = 4$ *Ans.*

EXERCISES

Determine the roots of the following problems.

1. $\sqrt{4}$	4. $\sqrt{100}$	7. $\sqrt[3]{64}$
2. $\sqrt{16}$	5. $\sqrt{144}$	8. $\sqrt[3]{1,000}$
3. $\sqrt{49}$	6. $\sqrt[3]{27}$	9. $\sqrt[4]{10,000}$

10. $\sqrt[5]{32}$ 14. $\sqrt[3]{8/27}$ 18. $\sqrt{0.5 \times 8}$

11. $\sqrt{4/25}$ 15. $\sqrt[3]{64/125}$ 19. $\sqrt[3]{428.8/6.7}$

12. $\sqrt{9/100}$ 16. $\sqrt{21 + 79}$ 20. $\sqrt{3/4 \times 3/4}$

13. $\sqrt{16/81}$ 17. $\sqrt{16.4 - 7.4}$

General method of computing square roots

The square root examples shown have all consisted of perfect squares or fractions having numerators and/or denominators which are perfect squares. *Perfect squares* are numbers which have whole number square roots. These roots are relatively easy to determine by observation.

Most numbers do not have whole number square roots; therefore, a definite procedure must be used in computing square roots of most numbers. The procedure has the advantage of eliminating estimating, and proceeds directly to the required square root to any degree of accuracy required. The following examples demonstrate this procedure. Study them carefully.

Example 1. Find the square root of 3,969.

THINK	DO THIS
1. Begin at the decimal point and mark off the digits in groups of two. Here 69 and 39 are the "pairs." They are called *periods*.	1. 6 3 '39'69. 2. 36 3, 4. 123 $\overline{)\,369}$ 5. 369 6. 0
2. Determine the largest digit which, when squared, is contained in the first period. (The left-hand one. This period may sometimes contain only one digit.) Here 36, the square of 6, is the largest square of a digit that is contained in 39, the first period. Write 36 under the 39 and subtract. Put 6 in the answer as shown.	

3. Bring down the second period, 69, getting 369.

From this point forward the procedure is different from steps 1, 2, 3, but this new procedure is repeated over and over again. The new procedure involves the following three steps:

4. Multiply the digit in the answer, 6, by 2; place the product, 12, in a box to the left of 369.

5. Divide 12 into all but the last digit of 369.12 into 36 = 3. Put this result after the 12 to get 123, and also after the 6 in the answer, giving 63.

6. Multiply 123 by the last digit in the answer, 3; place the product under 369 and subtract.

7. Check by squaring 63.

7. Check 63 x 63 = 3,969

Here one application of the procedure of steps 4, 5, and 6 resulted in 0. The answer 63 is exactly the square root of 3,969.

Ans. 63 is $\sqrt{3,969}$

Example 2. Calculate the square root of 467.953 to two decimal places.

THINK

DO THIS

$$2\ 1.\ 6\ 3$$

1. Beginning at the decimal point, mark off the digits of the number into periods of two digits, going both to the left, ←, and to the right, ⟶ . Affix one 0 to the decimal part to complete the right-hand period.

Place a decimal point in the answer immediately above the decimal point of the number.

1.
2. 4'67.95'30
3. 4
4, 5. 41⟌ 67
       41
6, 7, 8. 426⟌2695
          2556
9. 4323⟌13930
    12969
    961

2. Determine the largest square contained in the first period. The first period is 4; the largest square contained in it is 4.

3. Write 4 under the first period, 4, and subtract. Write 2 above the first period. Bring down the second period, 67.

From this point forward the procedure differs from that of steps 1, 2 and 3, but this new procedure is repeated over and over again.

4. Multiply the digit in the answer, 2, by 2, get 4; make a box to the left of the last number brought down (67) and place this result in the box.

5. Cover the last digit of 6 ⑦ , divide the 6 by the factor 4. Put the resulting factor, 1, in the answer above the second period and also in the box beside the 4 to make 41.

6. Multiply the factor 41 by 1. Put the result under 67, subtract and bring down the next period, giving 2695.

From here on the procedures given in steps 4, 5 and 6 are repeated until the answer, the square root, is obtained to the required number of decimal places.

7. (Like Step 4.) Multiply the digits obtained thus far in the answer, 21, by 2, which gives 42. Make a box to the left of the last number brought down, 2695, and place this result, 42, in the box.

```
           2  1. 6  3
          ⎺⎺⎺⎺⎺⎺⎺⎺⎺⎺
 1.       ← ←  → →
 2.       4'67.95'30
 3.       4
 4, 5.    41⟌67
            41
 6, 7, 8. 426⟌2695
            2556
 9.       4323⟌13930
            12969
              961
```

8. Cover the last digit of 269⑤ , divide 42 into the 269. Put the result, 6, in the answer above the third period and in the box beside 42, making that factor 426.

9. Multiply the new factor 426 by 6. Put the result, 2556, under 2695; subtract and bring down the next period, getting 13930.

10. Repeat steps 4, 5, 6 as needed to obtain the square root to the required number of decimal places. If the last operation does not give zero, the square root is approximate.

Ans. 21.63 is the square root of 467.953 to two decimal places.

Summary

To obtain the square root of any number, follow steps 1 to 6 inclusive as shown in Example 2. Repeat steps 4, 5, and 6 as needed to obtain the square root to the required number of decimal places. The result when squared should approximately equal the given number.

EXERCISES

Find the square roots of the numbers in Exercises 1-15. Carry the square root as far as hundredths, if necessary, but no further.

1. 225	6. 18,769	11. 9,234
2. 629	7. 1,814.76	12. 6.1933
3. 512	8. 62.41	13. 81.077
4. 4,225	9. 0.1369	14. 0.0569
5. 2,601	10. 32.671	15. 1.7272

Table 4 in the Appendix can be conveniently used to find the squares, square roots, cubes, and cube roots of whole numbers from 1-100.

Evaluation of algebraic expressions

The numerical value of a formula or any algebraic expression is found by substituting given numerical values for literal values (letters) and solving the expressions by following the proper order of operations as in arithmetic.

Usually algebraic expressions are made up of two or more arithmetic operations. Therefore, it is essential that one applies the proper order of operations in solving mathematical expressions.

Perform operations in the following order:

1. All operations within parentheses. If there are parentheses within parentheses, perform the operations within the innermost parentheses first.
2. Powers and roots. Powers and roots are determined as they occur.
3. Multiplication and division. Perform multiplication and division operations in the order in which they occur from left to right.
4. Addition and subtraction. Addition and subtraction are performed in order from left to right.

Parentheses are used to indicate the quantity upon which some operation is to be performed. Expressions inside parentheses are treated as single quantities. If we write $2(5 + 4)$, the parentheses around the expression $5 + 4$ show that $5 + 4$ must be first added, then the sum "9" is multiplied by 2. $2(5 + 4) = 2 \times 9 = 18$.

The procedure for determining the numerical value of one letter of an algebraic expression when the numerical values of the other letters are known is as follows:

1. Write the algebraic expression.
2. Replace each letter in the expression by its numerical value.

Wherever multiplication is required, insert a multiplication sign.

3. Perform the operations in the proper order.
4. The result is the numerical value required.

Example 1. If $r = 240$ and $C = 2.5$, find the numerical value of S in the formula $S = rC$.

THINK	DO THIS
1. Write the formula.	1. $S = rC$
2. Replace each letter in the formula by its given numerical value. The only operation required is multiplication. Insert the multiplication sign.	2. $S = 240 \times 2.5$
3. Perform the multiplication operation.	3. $S = 600$ *Ans.*

Example 2. Determine the perimeter of the rectangle shown in Fig. 5-4. The perimeter of a rectangle $= 2\ell + 2w$; $P = 2\ell + 2w$.

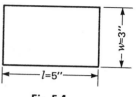

Fig. 5-4.

THINK	DO THIS
1. Write the formula.	1. $P = 2\ell + 2w$
2. Replace each letter in the formula by its given numerical value. Insert the multiplication sign where necessary.	2. $P = 2 \times 5 + 2 \times 3$
3. Perform the operations in the proper order. a. First: multiply. b. Then: add.	3. a. $P = 10 + 6$ b. $P = 16''$ *Ans.*

Example 3. If $B = 10$, $b = 8$ and $a = 4.5$, find the value of A, if $A = 1/2a$ $(B + b)$

THINK	DO THIS
1. Write the formula.	1. $A = 1/2a (B + b)$
2. Replace each letter in the formula by its given numerical value. Insert the multiplication sign where necessary.	2. $A = 1/2 \times 4.5 \times (10 + 8)$
3. Perform the operations in the proper order. a. First: perform addition within parentheses. b. Then: multiply.	3. a. $A = 1/2 \times 4.5 \times 18$ b. $A = 2.25 \times 18 = 40.5$ *Ans.*

Example 4. Determine the area of the ring shown in Fig. 5-5. The area of a ring $= \pi R^2 - \pi r^2$; $A = \pi R^2 - \pi r^2$; ($\pi = 3.14$).

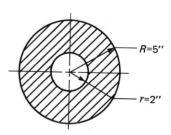

Fig. 5-5.

THINK	DO THIS
1. Write the formula.	1. $A = \pi R^2 - \pi r^2$
2. Replace each letter in the formula by its given numerical value. Insert the multiplication sign where necessary.	2. $A = 3.14 \times 5^2 - 3.14 \times 2^2$
3. Perform the operations in the proper order.	3. a. $A = 3.14 \times 25 - 3.14 \times 4$
a. First: perform the squaring operations.	b. $A = 78.50 - 12.56$
b. Second: multiply.	c. $A = 65.94$ sq. in. *Ans.*
c. Third: subtract.	

Example 5. Determine the perimeter of the ellipse shown in Fig. 5-6. The approximate perimeter of an ellipse equals $\pi\sqrt{2(a^2 + b^2)}$;
$P = \pi\sqrt{2(a^2 + b^2)}$,
$(\pi = 3.14)$

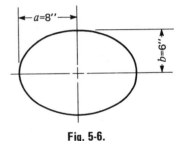

Fig. 5-6.

THINK	DO THIS
1. Write the formula.	1. $P = \pi\sqrt{2(a^2 + b^2)}$
2. Replace each letter in the formula by its given numerical value. Insert the multiplication sign where necessary.	2. $P = 3.14 \times \sqrt{2 \times (8^2 + 6^2)}$
3. Perform the operations in the proper order. The operations within the radical sign must be performed before multiplying by π.	3. a. $P = 3.14 \times \sqrt{2 \times (64 + 36)}$
	b. $P = 3.14 \times \sqrt{2 \times 100}$
	c. $P = 3.14 \times \sqrt{200}$
	d. $P = 3.14 \times 14.14$
a. First: perform the squaring operations within the parentheses.	e. $P = 44.40$ in. *Ans.*
b. Second: and the values within the parentheses.	
c. Third: multiply the values within the radical sign.	
d. Fourth: perform the square root operation.	
e. Fifth: multiply.	

EXERCISES

Substitute the given numbers for letters and find the value of the unknown in each of the following exercises:

1. Find p, when $S = 8$ in $p = 4S$.
2. Find C, when $\pi = 3.14$ and $r = 10$ in $C = 2\pi r$.
3. Find S, when $C = 20$ and $g = 8.5$ in $S = c + g$.
4. Find F, when $C = 80$ in $F = 1.8C + 32$.
5. Find B, when $A = 65$ in $B = 90 - A$.
6. Find b, when $p = 72$ and $e = 18$ in $b = p - 2e$.
7. Find a, when $n = 12$ in $a = 360/n$.
8. Find V, when $B = 30$ and $h = 2.5$ in $V = (Bh)/3$.
9. Find V, when $e = 7$ in $V = e^3$.
10. Find A, when $\pi = 3.14$ and $r = 2$ in $A = \pi r^2$.
11. Find F, when $W = 40, v = 30, g = 32$, and $r = 12$ in $F = (wv^2)/gr$.
12. Find p, when $\ell = 25$ and $w = 15$ in $p = 2(\ell + w)$.
13. Find A, when $\pi = 3.14, r = 3$, and $h = 10$ in $A = 2\pi r(r + h)$.
14. Find A, when $h = 6, b = 12$, and $b' = 8$ in $A = (h/2)(b + b')$.
15. Find c, when $a = 12$ and $b = 5$ in $c = \sqrt{a^2 + b^2}$.
16. Find R, when $V = 6{,}280, \pi = 3.14$, and $H = 2$ in $R = \sqrt{V/(\pi \times H)}$.
17. Find C, when $H = 15$ and $R = 3.75$ in $C = 2\sqrt{H(2R + H)}$.

PROBLEMS

For problems 6-10, the formula to be used is given at the right. The letters used in each formula are given in parentheses in the problem.

Problems	*Formula*
6. How much current (I) does a heater having a resistance (R) of 30 ohms take if the applied voltage (E) is 120 volts?	$I = \dfrac{E}{R}$
7. How much power (P) is lost in a transmission line having a resistance (R) of 0.6 ohms and carrying a current (I) of 40 amperes?	$P = I^2 R$

8. Find the horsepower developed by an engine having a piston diameter (d) of 4″, a stroke length (L) of 5″ and a mean effective pressure (P) of 42 lb. per square inch, and turning at the rate of 275 rpm (N). In this formula (A) represents the area of the piston and is equal to $0.7854d^2$.

$$h.p. = \frac{PLAN}{33,000}$$

9. Castings weighing 15 lb. each (f) need to be moved to a new location 25 feet away (s). How many foot-pounds of work (W) are done in moving one casting? How much work is done in moving 1500 castings?

$$W = f \times s$$

10. The diameter (D) of the large end of a taper plug is 1 7/8″, and that of the small end (d) is 1 5/8″. The length of the taper (l) is 4″. What is the taper per inch (T)?

$$T = \frac{D - d}{l}$$

For problems 11-22, refer to the illustrations, substitute given numerical values, and find the value of each of the following problems.

11. a. Area = $(1/2)d^2$, Area = ?
    b. $S = 0.7071\ d$, S = ?

12. a. $l = (\pi R\alpha)/180$, l = ?
    b. Area = $(1/2)Rl$, Area = ?

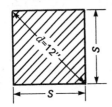

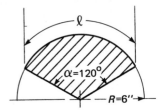

13. a. $S = (1/2)(a + b + c)$, S = ?
    b. Area = $\sqrt{S\ (S - a)\ (S - b)\ (S - C)}$, Area = ?

14. a. $r = (c^2 + 4h^2)/8h$, r = ?
    b. $l = 0.0175\ r\alpha$, l = ?

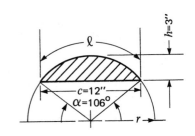

15. Area = $\dfrac{(H + h)\, b + ch + aH}{2}$

    Area = ?

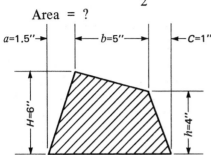

16. Length of belt on pulleys =

    $2C + \dfrac{11D + 11d}{7} + \dfrac{(D - d)^2}{4C}$

    Length of belt = ?

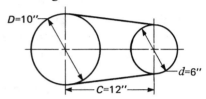

17. Area = $dt + 2a(s + n)$

    Area = ?

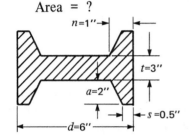

18. Area = $\pi(ab - cd)$

    Area = ?

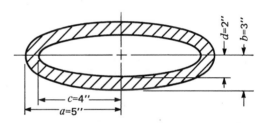

19. Area = $\dfrac{\pi(R^2 - r^2)}{2}$

    Area = ?

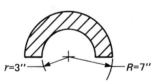

20. Area = $t[b + 2(a - t)]$

    Area = ?

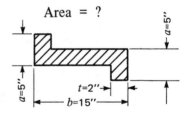

21. a. $S = \sqrt{(R - r)^2 + h^2}$, $S = ?$

    b. Volume =
       $1.05\, h\, (R^2 + Rr + r^2)$
       Volume = ?

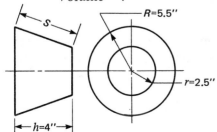

22. Volume = $\dfrac{(2a + c)bh}{6}$

    Volume = ?

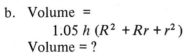

Signed numbers

All the numbers which have been used in this book have been numbers larger than zero. Numbers less than zero are now introduced so that many other problems can be solved.

Numbers that are preceded by a plus or a minus sign are called *signed numbers.* Signed numbers are sometimes required for solving problems in mechanics and trigonometry. A plus sign indicates that the number is greater than zero, and a minus sign indicates that the number is less than zero.

A number which has no sign, or one which is preceded by a plus sign, is a *positive number.* The numbers used in arithmetic are all positive numbers. For example, +8 is the same as 8. They both mean 8 units greater than 0. A number which is preceded by a minus sign is a *negative number.* For example, –8 is a negative number and means 8 units less than 0.

The number scale

The *number scale* in Fig. 5-7 shows the relationship of positive and negative numbers. It shows distance and direction between numbers. Considering a number as a starting point and counting to a second number to the right represents positive (+) direction. Counting to the left represents negative (–) direction.

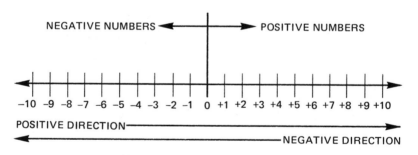

Fig. 5-7.

Examples of using the number scale.

Example 1. Starting at –2 and counting to the right to +6 represents 8 units in a positive (+) direction; +6 is 8 units greater than –2.

Example 2. Starting at +6 and counting to the left to –2 represents 8 units in a negative (–) direction; –2 is 8 units less than +6.

Example 3. Starting at –3 and counting to the left to –10 represents 7 units in a (–) direction; –10 is 7 units less than –3.

Example 4. Starting at –9 and counting to the right to 0 represents 9 units in a (+) direction; 0 is 9 units greater than –9.

EXERCISES

For exercises 1-20, refer to the number scale below and give the direction (+ or –) and the number of units counted going from the first to the second number.

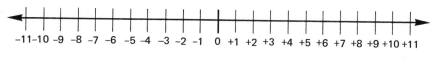

1. –11 to –2
2. –8 to –1
3. –6 to 0
4. –2 to –8
5. +2 to –8
6. +3 to +9
7. +10 to –10

8. +10 to 0
9. +2 to +7
10. +9 to +1
11. +11 to 0
12. 0 to –4
13. –7.5 to +10
14. +10 to –7.5

15. –10.8 to 2.3
16. –2.3 to 0.8
17. 2.3 to 0.8
18. +7 1/2 to 2 1/4
19. 6 7/8 to 0
20. 4 1/8 to 1/8

For exercises 21-32, select the greater of the two signed values, and indicate the number of units by which it is greater.

21. +5, –14
22. +7, –3
23. –6, –2
24. +8, +13

25. +20, –22
26. –18, –4
27. +14.3, +21
28. –1.8, +1.8

29. +17.6, –21.9
30. –2 1/4, –5
31. 6 3/8, –1 1/4
32. –32 1/16, 30

For exercises 33-38, list the signed numbers in order of increasing value starting with the smallest number.

33. +17, –1, +2, 0, –18, +4, –22
34. –5, +5, 0, +13, +27, –21, –2, –19
35. +10, –10, –7, +7, 0, +25, –25, +14
36. 0, 15, –3.6, –2.5, –14.9, +17, +0.3
37. 13, –13, –1, +1, –3.2, +18.7
38. –16, 14 1/8, –13 7/8, +6, –4 3/8

Operations using signed numbers

In order to solve problems in algebra, one must be able to perform the basic operations of addition, subtraction, multiplication, division, roots, and powers. The procedures for performing certain operations of signed numbers are based on an understanding of absolute value. The *absolute value* of a number is the number without a sign. For example, the absolute value of +7 is 7, the absolute value of –7 is also 7. Therefore, the absolute value of +7 and –7 is the same value, 7. The absolute value of –10 is 8 greater than the absolute value of +2; 10 is 8 greater than 2.

The following procedures and examples show how to perform the six basic operations of signed numbers. It is essential that one learn and apply these procedures.

Addition of signed numbers

Procedure for adding two or more positive numbers:

Add the numbers as in arithmetic.

Examples: Add the following numbers.

1. $+2$
   $+8$
   $+10$ or 10 *Ans.*

2. 14
   6
   $+20$ or 20 *Ans.*

3. $4 + 8 + 10 = +22$
   or 22 *Ans.*

4. $+12 + (+5) = +17$
   or 17 *Ans.*

Procedure for adding two or more negative numbers:

1. Add the absolute values of the numbers.
2. Prefix a minus sign.

Example 1. Add –4 and –12.

THINK

1. The absolute values of –4 and –12 are 4 and 12. Add 4 and 12.

2. Prefix a minus sign.

DO THIS

1. 4
   $+12$
   16

2. –16 *Ans.*

Example 2. Add –9, –4, and –12. Procedure: add absolute values 9 + 4 + 12 = 25; prefix a minus sign, –25. *Ans.*

Example 3. $-6 + (-5) = -11$ *Ans.*

Example 4. $-8 + (-10) + (-4) + (-1) = -23$ *Ans.*

Procedure for adding a positive and a negative number:

1. Subtract the smaller absolute value from the larger absolute value.
2. Prefix the sign of the number having the larger absolute value.

Example 1. Add +6 and –2.

THINK

1. The absolute values of +6 and –2 are 6 and 2. The absolute value of 2 is smaller than the absolute value of 6. Subtract 2 from 6.

2. The larger absolute value, 6, originally was +6. Prefix the + sign.

DO THIS

1. 6
   –2
   ──
    4

2. +4 or 4 *Ans.*

Example 2. Add –8 and +3.

THINK

1. The absolute values of –8 and +3 are 8 and 3. The absolute value of 3 is smaller than the absolute value of 8. Subtract 3 from 8.

2. The larger absolute value, 8, originally was –8. Prefix the – sign.

DO THIS

1. 8
   –3
   ──
    5

2. –5 *Ans.*

Example 3. +12 + (–8) = +4 or 4 *Ans.*

Example 4. –12 + (+8) = –4 *Ans.*

Example 5. –25 + (+25) = 0 *Ans.*

Procedure for adding more than two positive and negative numbers:
1. Add all the positive numbers.
2. Add all the negative numbers.
3. Add their sums following the procedure for adding signed numbers.

Example 1. Add: –3 + 8 + (–12) + 2.

THINK

1. Add all the positive numbers.

2. Add all the negative numbers.

3. Add the sums: add +10 and –15. The absolute values of +10 and –15 are 10 and 15.

   a. Subtract 10 from 15.
   b. The larger absolute value 15 originally was –15. Prefix the – sign.

DO THIS

1. +8 + (+2) = +10

2. –3 + (–12) = –15

3. a. 15 – 10 = 5
   b. –5 *Ans.*

Example 2. Add: $8 + 7 + (-6) + 4 + (-3) + (-5) + 10$. Add all positive numbers: $8 + 7 + 4 + 10 = 29$. Add all the negative numbers: $-6 + (-3) + (-5) = -14$. Add the sums: $+29 + (-14) = 29 -14 = 15$. *Ans.*

Example 3. $4 + (-6) + 12 + 3 + (-7) + 1 + (-5) + (-2) = [(4 + 12 + 3 + 1)] + [(-6) + (-7) + (-5) + (-2)] = 20 + (-20) = 0$ *Ans.*

Subtraction of signed numbers

Procedure for subtracting signed numbers:

1. Change the sign of the number subtracted (subtrahend) to the opposite sign.
2. Follow the procedure for addition of signed numbers.

Note: When the sign of the subtrahend is changed, the problem becomes one in addition. Therefore, subtracting a negative number is the same as adding a positive number, and subtracting a positive number is the same as adding a negative number.

Example 1. Subtract 7 from 3.

THINK	DO THIS
1. Change the sign of the subtrahend to the opposite sign.	1. -7
2. Follow the procedure for addition of signed numbers.	2. $3 + (-7) = -4$ *Ans.*

Example 2. Subtract -7 from 3.

THINK	DO THIS
1. Change the sign of the subtrahend to the opposite sign.	1. $+7$
2. Follow the procedure for addition of signed numbers.	2. $3 + (+7) = 10$ *Ans.*

Example 3. Subtract -7 from -3: $-3 - (-7) = -3 + (+7) = 4$. *Ans.*

Example 4. $7 - (-3) = 7 + (+3) = 10$ *Ans.*

Example 5. $-7 - (+3) = -7 + (-3) = -10$ *Ans.*

Example 6. $0 - (-16) = 0 + (+16) = 16$ *Ans.*

Example 7. $0 - (+16) = 0 + (-16) = -16$ *Ans.*

Example 8. $-16 - (-16) = -16 + (+16) = 0$ *Ans.*

Example 9. $(12 - 5) - (-20 + 7)$. Note: The operations enclosed within the parentheses must be done first: $(12 - 5) - (-20 + 7)$ $= 7 - (-13) = 7 + 13 = 20$. *Ans.*

EXERCISES

For exercises 1-12, convert the pairs of signed numbers to absolute values and subtract the smaller absolute value from the larger absolute value.

1. +18, -10
2. -18, +10
3. +15, +15
4. -15, +15
5. -15, -15
6. 0, -23

7. +7.2, -3.8
8. -7.2, +3.8
9. -10.7, -12.5
10. +2 1/8, -1 5/8
11. -37 3/16, -28 1/16
12. +14 5/32, +9 9/32

For exercises 13-33, add the signed numbers as indicated.

13. $+3 + (+12)$
14. $+12 + (+3)$
15. $+4 + (-12)$
16. $-12 + (+4)$
17. $-12 + (-4)$
18. $+12 + (-4)$
19. $-18 + (-18)$
20. $-40 + (-7.3)$
21. $-12.8 + (+4.2)$
22. $+23.5 + (-36.2)$
23. $+8.6 + (-8.6)$

24. $-3/4 + (+1/4)$
25. $-7/8 + (-15/16)$
26. $+17/64 + (+3/32)$
27. $-15 + 7 + 8 + (-3)$
28. $18 + (-7) + (-13) + (-6)$
29. $4 + 12 + (-14) + (18) + (-1)$
30. $-9.2 + (-10) + (-3.5) + 2.1$
31. $-3.4 + (-7.3) + (-1.5) + (-7.6)$
32. $-1/8 + (-5/8) + 3/8 + (-5 7/8)$
33. $7/32 + 10 1/2 + (-19) + (- 5/16)$

For exercises 34-54, subtract the signed numbers as indicated.

34. $-12 - (-9)$
35. $+12 - (+9)$
36. $+4 - (-9)$
37. $+4 - (+8)$
38. $+35 - (+35)$
39. $-35 - (-35)$
40. $-17 - 0$
41. $0 - (+1)$
42. $0 - (-1)$
43. $30 - (+42)$
44. $1.6 - (+7.8)$

45. $-3.9 - (+12)$
46. $9/32 - (-11/32)$
47. $-3 1/8 - (-15/16)$
48. $(5 + 9) - (3 + 8)$
49. $(7 + 6) - (12 + 1)$
50. $4 - (6 + 1)$
51. $(7.2 - 5) - (10.2 - 1.7)$
52. $(16 - 7) - (-3 + 2)$
53. $(-1/8 + 3/8) - (7 + 5/8)$
54. $(3/16 - 1/16) - (15/32 + 1/2)$

Multiplication of signed numbers

Procedure for multiplying two or more signed numbers:

1. Find the product of the numbers as in arithmetic, disregarding their signs.
2. Count the number of negative signs.
   a. If there is an odd number of negative signs, the product is negative.
   b. If there is an even number of negative signs, the product is positive.
   c. If the problem consists of all positive numbers, the product is positive.

It is not necessary to count the number of positive values in a problem consisting of both positive and negative numbers. Only count the number of negative values to determine the sign of the product.

Remember that factors can be enclosed within parentheses to eliminate the possibility of the multiplication sign (x) being mistaken for the letter x. For example, 5 x 7 can be written 5(7).

Example 1. Multiply: 2(–4).

THINK	DO THIS
1. Disregard the signs and multiply.	1. 2 x 4 = 8
2. Count the number of negative signs. There is one negative sign. One is an odd number, therefore, the product is negative.	2. –8 *Ans.*

Example 2. Multiply: –2(–4). Procedure: 2 x 4 = 8; there are 2 negative signs; 2 is an even number, therefore, –2(–4) = +8.

Example 3. (–2) (–4) (–3) (–1) (–2) (–1) = +48; (6 negatives, even number).

Example 4. (–2) (–4) (–3) (–1) (–2) (+1) = –48; (5 negatives, odd number).

Example 5. (2) (–4) (–3) (1) (–2) (1) = –48; (3 negatives, odd number)

Example 6. (–2) (4) (–3) (–1) (–2) (1) = 48; (4 negatives, even number)

Example 7. (2) (4) (3) (1) (2) (1) = +48; (all positive numbers)

The product of any number and 0 = 0; for example, 0 x 7 = 0, 0 x (–7) = 0.

Division of signed numbers

Procedure for dividing two numbers that have the same sign (both positive or both negative):

1. Divide the numbers as in arithmetic, disregarding their signs.
2. Prefix a plus sign.

Example 1. Divide –18 by –3.

THINK	DO THIS
1. Divide as in arithmetic, disregarding signs.	1. $18 \div 3 = 6$
2. Both values are negative; therefore, the quotient is positive. Prefix a plus sign.	2. +6 *Ans.*

Example 2. $+18 \div (+3) = +6$ or 6 (both values are positive; therefore, the quotient is positive.)

Procedure for dividing two numbers with unlike signs (one positive and one negative):

1. Divide the numbers as in arithmetic, disregarding their signs.
2. Prefix a minus sign.

Example 1. Divide –35 by +7.

THINK	DO THIS
1. Divide as in arithmetic, disregarding signs.	1. $35 \div 7 = 5$
2. One value is positive, the other negative; therefore, the quotient is negative. Prefix a minus sign.	2. –5 *Ans.*

Example 2. $-6 \overline{)\ 24} = -4$ *Ans.* (One value is positive and the other negative; therefore, the quotient is negative.)

Note: Zero divided by any number equals 0. For example, $0 \div +7 = 0$, $0 \div -7 = 0$, and $0 \div y = 0$. A number divided by 0 is not allowed. For the purposes of this text, division by 0 should be considered as having no meaning. For example, $15 \div 0$, $-15 \div 0$, and $y \div 0$ have no meaning.

EXERCISES

For exercises 1-20, multiply the signed numbers as indicated.

1. (–7) (5)
2. (7) (5)
3. (7) (–5)
4. (+15) (–4)
5. (0) (17)
6. (0) (–17)

7. $(12.5)(-2)$
8. $(-0.01)(-100)$
9. $(+6)(-0.3)$
10. $(5)(0.7)$
11. $5(-1/2)$
12. $(-2\ 1/8)(-1\ 3/8)$
13. $(-1\ 5/16)(0)$

14. $(3\ 7/16)(2\ 1/4)$
15. $(-10)(-1/10)$
16. $(-3)(3)(-2)$
17. $(-3)(-3)(-2)$
18. $(5)(-1/2)(10)(-4)$
19. $(-0.6)(-0.5)(-8)(1)$
20. $(-1/2)(-1/2)(-1)(-1)$

For exercises 21-40, divide the following signed numbers as indicated.

21. $-21/3$
22. $21/-3$
23. $-21/-3$
24. $30 \div 5$
25. $0 \div 12$
26. $0 \div (-12)$
27. $-4\,\overline{)\,28}$
28. $13\,\overline{)\,-39}$
29. $-2\,\overline{)\,-28}$
30. $+15\,\overline{)\,+45}$

31. $-80 \div 0.5$
32. $-12.6 \div 0.3$
33. $0 \div 0.09$
34. $-21.2 \div (-0.1)$
35. $-0.38/380$
36. $28 \div 1/4$
37. $0 \div 7\ 7/8$
38. $-1/2\,\overline{)\,+1/2}$
39. $-4\ 1/8 \div -(24)$
40. $3\ 3/8 \div (-1/16)$

Powers of signed numbers

The same procedures for multiplying signed numbers apply to raising signed numbers to powers. A positive number raised to any power is positive. Examples: $3^2 = +9$; $3^3 = +27$; $2^4 = +16$; and $2^5 = +32$.

A negative number raised to an even power is positive, and raised to an odd power is negative.

Example 1. Square -4. The exponent 2 of -4^2 is an even number. Therefore, the square of -4 is positive; $-4^2 = (-4)(-4) = +16$. *Ans.*

Example 2. Cube -4. The exponent 3 of -4^3 is an odd number. Therefore, the cube of -4 is negative; $-4^3 = (-4)(-4)(-4) = -64$. *Ans.*

Example 3. $-2^4 = (-2)(-2)(-2)(-2) = +16$ *Ans.* (4 is even, answer is +)

Example 4. $-2^5 = (-2)(-2)(-2)(-2)(-2) = -32$ *Ans.* (5 is odd, answer is -)

Values in formulas are sometimes expressed as having negative or fractional exponents. The following procedures and examples show how to solve these types of problems. Recall that any number can be considered as having a denominator of 1. For example, 3 = 3/1; 2 = 2/1; –4 = –4/1.

Procedure for determining values with negative exponents:

1. Invert the number.
2. Change the negative exponent to a positive exponent and solve.

Example 1. Solve 3^{-2}.

THINK	DO THIS
1. Any number has a denominator 1. Invert 3.	1. $3 = \dfrac{3}{1}$ $\dfrac{3}{1}$ inverted is $\dfrac{1}{3}$
2. Change the negative exponent to a positive exponent and solve.	2. $3^{-2} = 1/3^2$ $\dfrac{1}{3^2} = \dfrac{1}{(3)(3)} = \dfrac{1}{9}$ *Ans.*

Example 2. $2^{-3} = 2^{-3}/1$ Invert and change to a positive exponent; $2^{-3/1} = 1/2^3$; $1/2^3 = 1/[(2)(2)(2)] = 1/8$. *Ans.*

Example 3. $-4^{-3} = -4^{-3}/1$; $-4^{-3}/1 = 1/-4^3 = 1/[(-4)(-4)(-4)] = 1/-64$ *Ans.*

Procedure for raising numbers to a fractional power:

1. a. Write the numerator of the fraction as the exponent of the number.
   b. Write the denominator of the fraction as the root index of the number.
2. Solve.

Example 1. Solve $8^{2/3}$.

THINK	DO THIS
1. Considering the fractional exponent $2/3$, write the numerator 2 as the exponent and the denominator 3 as the root index of the number 8.	1. $\sqrt[3]{8^2}$
2. Solve. Square 8, then take its cube root.	2. $\sqrt[3]{8^2} = \sqrt[3]{64}$ $= 4$ *Ans.*

Example 2. Solve $25^{1/2}$. The exponent is 1 and the root index is 2. Therefore, $25^{1/2} = \sqrt[2]{25^1} = \sqrt{25} = 5$. *Ans.*

Example 3. Solve $36^{-1/2}$. Since the exponent is negative, the number must first be inverted and the exponent made positive: $36^{-1/2} = 1/36^{1/2}$. Next apply the procedure for fractional powers: $1/36^{1/2} = 1/\sqrt{36} = 1/6$. *Ans.*

Roots

The values of the roots of signed numbers are determined by the following two principles:

Principle 1. A positive number which has an even index has both a positive and a negative root. Example: $\sqrt{25} = \sqrt{(+5)(+5)} = +5$; $\sqrt{25}$ also equals $\sqrt{(-5)(-5)}$. Therefore, $\sqrt{25} = +5$ and -5, or ± 5. *Ans.*

Note: In this text, disregard the negative root and use only the positive root. Although $\sqrt{25} = +5$ and -5, disregard the -5 and use only the $+5$; $\sqrt{25} = +5$. *Ans.*

Principle 2. If the index of a number is odd, the sign of the root is the same as the sign of the number.

Example 1. $\sqrt[3]{8} = 2$, both number (8) and root (2) are +.

Example 2. $\sqrt[3]{-8} = -2$, both number (-8) and root (-2) are -.

Example 3. $\sqrt[5]{32} = 2$.

Example 4. $\sqrt[5]{-32} = -2$.

Example 5. $\sqrt[3]{-8/27} = -2/3$ Both numerator (-8) and root (-2) are -. Both denominator (27) and root (3) are +.

Note: The square root of a negative number has no solution in the real number system.

For example: $\sqrt{-4}$ does not equal either +2 or -2; $\sqrt{(+2)(+2)} = \sqrt{+4}$ and $\sqrt{(-2)(-2)} = \sqrt{+4}$.

Combined operations of signed numbers

Expressions consisting of two or more operations of signed numbers are solved using the same order of operations as in arithmetic.

Perform operations in the following order:

1. All operations within parentheses.
2. Powers and roots.
3. Multiplication and division in order from left to right.
4. Addition and subtraction in order from left to right.

Example.   Solve: $100 - (+4) [-3 + (-2^3)(5)]$.

THINK	DO THIS
a. The power operation within parentheses must be done first.	a. $-2^3 = -8$. $100 - (+4) [-3 + (-8)(5)]$
b. Multiply within parentheses.	b. $(-8)(5) = -40$ $100 - (+4) [-3 + (-40)]$
c. Add within parentheses. The operations within parentheses have been completed.	c. $-3 + (-40) = -43$ $100 - (+4) (-43)$
d. Multiply.	d. $(+4) (-43) = -172$ $100 - (-172)$
e. Subtract.	e. $100 - (-172) =$ $100 + (+172) = 272$ *Ans.*

EXERCISES

For exercises 1-25, raise each of the signed numbers to the indicated powers.

1. -3^3
2. -3^2
3. $+3^3$
4. -2^3
5. -2^4
6. $+2^4$
7. -5^3
8. -5^1
9. -1.5^2

10. -2.1^3
11. -0.3^4
12. $(+1/2)^2$
13. $(-1/2)^2$
14. $(-1/4)^3$
15. $(-3/4)^3$
16. $+4^{-2}$
17. $+3^{-3}$
18. $+1^{-5}$

19. -5^{-2}
20. -2^{-3}
21. $+4^{1/2}$
22. $+16^{1/2}$
23. $+27^{1/3}$
24. $9^{-1/2}$
25. $8^{-2/3}$

For exercises 26-45, determine the indicated root of each of the signed numbers.

26. $\sqrt[3]{-64}$
27. $\sqrt[3]{+64}$
28. $\sqrt{+64}$

29. $\sqrt[3]{-1000}$
30. $\sqrt[5]{-32}$
31. $\sqrt[4]{+81}$

32. $\sqrt[3]{-125}$
33. $\sqrt[3]{125}$
34. $\sqrt[7]{-128}$

35. $\sqrt[3]{-27}$ 40. $\sqrt[9]{-1}$ 44. $\dfrac{\sqrt[5]{-32}}{9}$

36. $\sqrt[5]{+32}$ 41. $\sqrt[3]{-8/27}$

37. $\sqrt[3]{+1}$ 42. $\sqrt[3]{8/-27}$ 45. $\dfrac{-1}{\sqrt[3]{-27}}$

38. $\sqrt[3]{-1}$ 43. $\sqrt[3]{-8/-27}$

39. $\sqrt[5]{-1}$

Solve each of the combined operation exercises, 46-55, using the proper order of operations.

46. $17 - (+3)(-2) + (-5)^2$
47. $4 - (+5)(8 - 10)$
48. $(-2)(4 + 2) + 3(5 - 7)$
49. $5 - (+3)(8 - 6) - (1 - 4)$
50. $(-1)(8 - 3)(3 - 8)(1 - 2)$

51. $8 + (-4)(19 - 5+3 - 6) +10$
52. $-3^3 + 3^3 - (-6)(+3) - (-6/2)$
53. $4^2 + \sqrt[3]{-8} + (-4)(0)(-3)$
54. $\sqrt[3]{(-21)(2) - (-15)} + (16)^{1/2}$
55. $[(15 - 40)(5)]^{-1/3} + 10^{-2}$

Basic algebraic operations

A knowledge of basic algebraic operations is required in order to solve certain algebraic expressions. Formulas given in trade handbooks cannot always be used directly as given, but must be rearranged. The following algebraic operations procedures will enable one to solve more complex expressions.

It is important to understand the following definitions in order to apply procedures which are required for solving problems involving basic operations.

A *term* of an algebraic expression is that part of the expression which is separated from the rest by a plus or a minus sign. For example, $4x + ab/2x - 12 + 3ab^2x - 8a\sqrt{b}$ is an expression that consists of five terms: $4x$, $ab/2x$, 12, $3ab^2x$, and $8a\sqrt{b}$.

A *factor* is one of two or more literal and numerical values of a term that are multiplied. For example, 4 and x are each factors of $4x$; 3, a, b^2 and x are each factors of $3ab^2x$; 8, a, and $\sqrt{b}$ are each factors of $8a\sqrt{b}$.

Note: It is of primary importance to distinguish between factors and terms. A *numerical coefficient* is the number factor of a term. The letter factors of a term are the *literal factors*. For example, in the term $5x$, 5 is the numerical coefficient; x is the literal factor.

In the term $(1/3)ab^2c^3$, 1/3 is the numerical coefficient; a, b^2, and c^3 are the literal factors.

Like terms are terms that have identical literal factors, including exponents; the numerical coefficients do not have to be the same. For example, $6x$ and $13x$ are like terms; $15ab^2c^3$, $3.2ab^2c^3$, and $(1/8)ab^2c^3$ are like terms.

Unlike terms are terms which have different literal factors. For example, $12x$ and $12y$ are unlike terms. The values of $15xy$, $3x^2y$, and $4x^2y^2$ are also unlike terms; although the literal factors are x and y, they are raised to different powers.

Addition

Like terms can be added; the addition of unlike terms can only be indicated. As in arithmetic, like things can be added, but unlike things cannot be added. For example, 4 inches + 5 inches = 9 inches. Both values are inches; therefore, they can be added. But 4 inches + 5 pounds cannot be added because they are unlike things.

Procedure for adding like terms:

Add the numerical coefficients applying the procedure for addition of signed numbers and leave the literal factors unchanged. If a term does not have a numerical coefficient, the coefficient 1 is understood; $x = 1x$, $abc = 1abc$, and $n^2rs^3 = 1n^2rs^3$.

Examples: Add the following like terms.

Example 1. Add $3x$ and $12x$.

THINK

DO THIS

Both terms have identical literal factors, x, therefore, they are like terms. Add the numerical coefficients applying the procedure for addition of signed numbers. Leave the literal factors unchanged.

$$\begin{array}{r} 3\,x \\ +\,12\,x \\ \hline 15\,x \ \ Ans. \end{array}$$

Example 2.

$$\begin{array}{r} x \\ +\,(-14x) \\ \hline -13x \ Ans. \end{array}$$

Example 3.

$$\begin{array}{r} -5xy \\ +(+5xy\,) \\ \hline 0 \ \ Ans. \end{array}$$

Example 4.

$$\begin{array}{r} -13x^2\,y^3 \\ +\ \ 6x^2\,y^3 \\ \hline -\,7x^2\,y^3 \ \ Ans. \end{array}$$

Example 5.

$$\begin{array}{r} 2\,ab \\ -3\,ab \\ +\ 7\,ab \\ \hline 6\,ab \ \ Ans. \end{array}$$

Unlike terms. The addition of unlike terms can only be indicated. Examples: Add the following unlike terms.

Example 1	Example 2	Example 3	Example 4
15	$7x$	$3x$	$8a$
$+\ \ x$	$+\ 8y$	$+\ (-7x^2)$	$-6b$
$15 + x$ *Ans.*	$7x + 8y$ *Ans.*	$3x + (-7x^2)$	$+\ 2c$
		Ans.	$8a + (-6b) + 2c$
			Ans.

Procedure for adding expressions that consist of more than one term:
1. Group like terms in the same column.
2. Add like terms and indicate unlike terms.

Example 1. Add: $12x - 2xy + 6x^2y^3$ and $-4x - 7xy + 5x^2y^3$.

THINK	DO THIS
1. Group like terms in the same column.	1. $12x - 2xy + 6x^2\ y^3$
	$\quad -4x - 7xy + 5x^2\ y^3$
2. Add like terms and indicate unlike terms.	2. $\overline{8x - 9xy + 11x^2y^3}$ *Ans.*

Example 2. Add $6a - 7b$ and $a + 18b - 3ab$ and $-14a - 5ab - ab^2$.

THINK	DO THIS
1. Group like terms in the same column.	1. $\quad 6a - \ \ 7b$
	$\quad\quad a + 18b - 3ab$
	$\quad -14a \quad\quad\quad - 5ab - ab^2$
2. Add like terms and indicate unlike terms.	2. $\overline{-7a + 11b - 8ab - ab^2}$ *Ans.*

EXERCISES

Exercises 1-21 consist of expressions with single terms. Add these terms.

1. $5x, 20x$
2. $3a, 16a$
3. $-10y, y$
4. $6r, -2r$
5. $8x^2y, 5x^2y$
6. $-7c^2d, -c^2d$
7. $d^2, 0$
8. $1.7P, 2.3P$
9. $-8.5m^2, .5m^2$
10. $-0.03c, -0.7c$
11. $(1/4)mn, (1\ 1/4)mn$

12. $(-1/2)xy, (3/8)xy$
13. $(-3/16)V^2, (-3/4)V^2$
14. $6xy, 3xy$
15. $25x^3, 4x^3$
16. $x, 3x, 15x$
17. $17H, -22H, 2H$
18. $7.8dt^2, 0.6dt^2, 5\ dt^2$
19. $0.05a, 0.5a, -5a$
20. $-15xy, -xy, (3/4)xy$
21. $(1/8)R, (-3\ 3/16)R, (-3/4)R$

Exercises 22-33 consist of expressions with two or more terms. Add these terms.

22. $(7a + 10ab + 5ab)$, $(11a + 4ab + 2b)$
23. $(5x + 7xy + 8y)$, $(9x + 12xy + 13y)$
24. $(-5x + 7xy - 8y)$, $(-9x - 12xy - 13y)$
25. $(3a - 11d - 8m)$, $(-a + 11d - 3m)$
26. $(-6ab - 7a^2b^2 - 3a^3b)$, $(ab - 2a^3b)$
27. $(-5cd - 12c^2d)$, $(9cd + 8cd^2)$
28. $(3xy^2 + x^2y - x^2y^2)$, $(2x^2y + x^2y^2)$
29. $(10a - 5b)$, $(-12a - 7b)$, $(17a + b)$
30. $(x - 3x^3)$, $(3x - 7x^3 + 7)$, $(x - x^3)$
31. $(b^4 + 4b^3c - 2b^2c)$, $(4b^3c - 7bc)$
32. $(n^2 - 4nm)$, $(4nm - m^2)$, $(-n^2 + m^2)$
33. $(0.3T - 8.5T^2)$, $(6T^2 + 0.3T^3)$

Subtraction

As in addition, only like terms can be subtracted; the subtraction of unlike terms can only be indicated. The same principles apply in arithmetic. For example, 8 feet - 3 feet = 5 feet, but 8 feet - 3 ounces cannot be subtracted.

Procedure for subtracting like terms:

Subtract the numerical coefficients applying the procedures for signed numbers and leave the literal factors unchanged.

Examples: Subtract the following like terms as indicated.

Example 1. $20ab - (-12ab)$

THINK	DO THIS
Both terms have identical literal factors, ab; therefore, they are like terms. Subtract the numerical coefficients applying the procedure for subtraction of signed numbers. Leave the literal factors unchanged.	$20\,ab$ $\underline{-\,(-12\,ab)}$ $32\,ab$ *Ans.*

Example 2. $18P - 7P = 11P$ *Ans.*

Example 3. $bx^2y^3 - 13bx^2y^3 = -12bx^2y^3$ *Ans.*

Example 4. $-5x^2y - 8x^2y = -13x^2y$ *Ans.*

Example 5. $-24dmr - (-24dmr) = 0$ *Ans.*

Unlike terms. The subtraction of unlike terms can only be indicated. Examples: Subtract the following unlike terms as indicated.

Example 1

$$3y$$
$$- (+2x)$$
$$\overline{3y \; - \; 2x \;\; Ans.}$$

Example 2

$$-13abc$$
$$- (+8abc^2)$$
$$\overline{-13abc \; - \; 8abc^2 \;\; Ans.}$$

Example 3

$$-2xy$$
$$- (-7y)$$
$$\overline{-2xy \; - \; (-7y) \text{ or}}$$
$$-2xy \; + \; 7y. \;\; Ans.$$

Procedure for subtracting expressions that consist of more than one term:

1. Group like terms in the same column.
2. Subtract like terms and indicate unlike terms.

Note that each term of the subtrahend is subtracted following the procedure for subtraction of signed numbers.

Example 1. Subtract $7a + 3b - 3d$ from $8a - 7b + 5d$.

THINK	DO THIS
1. Group like terms in the same column.	1. $\quad 8a - 7b + 5d$ $\quad -(7a + 3b - 3d)$
2. Subtract like terms. Each term, the $7a$, $3b$, and $-3d$ is subtracted following the procedure for subtraction of signed numbers. Change the sign of each term in the subtrahend and follow the rules of addition of signed numbers.	2. $\quad 8a - 7b + 5d$ $\quad +(-7a - 3b + 3d)$ $\quad \overline{\quad a - 10b + 8d \;\; Ans.}$

Example 2. Subtract as indicated, $(3x^2 + 5x - 12xy) - (7x^2 - x - 3x^3 + 6y)$.

$$3x^2 + 5x - 12xy$$
$$-(7x^2 - \; x \qquad - 3x^3 + 6y)$$

$$= \qquad 3x^2 + 5x - 12xy$$
$$+(-7x^2 + \; x \qquad + 3x^3 - 6y)$$
$$\overline{-4x^2 + 6x - 12xy + 3x^3 - 6y}$$
$$Ans.$$

EXERCISES

Exercises 1-21 consist of expressions with single terms. Subtract terms as indicated.

1. $7xy^2 - (-13xy^2)$
2. $3xy - xy$
3. $-3xy - xy$
4. $-3xy - (-xy)$
5. $9ab - (-9ab)$
6. $-5a^2 - (5a^2)$
7. $0.7a^2b^2 - 1.5a^2b^2$
8. $0 - (-8mn^3)$
9. $-8mn^3 - 0$
10. $7/8\ x^2 - (-3/8\ x^2)$
11. $13a^2 - 7a^2$
12. $-13a^2 - (-7a^2)$
13. $0.6xy^3 - 0.9xy^3$
14. $-ax^2 - ax^2$
15. $1/2\ dt - (-3/8\ dt)$
16. $1/2\ d^2t^2 - (-1/2\ d^2t^2)$
17. $18 - 3x$
18. $3x - 18$
19. $-3.2d - 6.4d$
20. $-1.4xy - (-1.4xy)$
21. $-0.02M - 1.06M$

Exercises 22-33 consist of expressions with two or more terms. Subtract these terms as indicated.

22. $(2a^2 - 3a) - (7a^2 - 8a)$
23. $(4x^2 + 8xy) - (3x^2 + 5xy)$
24. $(9b^2 + 1) - (9b^2 - 1)$
25. $(9b^2 - 1) - (9b^2 - 1)$
26. $(xy^2 - x^2y^2 + x^3y^3) - 0$
27. $0 - (xy^2 - x^2y^2 + x^3y^3)$
28. $(2a^3 - .5a^2) - (-a^3 + a^2 - a)$
29. $(-a^3 + a^2 - a) - (2a^3 - 0.5a^2)$
30. $(5x + 3xy - 7y) - (3y^2 - x^2y)$
31. $(8L - 12P) - (-10L - 6P^2)$
32. $(-d^2 - dt + dt^2) - (dt - 4)$
33. $(13g - 15) - (23g - 17 + h)$

Multiplication

The term x^2 means $x \cdot x$ (x is used as a factor 2 times), and the term x^4 means $x \cdot x \cdot x \cdot x$ (x is used as a factor 4 times). Therefore, $(x^2)(x^4)$ means $(x \cdot x)(x \cdot x \cdot x \cdot x)$ which equals x^6, (x is used as a factor 6 times.)

Procedure for multiplying two or more terms:

1. Multiply the numerical coefficients following this procedure for multiplication of signed numbers.
2. Add the exponents of the same letter factors.
3. Show the product as a combination of all numerical and literal factors.

Example 1. Multiply: $(-3x^2)(6x^4)$

THINK	DO THIS
1. Multiply the numerical coefficients following the procedure for signed numbers.	1. $(-3)(6) = -18$
2. Add the exponents of the same letter factors; x^2 and x^4 have the same letter factor, x.	2. $(x^2)(x^4) = x^{2+4} = x^6$
3. Show the product as a combination of all numerical and literal factors.	3. $-18x^6$ *Ans.*

Example 2. $(3a^2b^3)(7ab^3) = (3)(7)(a^{2+1})(b^{3+3}) = 21a^3b^6$. *Ans.*

Example 3. $(-4a)(-7b^2c^2)(-2ac^3d^3) = (-4)(-7)(-2)(a^{1+1})(b^2)$ $(c^{2+3})(d^3) = -56a^2b^2c^5d^3$. *Ans.*

It is sometimes necessary to multiply expressions that consist of more than one term within an expression, such as $5x(10 + 2x)$ and $(-3m + 7n)$ $(2m - 3n)$.

Procedure for multiplying expressions that consist of more than one term within an expression:

1. Multiply each term of one expression by each term of the other expression.

2. Combine like terms.

Before applying the procedure to algebraic expressions, two examples will be given to show that the procedure is consistent with arithmetic.

Arithmetic: **Example 1.** Multiply: $3(4 + 2)$. From arithmetic, $3(4 + 2)$ $= 3(6) = 18$. The same answer is obtained by applying the above procedure:

THINK	DO THIS
1. Multiply each term of one expression by each term of the other expression.	1. $3(4 + 2) = 3(4) + 3(2) = 12 + 6$
2. Combine like terms.	2. $12 + 6 = 18$ *Ans.*

Arithmetic: **Example 2.** Multiply: $(5 + 3)(2 + 4)$. From arithmetic, $(5 + 3)(2 + 4) = (8)(6) = 48$. Applying the procedure the same answer is obtained:

THINK

1. Multiply each term of one expression by each term of the other expression.

DO THIS

1.
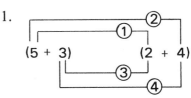

Step 1 Step 2 Step 3 Step 4

$5(2) + 5(4) + 3(2) + 3(4) =$
$10 + 20 + 6 + 12$

2. Combine like terms.

2. $10 + 20 + 6 + 12 = 48$ *Ans.*

Algebra: **Example 1.** Multiply: $3a(6 + 2a^2)$.

Multiply each term: $3a(6 + 2a^2) = (3a)(6) + (3a)(2a^2) = 18a + 6a^3$.

Note that $18a$ and $6a^3$ are unlike terms and cannot be combined. The answer is $18a + 6a^3$.

Algebra: **Example 2.** Multiply: $(3c + 5d^2)(4d^2 - 2c)$.

Multiply each term:

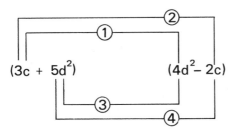

Step 1   Step 2   Step 3   Step 4

$3c(4d^2) + 3c(-2c) + 5d^2(4d^2) + 5d^2(-2c) =$
$12cd^2 - 6c^2 + 20d^4 - 10cd^2$

Combine like terms: Combine

$12cd^2 - 6c^2 + 20d^4 - 10cd^2 =$
$2cd^2 - 6c^2 + 20d^4$ *Ans.*

EXERCISES

Exercises 1-21 consist of expressions with single terms. Multiply these terms as indicated:

1. $(-5b^2c)(3b^3)$
2. $(x)(x^2)$
3. $(-3a^2)(-4a^3)$
4. $(8ab^2c)(7a^3bc^2)$
5. $(-x^3y^3)(3x^2y^4)$
6. $(-3xy)(0)$
7. $(7ab^4)(3a^4b)$
8. $(-3d^5r^4)(-d^3)$
9. $(-3d^5r^4)(-d^3)(-1)$
10. $(0.3x^2y^4)(0.4x^5)$
11. $(1/4\ a^3)(3/8\ a^2)$

12. $(-5x)(0)(-5x)$
13. $(m^2t)(st)$
14. $(-1.6bc)(2.1)$
15. $(abc^3)(c^3d)$
16. $(2x^6y^6)(-2x^2)$
17. $(-2/3\ mt)(t^4)$
18. $(7ab^3)(-7a^3b)$
19. $(-0.3a^3b^2)(-5b^4)$
20. $(-x^2y)(-xy)(-x)$
21. $(d^4m^2)(-1)(-m^3)$

Exercises 22-33 consist of expressions with two or more terms. Multiply these terms as indicated:

22. $-7x^2y^3\ (2xy^2 - 3x^4)$
23. $3a^2\ (-a^2 + a^3b)$
24. $-2a^3b^2\ (4ab^3 - b^2 - 4)$
25. $xy^2\ (x^2 + y^3 + xy)$
26. $-4(dt + t^2 - 1)$
27. $(m^2t^3S^4)(-m^4S^2 + m - S^5)$

28. $(a + b)(a + 3)$
29. $(3x + 7)(x^2 + 8)$
30. $(7x^2 - y^3)(-2x^3 + y^2)$
31. $(5ax^3 + bx)(2a^2x^3 + b^2x)$
32. $(-4a^2b^3 + 5xy^2)(4a^2b^3 - 5xy)$
33. $(10c - cd^2)(3c^2 + cd^2)$

Procedure for dividing two terms:

1. Divide the numerical coefficients, following the procedure for signed numbers.
2. Subtract the exponents of letter factors of the divisor from the exponents of the same letter factors of the dividend.
3. Combine numerical and literal factors.

Note that this division procedure is consistent with arithmetic. For example, $2^5/2^2 = (2 \cdot 2 \cdot 2 \cdot 2 \cdot 2)/(2 \cdot 2) = 2 \cdot 2 \cdot 2 = 2^3 = 8$. The same answer, 8, is obtained by applying the division procedure: $2^5/2^2 = 2^{5-2} = 2^3 = 8$. *Ans.*

Example 1. Divide $-16x^3$ by $8x$.

THINK	DO THIS
1. Divide the numerical coefficients, following the procedure for signed numbers.	1. $-16/8 = -2$

2. Subtract the exponents of the letter factors in the divisor from the exponent of the same letter factors in the dividend.	2. $x^3/x = x^{3-1} = x^2$
3. Combine numerical and literal factors.	3. $-2x^2$ *Ans.*

Example 2. Divide: $(-30a^3b^5c^2)/(-5a^2b^3)$. Divide numberical coefficients: $-30/-5 = 6$. Subtract exponents of like literal factors: $(a^{3-2})(b^{5-3})(c^2) = ab^2c^2$. Combine: $6ab^2c^2$. *Ans.*

In arithmetic, any number divided by itself equals 1. For example, $4/4 = 1$. Applying the division procedure, $4/4 = 4^{1-1} = 4^0$. Therefore, $4^0 = 1$. Any number raised to the zero power equals 1: $5^3/5^3 = 5^{3-3}$ $= 5^0 = 1$; $(a^3b^2c)/(a^3b^2c) = (a^{3-3})(b^{2-2})(c^{1-1}) = a^0b^0c^0 = (1)(1)(1)$ $= 1$; $3x/x = 3x^{1-1} = 3x^0 = 3(1) = 3$.

Procedure for dividing when the dividend consists of more than one term:

1. Divide each term of the dividend by the divisor, applying the procedure for dividing two terms.
2. Combine terms.

This procedure is consistent with arithmetic. For example, in arithmetic, $(6 + 8)/2 = 14/2 = 7$. Applying the division procedure: $(6 + 8)/2 = 6/2 + 8/2 = 3 + 4 = 7$. *Ans.*

Example 1. Divide: $\dfrac{-20xy^2 + 15x^2y^3 + 35x^3yz}{-5xy}$.

THINK	DO THIS
1. Divide each term of the dividend by the divisor, applying the procedure for dividing two terms.	1. $\dfrac{-20xy^2}{-5xy} = 4y$
	$\dfrac{15x^2y^3}{-5xy} = -3xy^2$
	$\dfrac{35x^3yz}{-5xy} = -7x^2z$
2. Combine terms.	2. $4y - 3xy^2 - 7x^2z$ *Ans.*

EXERCISES

Exercises 1-21 consist of expressions with single terms. Divide these terms as indicated.

1. $-18a^4b^5 \div 6ab^2$
2. $21x^3y^4 \div 3y^3$
3. $32c^3y^2 \div c^3y^2$
4. $m^4n^5 \div m^4n^5$
5. $-m^4n^5 \div m^4n^5$
6. $0 \div 13ab^2$
7. $-30a^5d^2 \div -6a^2d^2$
8. $-1.2x^2y \div 0.3xy$
9. $0.6PV^2 \div 0.3V$
10. $xy^2 \div -1$
11. $-xy^2 \div -1$
12. $1.8ab \div ab$
13. $-6.6gh^3 \div -3h$
14. $1\ 3/4\ a^2d^3 \div 1/4\ ad^3$
15. $-3e^3f^2 \div -1/2\ ef^2$
16. $-12x^2y^8 \div -3x^2y^5$
17. $-6d^3t^2 \div 6d^3t^2$
18. $x^2y^3z^4 \div xy^3z$
19. $9a^3bc^2y \div -a^3$
20. $4.5xy \div -0.5y$
21. $1/8\ P^2V \div 1/32$

Exercises 22-33 consist of expressions in which the dividends consist of two or more terms. Divide as indicated.

22. $(9x^6y^3 - 6x^2y^5) \div 3xy^2$
23. $(15a^2 + 25a^5) \div (-a)$
24. $(2x - 4y) \div 2$
25. $(-18a^2b^7 - 12a^5b^5) \div (-6a^2b^5)$
26. $(7cd^2 - 35c^2d - 7) \div (-7)$
27. $(0.8x^5y^6 + 0.2x^4y^7) \div 2x^2y^4$
28. $(-0.9a^2x - 0.3ax^2 + 0.6) \div (-0.3)$
29. $(5y^2 - 25xy^2 - 10y^5) \div 5y^2$
30. $(2.5c^2d + 0.5cd^2 - c^2d^2) \div 0.5d$
31. $(1/8\ P^2V - 1/2\ P^3V^2 - PV^3) \div 1/16\ PV$
32. $(-x^3y^3z^3 + x^2z^4 + x^3y) \div -x^2$
33. $(50MN^3P^2 - 20M^3N^3P) \div 0.1MN^2P$

Powers

Procedure for raising a single term to a power:

1. Raise the numerical coefficient to the indicated power, following the procedure for powers of signed numbers.
2. Multiply each of the literal factor exponents by the exponent of the power to which it is to be raised.
3. Combine numerical and literal factors.

This power procedure is consistent with arithmetic. For example, $(2^2)^3 = 4^3 = 64$ or $(2^2)^3 = (2 \cdot 2)(2 \cdot 2)(2 \cdot 2) = 2^6 = 64$ (2 is used as a factor 6 times). Applying the power procedure the same answer is obtained: $(2^2)^3 = 2^{2 \cdot 3} = 2^6 = 64$. *Ans.*

Example 1. Raise to the indicated power: $(3x^3)^2$.

THINK	DO THIS
1. Raise the numerical coefficient to the indicated power, following the procedure for powers of signed numbers.	1. $3^2 = 9$
2. Multiply each literal factor exponent by the exponent of the power to which it is to be raised.	2. $(x^3)^2 = x^{3 \cdot 2} = x^6$
3. Combine numerical and literal factors.	3. $9x^6$ *Ans.*

Note: $(x^3)^2$ is <u>not</u> the same as $x^3 x^2$; $x^3 x^2 = (x \cdot x \cdot x)(x \cdot x)$ $= x^5$ (x is used as a factor 5 times). The value of $(x^3)^2 = (x^3)(x^3)$ or $(x \cdot x \cdot x)(x \cdot x \cdot x) = x^6$ (x is used as a factor 6 times).

Example 2. Solve: $(-3a^2 b^4 c)^3$. Raise -3 to the third power: $(-3)^3$ $= (-3)(-3)(-3) = -27$. Multiply the exponents of $a^2 b^4 c$ by the exponent 3: $(a^2 b^4 c)^3 = a^{2 \cdot 3} b^{4 \cdot 3} c^{1 \cdot 3} = a^6 b^{12} c^3$. Combine: $-27a^6 b^{12} c^3$. *Ans.*

Example 3. Solve: $[-1/2\, x^3\, (yd^2)^3 r^4]^2$. Note: yd^2 is raised to the third power: $(yd^2)^3 = y^3 d^6$. Therefore, $[-1/2\, x^3\, (yd^2)^3 r^4]^2$ $= (-1/2\, x^3 y^3 d^6 r^4)^2$. Apply the power procedure: $(-1/2\, x^3 y^3 d^6 r^4)^2 = (-1/2)^2 x^{3 \cdot 2} y^{3 \cdot 2} d^{6 \cdot 2} r^{4 \cdot 2} =$ $1/4\, x^6 y^6 d^{12} r^8$. *Ans.*

If an expression consists of two or more terms raised to a power, solve as a multiplication problem, applying the procedure for multiplying expressions that consist of more than one term.

Example 1. Solve: $(2x + 4y^3)^2$. The expression $(2x + 4y^3)^2$ is the same as $(2x + 4y^3)(2x + 4y^3)$. Apply the multiplication procedure:

$$
\begin{aligned}
&\quad\text{Step 1}\quad\;\text{Step 2}\qquad\text{Step 3}\qquad\text{Step 4}\\
(2x+4y^3)(2x+4y^3) &= 2x(2x) + 2x(4y^3) + 4y^3(2x) + 4y^3(4y^3) =\\
&= 4x^2 + 8xy^3 + 8xy^3 + 16y^6 =
\end{aligned}
$$

$$4x^2 + 16xy^3 + 16y^6. \quad Ans.$$

EXERCISES

Exercises 1-18 consist of expressions with single terms. Raise these terms to the indicated power.

1. $(2xy)^2$
2. $(4a^4b^3)^2$
3. $(-4a^3b^2c^4)^3$
4. $(3de^2)^4$
5. $(-5x^4y^5)^2$
6. $(-2a^2b^2c^3)^4$
7. $(x^2yz^3)^3$
8. $(-d^4m^5x)^3$
9. $(4P^4V^2T)^2$
10. $(0.5x^3y)^3$
11. $(-0.7c^2d^3e)^3$
12. $(1.1m^3np^2)^2$
13. $(1/2\ a^3b^2c)^2$
14. $(2/3 \cdot xyz^3)^3$
15. $[-6(a^2b^3)^2c]^2$
16. $[-3b^2(m^3)^2x^3]^3$
17. $[0.4x^3\ (yz^2)^3]^2$
18. $[(-5e^2f)^2(ef^2)^2]^3$

Exercises 19-27 consist of expressions of more than one term. Raise these expressions to the indicated power and combine like terms where possible.

19. $(3x^2 - 5y^3)^2$
20. $(a^3 + b^4)^2$
21. $(5t^2 - 6x)^2$
22. $(a^2b^3 + ab^3)^2$
23. $(0.6d^3t^2 - 0.2t)^2$
24. $(-0.4x^2y - y^4)^2$
25. $(2/3\ c^2d + 3/4\ cd^2)^2$
26. $[(x^2)^3 - (y^3)^2]^2$
27. $[(-a^4b)^2 + (x^2y)^3]^2$

Roots

Procedure for extracting the root of a term:

1. Determine the root of the numerical coefficient, following the procedure for roots of signed numbers.

2. The roots of the literal factors are determined by dividing the exponent of each literal factor by the index of the root.

3. Combine the numerical and literal factors.

This root procedure is consistent with arithmetic. For example, $\sqrt[2]{2^6} = \sqrt[2]{2 \cdot 2 \cdot 2 \cdot 2 \cdot 2 \cdot 2} = \sqrt[2]{64} = 8$. Applying the power procedure the same answer is obtained: $\sqrt[2]{2^6} = 2^{6/2} = 2^3 = 2 \cdot 2 \cdot 2 = 8$. *Ans.*

Example 1. Solve: $\sqrt{25a^6 b^4 c^8}$.

THINK	DO THIS
1. Determine the root of the numerical coefficient.	1. $\sqrt{25} = 5$
2. Determine the roots of the literal factors. Divide the exponent of each literal factor by the index of the root.	2. $\sqrt{a^6} = a^{6/2} = a^3$ $\sqrt{b^4} = b^{4/2} = b^2$ $\sqrt{c^8} = c^{8/2} = c^4$
3. Combine the numerical and literal factors.	3. $5a^3 b^2 c^4$ *Ans.*

Example 2. Solve: $\sqrt[3]{-27dx^9 y^2}$. Determine the root of the numerical coefficient: $\sqrt[3]{-27} = -3$. Determine the roots of the literal factors: $\sqrt[3]{dx^9 y^2} = d^{1/3} x^{9/3} y^{2/3} = d^{1/3} x^3 y^{2/3}$. Combine: $-3d^{1/3} x^3 y^{2/3}$. *Ans.*

Example 3. $\sqrt[4]{16/81\, d^8 t^{12} y^2}$ $2/3\, d^{8/4} t^{12/4} y^{2/4}$
$= 2/3\, d^2 t^3 y^{1/2}$ *Ans.*

Roots of expressions that consist of two or more terms *cannot* be extracted by this procedure. For example, $\sqrt{a^2 + b^2}$, consists of two terms and does *not* equal $\sqrt{a^2} + \sqrt{b^2}$. Therefore, the root of $\sqrt{a^2 + b^2}$ does *not* equal $a + b$. This mistake, commonly made by students, must be avoided. This fact, is consistent with arithmetic. For example, $\sqrt{3^2 + 4^2} = \sqrt{9 + 16} = \sqrt{25} = 5$, which is the correct answer; but $\sqrt{3^2 + 4^2}$ does *not* equal $\sqrt{3^2} + \sqrt{4^2}$; $\sqrt{3^2} + \sqrt{4^2} = 3 + 4 = 7$, which is an incorrect answer.

EXERCISES

Determine the roots of the following terms:

1. $\sqrt{36a^6 b^2 c^2}$

2. $\sqrt{e^4 b^2 d^6}$

3. $\sqrt{4x^2 y^4}$

4. $\sqrt{25x^8 y^6}$

5. $\sqrt{81c^{10}d^2x^6}$ 12. $\sqrt{100ab^4}$

6. $\sqrt[3]{64d^3t^9}$ 13. $\sqrt[3]{c^2dt^9}$

7. $\sqrt[3]{-64d^3t^9}$ 14. $\sqrt[3]{-x^2d^3t}$

8. $\sqrt[3]{-8cd}$ 15. $\sqrt[5]{-m^5y^{10}t^{15}}$

9. $\sqrt[3]{-64dt^9}$ 16. $\sqrt{1/16\ xy^6}$

10. $\sqrt[5]{-32x^{10}}$ 17. $\sqrt{4/9\ ab^2c^3}$

11. $\sqrt[5]{32x^{10}}$ 18. $\sqrt{64/81\ xy^6}$

Removal of parentheses

Procedure for removal of parentheses preceded by a plus (+) or a negative (−) sign:

If parentheses are preceded by a plus (+) sign, the parentheses are removed without changing the signs of any terms within the parentheses. For example, $5a + (4b + 7 - 3d) = 5a + 4b + 7 - 3d$.

If parentheses are preceded by a negative (−) sign remove the parentheses and change the sign of each term within the parentheses to the opposite sign. For example, $3c - (8x + 3y - 7) = 3c - 8x - 3y + 7$, and $3x - (-4y - y^2 + 5) = 3x + 4y + y^2 - 5$.

EXERCISES

Remove parentheses and combine like terms where possible.

1. $7x + (2x - 3x^2 - x^3)$ 7. $-15a^2 - (-12a^2 - b)$

2. $7x - (2x - 3x^2 - x^3)$ 8. $- (x^2 + y^2 - xy)$

3. $8x - (3d + c)$ 9. $+ (x^2 + y^2 - xy)$

4. $13 + (m^2 - 4)$ 10. $- (18 + ab) + ab$

5. $-2a^2 + (3 - 6a^2)$ 11. $- (x^2 + y^2) + (x^2 + y^2)$

6. $x^2y - (xy^2 - xy)$ 12. $- (x^2 + y^2) - (x^2 + y^2)$

Combined operations

Expressions which consist of two or more different operations are solved by applying the proper order of operations, which is discussed earlier in this unit.

Example 1. Simplify: $10x - 3x(2 + x - 4x^2)$.

Multiply: $-3x(2 + x - 4x^2) = -6x - 3x^2 + 12x^3$.

Add and subtract like terms (combine):

$10x - 6x - 3x^2 + 12x^3 = 4x - 3x^2 + 12x^3$. *Ans.*

Example 2. Simplify: $15a^6b^3 + (2a^2b)^3 - [a^7(b^3)^2]/(ab^3)$. Raise to indicated powers: $(2a^2b)^3 = 8a^6b^3$; $[-a^7(b^3)^2]/(ab^3) = (-a^7b^6)/(ab^3)$. Divide $(-a^7b^6)/(ab^3) = -a^6b^3$. Add and subtract like terms (combine): $15a^6b^3 + 8a^6b^3 - a^6b^3 = 22a^6b^3$. *Ans.*

EXERCISES

The following expressions consist of combined operations. Simplify and combine like terms where possible.

1. $12 - 3(-5a) - a$
2. $15 - 2(3xy)^2 + x^2y^2 - 3$
3. $5(a^2 - b) + a^2 - b$
4. $(2 - c^2)(2 + c^2) + c$
5. $-10(m^2 + 5) - (mn^2)^2$
6. $(ab)/a - [(-a^2b)/a - (a^3b)/a^3]$
7. $(4 -8x + 16x^2)/2 + 3x^4/x^2$

8. $(16xy^8)/(2xy^2) - (y^2)^3 + 15$
9. $\sqrt{(25x^2)/-5}\ (3xy^3) - (-4)$
10. $\sqrt{(64d^6)/9} \div d^2$
11. $(12x^6 + 16x^4y)/(2x)^2$
    $- \sqrt{16x^4y^2}$
12. $-4a[-8 + (ab^2)^3 - 12]$
13. $(c + 3d)^2 + (c - 3d)^2 - 10c^2$

Equations

An *equation* is a mathematical statement of equality between two or more quantities and always contain the equal sign (=). The value of all of the quantities on the left side of the equal sign equals the value of all quantities on the right side of the equal sign. A *formula* is a particular type of an equation which states a mathematical rule. The following expressions are examples of equations:

$$10 + 3 = 6 + 7 \qquad x + y = m - n \qquad 90° = 3 \times 20° + 30°$$
$$2 \times 6\ 1/4'' = 12\ 1/2'' \qquad A = \ell w \qquad C = 2\pi r$$

Because it expresses the equality of the quantities on the left and on the right of the equal sign, an equation is a balanced mathematical statement. An equation may be considered similar to a balanced scale.

In general, an equation is used to determine the numerical value of an unknown quantity. Although any letter or symbol can be used to represent the unknown quantity, the letter x is commonly used.

The first letter of the unknown quantity is often used. Some common letter designations are:

ℓ to represent length p to represent pressure
A to represent area f to represent feed of cutter
t to represent time w to represent weight

An equation asks a question. It asks for the value of the unknown which makes the left side of the equation equal to the right side. For example, $3 + x = 12$ asks, "What number added to 3 equals 12?" and

$6x = 24$ asks "What number multiplied by 6 equals 24?". These are simple equations which can be solved by common sense, but in actual practice equations are usually more complex. For example, an equation such as $1/8\ x - 3(x - 7) = 5\ 1/8\ x - 3$ is difficult and time consuming to solve unless a definite logical procedure is applied in its solution.

Principles of equality

There are specific procedures for solving equations using the fundamental principles of equality. The principles of equality which will be presented are those of addition, subtraction, multiplication, division, powers, and roots. Equations are solved directly and efficiently by the application of these principles.

Checking equations

It is important to check an equation after it is solved. An equation is checked by substituting in the original equation the value which was obtained for the unknown. An equation is correctly solved if the values on the left and right of the equal sign are equal. For example, for the equation $x + 8 = 15$, the value of $x = 7$. Substituting 7 in place of x, $7 + 8 = 15$ or $15 = 15$ shows that both sides of the equation are equal. The equation is checked.

Solution of equations by the subtraction principle of equality

The subtraction principle of equality states that if the same number is subtracted from both sides of the equation, the sides remain equal, or the equation remains balanced.

The subtraction principle is illustrated in Figs. 5-8, 5-9, and 5-10.

Figure 5-8 shows a balanced scale. Expressed as an equation, 10 lb. + 5 lb. = 15 lb.

If 5 pounds are removed from the left side only, the scale is not in balance, as shown in Fig. 5-9; 10 pounds does not equal 15 pounds.

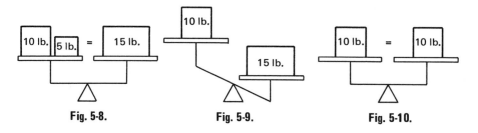

Fig. 5-8.   Fig. 5-9.   Fig. 5-10.

If 5 pounds are removed from both the left and right side, the scale remains balanced, as shown in Fig. 5-10;   10 lb. + 5 lb. - 5 lb. = 15 lb. - 5 lb.; 10 lb. = 10 lb.

The subtraction principle is used to solve an equation in which a number is added to the unknown, such as $y + 9 = 25$.

Procedure for solving an equation in which a number is added to the unknown:

Subtract the number which is added to the unknown from both sides of the equation. Check.

Example. 1. $x + 7 = 15$. Solve for x.

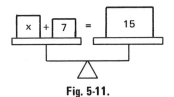

Fig. 5-11.

THINK	DO THIS
1. In the equation $x + 7 = 15$, 7 is added to x. Therefore, 7 must be subtracted from both sides of the equation to find the value of x.	1. $x + 7 = 15$ $\underline{\quad -7 \quad -7}$ $x \quad = \quad 8$ *Ans.*
2. Check. Substitute in the original equation the value obtained for the unknown. The equation must balance.	2. $x + 7 = 15$ $8 + 7 = 15$ $15 = 15$

Example 2. Figure 5-12 shows a dimensioned plate with a drilled hole. The unknown distance, d, is expressed in the equation, $d + 6 = 20$. Solve for d. Subtract 6 from both sides of the equation:

$$d + 6 = 20$$
$$\underline{\quad -6 \quad -6}$$
$$d \quad = 14'' \; Ans.$$

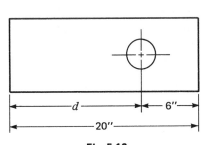

Fig. 5-12

Check: Substitute 14 for d in the original equation: $d + 6 = 20$; $14 + 6 = 20$; $20 = 20$. The equation is balanced.

Example 3. $-28 = M + 20$. Solve for M.

Subtract 20 from both sides of the equation:

$$-28 = M + 20$$
$$\underline{-20 = -20}$$
$$-48 = M \quad Ans.$$

Check: Substitute −48 for M in the original equation:

$$-28 = M + 20$$
$$-28 = -48 + 20$$
$$-28 = -28$$

Transposition

Transposing a term means transferring, or moving, it from one side of the equation to the other. Transposition is an alternate method of solving equations.

Procedure for transposing a term from one side of an equation to the other:

Change the operation of the term to the opposite operation.

Examples: The following equations are solved using transposition. Observe that solving these problems by transposition is actually a shortcut application of the subtraction principle of equality.

Example 1. $x + 12 = 28$. Solve for x.

THINK

1. To find the value of x (to get x to stand alone), 12 must be transposed to the right side of the equation. Twelve is added to x. Subtraction is the opposite operation of addition. Therefore, when 12 is moved to the right side of the equation, it is subtracted.

2. Check. The equation must balance.

DO THIS

1. $x + 12 = 28$
$x = 28 - 12$
$x = 16$ $Ans.$

2. $x + 12 = 28$, $16 + 12 = 28$,
$28 = 28$

Example 2. $a + 3.6 = 10$. Solve for a. Move 3.6 from the left side of the equation to the right side and subtract:

$$a + 3.6 = 10 \qquad \text{Check: } a + 3.6 = 10$$
$$a = 10 - 3.6 \qquad 6.4 + 3.6 = 10$$
$$a = 6.4 \text{ } Ans. \qquad 10 = 10$$

EXERCISES

Solve each of the following equations using both the subtraction principle of equality and transposition. Check all answers.

1. $x + 10 = 30$
2. $x + 37 = 56$
3. $y + 19 = 43$
4. $17 = a + 12$
5. $T + 50 = -33$
6. $P + 13.8 = 21.7$
7. $-33.6 = b + 16$
8. $54.4 = C + 18.6$
9. $d + 17.7 = 0$

10. $m + 0.03 = 0.72$
11. $15.87 = x + 19$
12. $V + 23 = -105$
13. $R + 1/8 = 7$
14. $S + 7/16 = 11/16$
15. $M + 1/4 = 3 \, 3/4$
16. $5/16 = B + 1/8$
17. $-3/32 = D + 5/16$
18. $x + 2 \, 3/32 = 1 \, 1/2$

Solution of equations by the addition principle of equality

The addition principle of equality states that if the same number is added to both sides of an equation, the sides remain equal, or the equation remains balanced.

The addition principle is used to solve an equation in which a number is subtracted from the unknown, such as $x - 3 = 9$.

Procedure for solving an equation in which a number is subtracted from the unknown:

Add the number which is subtracted from the unknown to both sides of the equation. Check.

Example 1. $y - 4 = 18$. Solve for y.

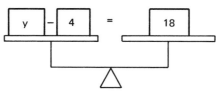

Fig. 5-13.

THINK	DO THIS
1. In the equation $y - 4 = 18$, 4 is subtracted from y. Therefore, 4 must be added to both sides of the equation to find the value of y.	1. $\begin{array}{rcl} y - 4 &=& 18 \\ + 4 &=& +4 \\ \hline y &=& 22 \quad Ans. \end{array}$
2. Check. Substitute in the original equation the value obtained for the unknown. The equation must balance.	2. $\begin{array}{rcl} y - 4 &=& 18 \\ 22 - 4 &=& 18 \\ 18 &=& 18 \end{array}$

Example 2. Three inches are cut from a block as shown in Fig. 5-14. The remaining block is 5 inches high. The original height, h, is expressed in the equation, $h - 3 = 5$. Solve for h. Add 3 to both sides of the equation:

$\begin{array}{rcl} h - 3 &=& 5 \\ + 3 &=& + 3 \\ \hline h &=& 8'' \quad Ans. \end{array}$ Check: $\begin{array}{rcl} h - 3 &=& 5 \\ 8 - 3 &=& 5 \\ 5 &=& 5 \end{array}$

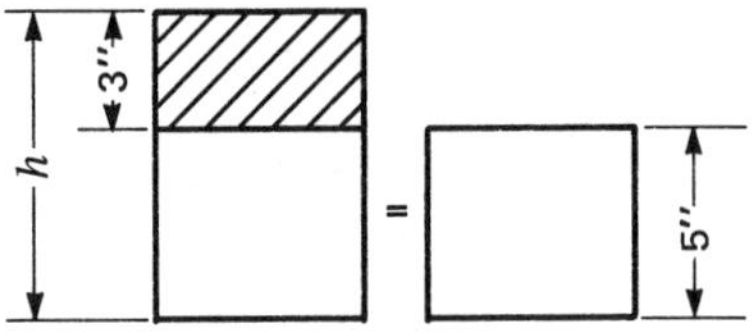

Fig. 5-14.

Transposition

Examples. The following examples are solved using transposition. Observe that solving these problems by transposition is actually a shortcut application of the addition principle of equality.

Example 1. $x - 10 = 22.8$. Solve for x.

THINK	DO THIS
1. To find the value of x (to get x to stand alone), 10 must be transposed to the right side of the equation. Ten is subtracted from x. Addition is the opposite operation of subtraction. Therefore, when 10 is moved to the right side of the equation, it is added.	1. $\begin{array}{rcl} x - 10 &=& 22.8 \\ x &=& 22.8 + 10 \\ x &=& 32.8 \quad Ans. \end{array}$
3. Check. The equation must balance.	2. $\begin{array}{rcl} x - 10 &=& 22.8 \\ 32.8 - 10 &=& 22.8 \\ 22.8 &=& 22.8 \end{array}$

Example 2. $-16 = T - 11.3$. Solve for T. Move -11.3 from the right side of the equation to the left side and add: $-16 = T - 11.3$, $-16 + 11.3 = T$, -4.7 (*Ans.*) $= T$. Check: $-16 = T - 11.3$, $-16 = -4.7 - 11.3$, $-16 = -16$.

EXERCISES

Solve each of the following equations using both the addition principle of equality and transposition. Check all answers.

1. $y - 10 = 30$
2. $x - 13 = 21$
3. $A - 5 = 14$
4. $P - 6 = 24$
5. $9 = T - 13$
6. $a - 40 = -27$
7. $14 = C - 3$
8. $-39 = M - 14$
9. $x - 6.2 = -33$

10. $y - 13.3 = 15.9$
11. $R - 19.4 = 0$
12. $0.08 = N - 1.77$
13. $H - 2.6 = -2.6$
14. $y - 1/2 = 3$
15. $x - 3/8 = 5\ 1/8$
16. $-5/16 = D - 5/32$
17. $21\ 1/8 = L - 3/8$
18. $W - 53\ 9/64 = -20$

Solution of equation by the division principle of equality

The division principle of equality states that if both sides of an equation are divided by the same number, the sides remain equal, or the equation remains balanced.

The division principle is used to solve an equation in which a number is multiplied by the unknown, such as $5x = 30$.

Procedure for solving an equation in which a number is multiplied by the unknown:

Divide both sides of the equation by the number which is multiplied by the unknown.

Example 1. $4x = 20$. Solve for x.

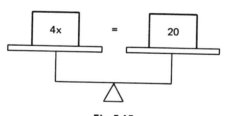

Fig. 5-15.

234 • Unit 5

THINK	DO THIS
1. In the equation $4x = 20$, x is multiplied by 4. Therefore, both sides of the equation must be divided by 4 to find the value of x.	1. $\dfrac{4x}{4} = \dfrac{20}{4}$ $x = 5$ *Ans.*
2. Check. Substitute in the original equation the value obtained for the unknown. The equation must balance.	2. $4x = 20$ $4(5) = 20$ $20 = 20$

Example 2. Figure 5-16 shows a plate with holes equally spaced at y'' apart which can be expressed by the equation $4y = 28$. Solve for y. Divide both sides of the equation by 4:

$$\frac{4y}{4} = \frac{28}{4}, \quad y = 7 \quad Ans.$$

Check: $4y = 28$
$4(7) = 28$
$28 = 28$

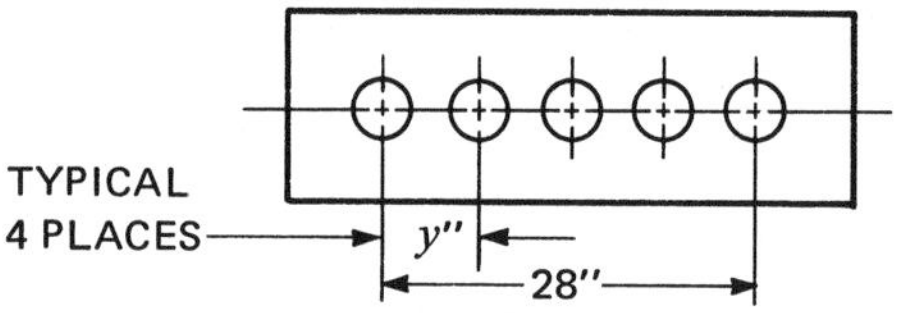

TYPICAL
4 PLACES

y''

28"

Fig. 5-16.

Transposition

The following examples are solved using transposition. Observe that solving these problems by transposition is actually a shortcut application of the division principle.

Example 1. $-18.6 = 3C$. Solve for C.

THINK	DO THIS
1. To find the value of C (to get C to stand alone), 3 must be transposed to the left side of the equation. Three is multiplied by C. Division is the opposite operation of multiplication. Therefore, when 3 is moved to the left side of the equation, it is used as a divisor.	1. $-18.6 = 3C$ $\dfrac{-18.6}{3} = C$ $-6.2\ (Ans.) = C$
2. Check. The equation must balance.	2. $-18.6 = 3C$ $-18.6 = 3(-6.2)$ $-18.6 = -18.6$

Example 2. $-1/4\ S = -2$. Solve for S. Move $-1/4$ from the left side of the equation to the right side and divide: $-1/4\ S = -2$, $S = -2 \div (-1/4)$, $S = 8$ *Ans.*

Check: $-1/4\ S = -2$, $-1/4\ (8) = -2$, $-2 = -2$

EXERCISES

Solve each of the following equations using both the division principle of equality and transposition. Check all answers.

1.	$8x = 56$	10.	$-0.5y = 18$
2.	$10P = 87$	11.	$-0.09x = -99$
3.	$32 = 4T$	12.	$100d = 0.1$
4.	$5P = -20$	13.	$1/2\ y = 13$
5.	$-12L = 48$	14.	$3/8\ V = 9$
6.	$-12L = -48$	15.	$3/32\ C = -6$
7.	$-55 = 2.5x$	16.	$1\ 1/4\ N = 12\ 1/2$
8.	$10.7S = 10.7$	17.	$3/16\ D = -3/8$
9.	$0 = 14M$	18.	$3/8\ x = 3/16$

Solution of equations by the multiplication principle of equality

The multiplication principle of equality states that if both sides of an equation are multiplied by the same number, the sides remain equal, or the equation remains balanced.

The multiplication principle is used to solve an equation in which the unknown is divided by a number, such as $y/6 = 5$.

Procedure for solving an equation in which the unknown is divided by a number:

Multiply both sides of the equation by the number which is divided into the unknown.

Example 1. $x/6 = 7$. Solve for x.

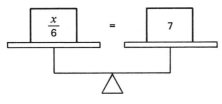

Fig. 5-17.

THINK	DO THIS
1. In the equation $x/6 = 7$, x is divided by 6. Therefore, both sides of the equation are multiplied by 6 to find the value of x.	1. $6\left(\dfrac{x}{6}\right) = 6(7)$ $\quad\quad x = 42$ *Ans.*
2. Check. Substitute in the original equation the value obtained for the unknown. The equation must balance.	2. $\dfrac{x}{6} = 7$ $\dfrac{42}{6} = 7$ $7 = 7$

Example 2. The pipe shown in Fig. 5-18 is cut into 3 equal pieces. Each piece is 8 1/4 inches long. The original length of pipe (ℓ) can be expressed in the equation $\ell/3 = 8\ 1/4$. Solve for ℓ. Multiply both sides of the equation by 3:

$$3\,\frac{(\ell)}{(3)} = 3\,(8\ 1/4),\ \ell = 24\ 3/4\quad Ans.$$

Check: $\dfrac{\ell}{3} = 8\ 1/4;\ \dfrac{24\ 3/4}{3} = 8\ 1/4;$
$8\ 1/4 = 8\ 1/4$

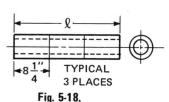

Fig. 5-18.

Transposition

Examples: The following examples are solved using transposition. Observe that solving these problems by transposition is actually a shortcut application of the multiplication principle.

Example 1. $4 = \dfrac{P}{2.4}$. Solve for P.

THINK	DO THIS
1. To find the value of P (to get P to stand alone), 2.4 must be transposed to the left side of the equation. The unknown P is divided by 2.4. Multiplication is the opposite operation of division. Therefore, when 2.4 is moved to the left side of the equation, it is used as a multiplier.	1. $4 = \dfrac{P}{2.4}$ $(2.4)(4) = \dfrac{(2.4)\,P}{2.4}$ $9.6\ (Ans.) = P$
2. Check. The equation must balance.	2. $4 = \dfrac{P}{2.4},\ 4 = \dfrac{9.6}{2.4},$ $4 = 4$

Example 2. $\dfrac{F}{7\,1/2} = -4$. Solve for F.

Move 7 1/2 to the right side of the equation and multiply:

$\dfrac{F}{7\,1/2} = -4$; $F = (7\,1/2)(-4)$; $F = -30$ *Ans.*

Check:

$\dfrac{F}{7\,1/2} = -4$; $\dfrac{-30}{7\,1/2} = -4$; $-4 = -4$.

EXERCISES

Solve each of the following equations using both the multiplication principle of equality and transposition. Check all answers.

1. $x/4 = 8$
2. $c/2 = 14$
3. $P \div 7 = 9$
4. $-10 = A \div 9$
5. $0 = P/3$
6. $x/(-8) = -3$
7. $V/1.5 = 16$
8. $y/0.6 = -20$
9. $S \div 0.01 = 1$

10. $T/-0.9 = -2.1$
11. $M \div 0.6 = 12.8$
12. $x \div (-15) = 2.5$
13. $R/1000 = -0.09$
14. $N \div 3/8 = 1/4$
15. $D \div 7/16 = -5/8$
16. $x \div (-1/2) = -3/4$
17. $S \div 3\,1/2 = 4$
18. $15\,3/4 = y \div 8$

Solution of equations by the root principle of equality

The root principle of equality states that if the same root of both sides of an equation is taken, the sides remain equal, or the equation remains balanced.

The root principle is used to solve an equation that contains an unknown which is raised to a power, such as $x^2 = 9$.

Procedure for solving an equation in which the unknown is raised to a power:

Extract the root of both sides of the equation which leaves the unknown with an exponent of one.

Example 1. $x^2 = 16$. Solve for x.

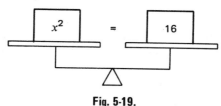

Fig. 5-19.

THINK	DO THIS
1. In the equation $x^2 = 16$, x is squared. The square root of $x^2 = x' = x$. Therefore, extract the square root of both sides of the equation.	1. $x^2 = 16$ $\sqrt{x^2} = \sqrt{16}$ $x = 4$ *Ans.*
2. Check. Substitute in the original equation the value obtained for the unknown. The equation must balance.	2. $x^2 = 16$ $4^2 = 16$ $16 = 16$

Example 2. The area of the square shown in Fig. 5-20 equals 9 square feet. The length of each side = S feet. Expressed as an equation, S^2 = area or $S^2 = 9$. Solve for S. The square root of $S^2 = S$; therefore, extract the square root of both sides of the equation: $S^2 = 9; \sqrt{S^2} = \sqrt{9}; S = 3$. *Ans.*
Check: $S^2 = 9; 3^2 = 9; 9 = 9$.

Fig. 5-20.

Example 3. $M^3 = -27$. Solve for M. The cube root of $M^3 = M$; therefore, extract the cube root of both sides of the equation: $M^3 = -27; \sqrt[3]{M^3} = \sqrt[3]{-27}; M = -3$. *Ans.*
Check: $M^3 = -27; -3^3 = -27; -27 = -27$.

Transposition

Examples: The following examples are solved using transposition. Observe that solving these problems by transposition is actually a shortcut application of the root principle.

Example 1. $y^2 = 25$. Solve for y.

THINK	DO THIS
1. The unknown y is squared. Extracting a square root is the opposite operation of squaring, therefore, when the squaring operation is moved to the right side of the equation, it is changed to a square root operation.	1. $y^2 = 25$ $y = \sqrt{25}$ $y = 5$ *Ans.*

2. Check.   The equation must   2. $y^2 = 25$, $5^2 = 25$,
balance. $25 = 25$.

Example 2. $64 = T^3$. Solve for T. Move the cubing operation from the right side of the equation to the left and change it to a cube root operation: $64 = T^3$; $\sqrt[3]{64} = T$; 4 *(Ans.)* $= T$. Check: $64 = T^3$; $64 = 4^3$; $64 = 64$.

EXERCISES

Solve each of the following equations using both the root principle of equality and transposition. Check all answers.

1. $x^2 = 36$
2. $L^2 = 81$
3. $y^3 = 8$
4. $y^3 = -8$
5. $S^2 = 100$
6. $T^3 = 125$
7. $a^2 = 144$
8. $M^3 = -125$
9. $H^2 = 10,000$

10. $x^2 = 4/81$
11. $L^2 = 49/121$
12. $N^3 = 8/27$
13. $C^3 = -64/125$
14. $D^2 = 1/36$
15. $F^2 = 1.44$
16. $W^3 = 0.064$
17. $x^2 = 0.0049$
18. $H^3 = -0.027$

Solution of Equations by the Power Principle of Equality

The power principle of equality states that if both sides of an equation are raised to the same power, the sides remain equal, or the equation remains balanced.

The power principle is used to solve an equation that contains a root of the unknown, such as $\sqrt{x} = 12$.

Procedure for solving an equation which contains a root of the unknown:

Raise both sides of the equation to the power which leaves the unknown with an exponent of one.

Example 1. $\sqrt{x} = 6$. Solve for x.

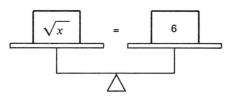

Fig. 5-21.

THINK	DO THIS
1. In the equation $\sqrt{x} = 6$, x is expressed as a square root. The square of $\sqrt{x} = x' = x$. Therefore, square both sides of the equation.	1. $\sqrt{x} = 6$ $(\sqrt{x})^2 = 6^2$ $x = 36$ *Ans.*
2. Check. Substitute in the original equation the value obtained for the unknown. The equation must balance.	2. $\sqrt{x} = 6$ $\sqrt{36} = 6$ $6 = 6$

Example 2. The length of each side of the cube shown in Fig. 5-22 equals 2 inches. The volume of the cube (V) can be expressed as an equation $\sqrt[3]{V} = 2$. Solve for V. The cube of $\sqrt[3]{V} = V$; therefore, cube both sides of the equation: $\sqrt[3]{V} = 2$; $(\sqrt[3]{V})^3 = 2^3$; $V = 8$. *Ans.* Check: $\sqrt[3]{V} = 2$; $\sqrt[3]{8} = 2$; $2 = 2$.

Fig. 5-22.

Transposition

Examples: The following examples are solved using transposition. Observe that solving these problems by transposition is actually a shortcut application of the power principle.

Example 1. $\sqrt{x} = 10$. Solve for x.

THINK	DO THIS
1. The square root of the unknown is given. Squaring is the opposite operation of extracting the square root. Therefore, when the square root operation is moved to the right side of the equation, it is changed to a squaring operation.	1. $\sqrt{x} = 10$ $x = 10^2$ $x = 100$ *Ans.*
2. Check. The equation must balance.	2. $\sqrt{x} = 10$ $\sqrt{100} = 10$ $10 = 10$

Example 2. $-3 = \sqrt[3]{y}$. Solve for y. Move the cube root operation from the right side of the equation to the left and change it to a cubing operation. $-3 = \sqrt[3]{y}$, $(-3)^3 = y$; -27 (*Ans.*) $= y$ Check: $-3 = \sqrt[3]{y}$; $-3 = \sqrt[3]{-27}$; $-3 = -3$.

EXERCISES

Solve each of the following equations using both the power principle of equality and transposition. Check all answers.

1. $\sqrt{T} = 9$

2. $\sqrt{x} = 3.2$

3. $\sqrt{y} = 0.06$

4. $\sqrt[3]{V} = 4$

5. $\sqrt[3]{b} = -4$

6. $0.1 = \sqrt{T}$

7. $\sqrt{P} = 0$

8. $\sqrt[4]{M} = 0.1$

9. $\sqrt[3]{D} = 2.2$

10. $\sqrt[3]{D} = -2.2$

11. $\sqrt[5]{N} = 1$

12. $2 = \sqrt[4]{F}$

13. $-2 = \sqrt[5]{G}$

14. $\sqrt{x} = 5/8$

15. $\sqrt{y} = 3\,1/4$

16. $\sqrt[3]{a} = -1\,1/2$

17. $\sqrt[3]{b} = 3/4$

18. $\sqrt[4]{c} = 1/2$

Solution of equations consisting of combined operations

The purpose of solving an equation is to find the value of the unknown. In order to solve an equation which requires more than one operation between the unknown and known values, such as $2x - 3 = x + 5$, a definite step-by-step procedure must be followed. Use of the proper procedure results in the unknown standing alone on one side of the equation with its value on the other.

Always solve for a positive unknown. A positive unknown may equal a negative value, but a negative unknown is not permitted. For example, $x = -7$ is correct, but $-x = 7$ is incorrect.

Procedure for solving equations consisting of combined operations.

It is essential that the steps used in solving an equation be taken in the following order. Some or all of these steps may be used, depending upon the particular equation.

1. Remove parentheses.
2. Combine like terms on each side of the equation.
   a. Apply the addition and subtraction principles of equality to get all unknown terms on one side of the equation and all known terms on the other side.
   b. Combine like terms.
3. Apply the multiplication and division principles of equality.
4. Apply the power and root principles of equality.

Example 1. $6y + 8 = 32$. Solve for y.

THINK	DO THIS
The operations involved are multiplication and addition; y is multiplied by 6 and 8 is added. Follow the proper order for solving an equation consisting of combined operations:	
1. Apply the subtraction principle. Subtract 8 from both sides of the equation.	1. $\quad \begin{aligned} 6y + 8 &= 32 \\ -8 &= -8 \\ \hline 6y \quad &= 24 \end{aligned}$
2. Apply the division principle. Divide both sides of the equation which resulted from Step 1 by 6.	2. $\quad \dfrac{6y}{6} = \dfrac{24}{6}$ $\quad y = 4$ *Ans.*
3. Check. Substitute 4 for y in the original equation. The equation must balance.	3. $\quad \begin{aligned} 6y + 8 &= 32 \\ 6(4) + 8 &= 32 \\ 24 + 8 &= 32 \\ 32 &= 32 \end{aligned}$

Example 2. $2x + 3x = 6x - 8x + 12 + 9$. Solve for x.

THINK	DO THIS
1. Combine like terms on the same side of the equation.	1. $5x = -2x + 21$
2. Apply the addition principle. Add $2x$ to both sides of the equation.	2. $\begin{aligned} 5x &= -2x + 21 \\ +2x &= +2x \\ \hline 7x &= \qquad 21 \end{aligned}$
3. Apply the division principle. Divide both sides of the equation by 7.	3. $\dfrac{7x}{7} = \dfrac{21}{7}; x = 3$ *Ans.*
4. Check. Substitute 3 wherever x appears in the original equation.	4. $\begin{aligned} 2x + 3x &= 6x - 8x + 12 + 9 \\ 2(3) + 3(3) &= 6(3) - 8(3) + 12 + 9 \\ 6 + 9 &= 18 - 24 + 12 + 9 \\ 15 &= 15 \end{aligned}$

Example 3. $12D + 6(D + 2) = 21$. Solve for D.

Step 1. Remove parentheses. $12D + 6D + 12 = 21$

Step 2.   Combine like terms on the same side of the equation.
$18D + 12 = 21$

Step 3.   Apply the subtraction principle.

$$
\begin{array}{rcl}
18D + 12 &=& 21 \\
- 12 &=& -12 \\
\hline
18D &=& 9
\end{array}
$$

Step 4.   Apply the division principle. $\dfrac{18D}{18} = \dfrac{9}{18}$; $x = 1/2$ *Ans.*

Check: $12D + 6(D + 2) = 21$; $12(1/2) + 6(1/2 + 2) = 21$; $6 + 15 = 21$; $21 = 21$.

The tradesman must often solve formulas in which all but one numerical value for letter values is known. The unknown letter value can appear anywhere within the formula. To determine the numerical value of the unknown, write the original formula, substitute the known number values for their respective letter values, and simplify. Then follow the procedure given for solving equations consisting of combined operations.

Example: The formula for the power factor of an alternating current circuit is

$$
P.F. = \frac{P}{IE} .
$$

Find the power (P) if the power factor ($P.F.$) is 0.8 when a current (I) of 25 amperes at a pressure (E) of 220 volts flows through the circuit.

THINK	DO THIS
1. Write the formula.	1. $P.F. = \dfrac{P}{IE}$
2. Substitute the known numerical values for their respective letter values and simplify.	2. $0.8 = \dfrac{P}{(25)(220)}$ $0.8 = \dfrac{P}{5,500}$
3. Apply the multiplication principle. Multiply both sides of the formula by 5,500.	3. $(5,500)(0.8) = \dfrac{(5,500)P}{5,500}$ $4,400 \,(Ans.) = P$
4. Check. Substitute 4,400 for P in the formula. The formula must balance.	4. $0.8 = \dfrac{4,400}{(25)(220)}$ $0.8 = 0.8$

One of the most important geometry formulas which has wide trade application is $c^2 = a^2 + b^2$. This formula, which is called the Pythagorean Theorem, states that in a right triangle, the square of the side (called *hypotenuse*) opposite the right angle is equal to the sum of the squares of the other two sides. See Fig. 5-23.

Using the formula, walls can be made vertical, corners can be made square, and certain distances between holes on machined parts can be easily computed. If the dimensions of a triangle are known, it can be determined if the triangle has a right angle. If two sides of a right triangle are known, the third side can be calculated.

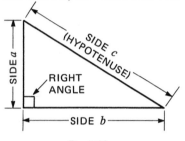

Fig. 5-23.

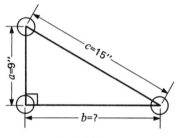

Fig. 5-24.

Example: Determine distance "b" between the holes shown in Fig. 5-24. The lengths of two sides of a right triangle are known; therefore, the Pythagorean Theorem, $c^2 = a^2 + b^2$ can be used to find the length of the third side.

THINK	DO THIS
1. Write the formula.	1. $c^2 = a^2 + b^2$
2. Substitute the known numerical values for their respective letter values in the formula and simplify. Note: side c is the hypotenuse.	2. $15^2 = 9^2 + b^2$ $225 = 81 + b^2$
3. Apply the subtraction principle. Subtract 81 from both sides of the formula.	3. $225 = 81 + b$ $\underline{-81 = -81}$ $144 = \quad b^2$
4. Apply the root principle. Take the square root of both sides of the formula.	4. $\sqrt{144} = \sqrt{b^2}$ $12''$ (*Ans.*) $= b$
5. Check. Substitute 12 for b in the formula. The formula must balance.	5. $c^2 = a^2 + b^2$ $15^2 = 9^2 + 12^2$ $225 = 81 + 144$ $225 = 225$

EXERCISES

Solve for the unknown in each of the following equations which consist of combined operations. Check.

1. $6x - 18 = 6$
2. $10y + 5 + 4y = 80$
3. $4D - 7 = D + 14$
4. $21M - 42 = 0$
5. $6P + 4 = 84 - P$
6. $2.5A + 8 = 15 - 6.5$
7. $6 - (-b + 4) = 9$
8. $3F + (2 - F) = 20$
9. $10(x - 2) = -5(x + 6)$
10. $-(2 + T) - (4 + 2T) = -6$

11. $9.39 - 0.01y = 0.29y$
12. $x/14 + 16 = 11.8$
13. $1/8\, d - 3(d - 7) = 5\ 1/8\ d - 3$
14. $(H + 4)(H - 4) = -20 + 2H^2$
15. $(a/2)^3 + 34 = 42$
16. $\sqrt{x + 2} = 6$
17. $16V^2 + 6V = 8V^2 + 6V + 72$
18. $-2(x - 3) = \sqrt{x^2} - 2x$
19. $14\sqrt{x} = 6(\sqrt{x} + 8) + 16$
20. $2\sqrt[3]{y} + 10\sqrt[3]{y} - 6\sqrt[3]{y} = 30$

PROBLEMS

Substitute given numerical values for letter values in each of the following formulas and solve. Check.

23. Solve each of the following electrical formulas for the quantity indicated. Calculate to three decimal places where necessary.

a. $f = (PS)/120$. Solve for P when $f = 25$ and $S = 1{,}250$.

b. $X_L = 2\pi fL$. Solve for L when $f = 60$ and $X_L = 82$.

c. $E_P T_S = E_S T_P$. Solve for T_S when $E_P = 440$, $E_S = 2{,}200$, and $T_P = 150$.

d. $H = 0.057I^2\, Rt$. Solve for I when $H = 720$, $t = 42$, and $R = 12$.

e. $R_t = R_{32}(1 + at)$. Solve for a when $R_t = 10.51$, $R_{32} = 9.59$, and $t = 43$.

f. $Z = \sqrt{R^2 + X_L^2}$. Solve for X_L when $R = 8$ and $Z = 60$.

24. Solve each of the following machine shop formulas for the quantity indicated. Calculate to four decimal places where necessary.

a. $F = 2.38P + 0.25$. Solve for P when $F = 1.750$.

b. $S = T - (1.732/N)$. Solve for T when $S = 0.4134$ and $N = 20$.

c. $S = [L/\ell][0.5(D - d)]$. Solve for D when $S = 0.25$, $L = 16$, and $\ell = 4$, and $d = 2.5$.

d. $W = S_t(0.55d^2 - 0.25d)$. Solve for S_t when $W = 1150$ and $d = 0.750$.

25. In the formula $C = (5/9)(F - 32)$, C is the temperature in degrees Celsius and F the corresponding temperature in degrees Fahrenheit. Find the Fahrenheit temperature to the closest whole number that is equivalent to: a. 5°C, b. 10°C, c. 25°C, and d. 30°C.

26. Use the Pythagorean Theorem, $c^2 = a^2 + b^2$, to solve the following problems. Calculate to two decimal places where necessary.

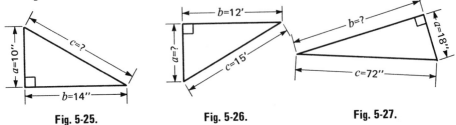

| Fig. 5-25. | Fig. 5-26. | Fig. 5-27. |

d. Determine the distance between centers of the pulleys shown in Fig. 5-28.

e. Find the distance between the centers of gears A and C as shown in Fig. 5-29. Line AC and line BC form a 90° angle.

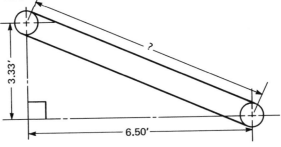

Fig. 5-28.

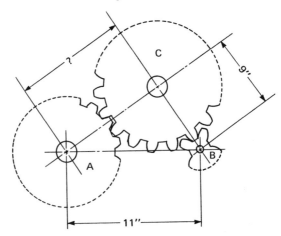

Fig. 5-29.

Rearranging formulas

A formula that is given in terms of a particular value must sometimes be rearranged in terms of another value. For example, without any numbers given for letter values, the formula $A = 1/2\ bh$, which is given in terms of A may have to be rearranged in terms of h.

Consider the letter to be solved for as the unknown term and the other letters in the formula as the known values. The formula must be rearranged so that the unknown term is on one side of the equation and all other values are on the other side. The formula is rearranged using the same procedure that is used for solving equations consisting of combined operations.

Example 1. $A = bh$. Solve for h. Apply the division principle: (Divide both sides by b.)

$$\frac{A}{b} = \frac{bh}{b}; Ans.\ \frac{A}{b} = h.$$

Example 2. $\ell = a + b$. Solve for a. Apply the subtraction principle: (Subtract b from both sides.)

$$\ell = a + b$$
$$\underline{-b = \quad - b}$$
$$Ans.\ \ell - b = a$$

Example 3. $R = \dfrac{KL}{d^2}$. Solve for L.

Step 1. Apply the multiplication principle: (Multiply both sides by d^2.)

$$d^2(R) = d^2\left(\frac{KL}{d^2}\right)$$
$$d^2 R = KL$$

Step 2. Apply the division principle: (Divide both sides by K.)

$$\frac{d^2 R}{K} = \frac{KL}{K}$$
$$Ans.\ \frac{d^2 R}{K} = L$$

EXERCISES

The following formulas are found in various trade handbooks. Rearrange and solve each for the designated letter.

1. $I = E/R, E = ?$
2. $d = rt, r = ?$
3. $V = lwh, h = ?$
4. $l = c - s, c = ?$
5. $A + B + C = 180, B = ?$
6. $S = c + g, g = ?$

7. $F = 2.38P + 0.25, P = ?$
8. $E = I(R + r), R = ?$
9. $HP = 0.000016MN, N = ?$
10. $Ca = S(C - F), C = ?$
11. $P = 2\,EI\,P.F., P.F. = ?$
12. $KW = (2\,IE\,P.F.)/1000, I = ?$
13. $Do = 2C - d + 2a, C = ?$

14. $M = D - 1.5155P + 3W, P = ?$
15. $Cx = By\,(F - 1), F = ?$
16. $M = E - 0.866P + 3W, E = ?$
17. $H.P. = D^2N/2.5, D = ?$
18. $C = 3.1416\,DN/12, N = ?$
19. $C = 0.7854\,D^2L/231, L = ?$
20. $L = 3.14(0.5D + 0.5d) + 2x, d = ?$

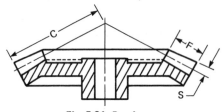

Fig. 5-31 Bevel gear.

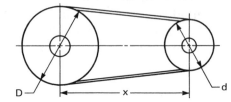

Fig. 5-32 Pulleys—open belt.

Ratio and proportion

The ability to solve applied problems using ratio and proportion is a requirement of most occupations. Compression ratios, transmission ratios, and rear axle ratios are commonly used by automobile mechanics. Ratios are used to determine pipe capacities and roof pitches in the building trades. Proportions are used to determine gear sizes and speeds, electrical resistance, wire sizes, and material requirements.

Description of ratios

Ratio is the comparison of two like quantities. For example, the compression ratio of an engine is the comparison between the amount of space in a cylinder when the piston is at the bottom of the stroke, and the amount of space when the piston is at the top of the stroke. See Fig. 5-33. The compression ratio is 8 to 1. The triangle shown in Fig. 5-34 has sides of 3, 4, and 5 feet. The comparison of side *a* to side *b* is expressed as the ratio of 3 to 4. The comparison of side *a* to side *c* is expressed as the ratio of 3 to 5. The comparison of side *b* to side *c* is expressed as the ratio of 4 to 5.

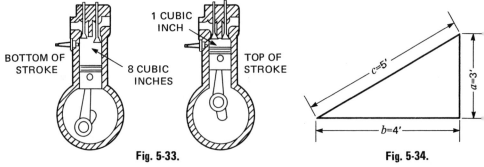

Fig. 5-33.

Fig. 5-34.

The terms of a ratio are the two numbers that are compared. Both terms must be expressed in the same units. For example, a length of pipe 11 inches long and a length of pipe 2 feet long cannot be compared as a ratio until the 2 foot length is converted to 24 inches. The two lengths are in the ratio of 11 to 24.

It is impossible to express two quantities as ratios if the terms have unlike units that cannot be converted to like units. For example, inches and pounds as shown in Fig. 5-35 cannot be compared as ratios.

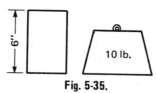

Fig. 5-35.

Ratios are expressed in the following two ways:
1. With a colon between the two terms, such as 5:8. The colon means "to." Therefore, the ratio 5:8 is read "5 to 8."
2. With a division sign separating the two numbers such as 5 ÷ 8 or as a fraction, 5/8.

Order of terms and reduction of ratios

The terms of a ratio must be compared in the order in which they are given. The first term is the numerator of a fraction and the second term is the denominator. Notice that 1 to $3 = 1 \div 3 = \frac{1}{3}$,

but 3 to $1 = 3 \div 1 = \frac{3}{1}$; and $x:y = x \div y = \frac{x}{y}$, but $y:x = y \div x = \frac{y}{x}$.

Generally, a ratio should be reduced to its lowest terms. The following examples illustrate the reduction of ratios.

$$3:9 = \frac{3}{9} = \frac{1}{3} \; ; \; 40 \text{ to } 15 = \frac{40}{15} = \frac{8}{3} \; ; \; \frac{3}{8} : \frac{9}{16} = \frac{3}{8} \div \frac{9}{16}$$

$$= \frac{3}{8} \times \frac{16}{9} = \frac{2}{3} \; ; \; x^3 \text{ to } x^2 = \frac{x^3}{x^2} = \frac{x}{1} \; ; \; 4ab:6a = \frac{4ab}{6a} = \frac{2b}{3}$$

EXERCISES

1. Express the following ratios in fractional form. Reduce where possible.

a. 6:15

b. 15:6

c. 2:11

d. 7:21

e. 12′ to 46′

f. 3 lb. to 18 lb.

g. 17 mi. to 9 mi.

h. 156′ to 200′

i. $3a^2b:6ab$

j. $xy:x^2y$

k. $\frac{2}{3}$ to $\frac{1}{2}$

l. $\frac{1}{2} : \frac{2}{3}$

2. Write the following comparisons as ratios. Convert units and reduce where possible.

a. 6" to 3'
b. 5' to 5"
c. 2 ft. to 3 yd.
d. 18 in. to 1 yd.

e. 1 sq. ft. to 48 sq. in.
f. 18 sq. ft. to 2 sq. yd.
g. 20 min. to 1 hr.
h. 40 sec. to 3 min.

PROBLEMS

27. What is the compression ratio of an engine if there are 28 cubic inches of space when the piston is at the bottom of the stroke and 2.8 cubic inches when the piston is at the top of the stroke?

28. In the building trades the terms pitch, rise, run, and span are used in the layout and construction of roofs. In the gable roof shown in Fig. 5-36, the span is twice the run. Pitch is the ratio of the rise to the span; pitch = rise/span. Determine the pitch of the following gable roofs:

a. 8' rise, 24' span
b. 6' rise, 20' span
c. 10' rise, 24' span

d. 15' rise, 15' span
e. 22' rise, 33' run
f. 12' rise, 24' run

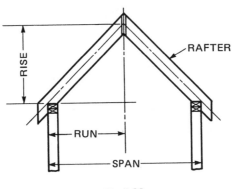

Fig. 5-36.

29. Refer to the hole locations given in Fig. 5-37 and determine the ratios of the following:

a. Dim. A to Dim. B.
b. Dim. A to Dim. C.
c. Dim. B to Dim. C.
d. Dim. B to Dim. D.

e. Dim. C to Dim. D.
f. Dim. D to Dim. A.
g. Dim. C to Dim. B.
h. Dim. D to Dim. C.

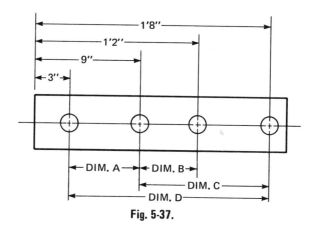

Fig. 5-37.

Description of proportions

A *proportion* is an expression that states the equality of two ratios. Proportions are expressed in the following two ways:

1. 5:8 :: 10:16, which is read as "5 is to 8 as 10 is to 16." Note: : means "is to," and :: means "as."

2. $\frac{5}{8} = \frac{10}{16}$. This is the equation form, and is the way that proportions are generally written for trade applications.

A proportion consists of four terms. The first and fourth terms are called *extremes* and the second and third terms are called *means*. In the proportion, 2:3 :: 4:6, 2 and 6 are the extremes, 3 and 4 are the means. In the proportion $\frac{3}{4} = \frac{6}{8}$, 3 and 8 are the extremes, 4 and 6 are the means.

The product of the means equals the product of the extremes. The following two examples illustrate this principle.

1. $\frac{3}{4} = \frac{6}{8}$. Cross multiply, $\frac{3}{4} \diagup\!\!\!\!\diagdown \frac{6}{8}$; 3 x 8 = 4 x 6; 24 = 24.

2. $\frac{a}{b} = \frac{c}{d}$. Cross multiply, $\frac{a}{b} \diagup\!\!\!\!\diagdown \frac{c}{d}$; a x d = b x c; ad = bc.

The method of cross multiplying is used in solving many practical trade applications for an unknown term when the values of three terms are known. Since a proportion is an equation, the principles used for solving equations are applied in determining the value of the unknown after the terms have been cross multiplied. The following examples illustrate the procedure for solving proportions.

Example 1. $\dfrac{5}{7} = \dfrac{x}{14}$. Solve for the value of x.

THINK	DO THIS
1. The product of the means equals the product of the extremes. Cross multiply.	1. $7x = 5(14)$ $7x = 70$
2. Apply the division principle of equality. Divide both sides of the equation by 7.	2. $\dfrac{7x}{7} = \dfrac{70}{7}$ $x = 10$ *Ans.*
3. Check. Substitute 10 for x in the original proportion and cross multiply.	3. $\dfrac{5}{7} = \dfrac{10}{14}$ $5(14) = 7(10),\ 70 = 70$

Example 2. $\dfrac{R_1}{R_2} = \dfrac{\ell_1}{\ell_2}$ is a proportion used in the electrical trades where R_1 and R_2 are the resistance of two wires and ℓ_1 and ℓ_2 are the lengths of the wires. Find R_1 when $R_2 = 10$ ohms, $\ell_1 = 30$ ft., and $\ell_2 = 75$ ft.

THINK	DO THIS
1. Substitute the known values for the letters.	1. $\dfrac{R_1}{10} = \dfrac{30}{75}$
2. Cross multiply.	2. $75\,R_1 = 10(30)$ $75\,R_1 = 300$
3. Apply the division principle of equality. Divide both sides of the equation by 75.	3. $\dfrac{75\,R_1}{75} = \dfrac{300}{75}$ $R_1 = 4$ ohms. *Ans.*
4. Check. Substitute 4 for R_1 in the original equation and cross multiply.	4. $\dfrac{4}{10} = \dfrac{30}{75}$ $4(75) = (10)(30),\ 300 = 300$

EXERCISES

1. Solve for the unknown value in each of the following proportions. Check all answers.

   a. $\dfrac{2}{3} = \dfrac{x}{21}$ c. $\dfrac{19}{a} = \dfrac{1}{2}$ e. $\dfrac{6}{7} = \dfrac{15}{m}$

   b. $\dfrac{7}{10} = \dfrac{y}{50}$ d. $\dfrac{C}{12} = \dfrac{25}{4}$ f. $\dfrac{x}{17.76} = \dfrac{19.3}{37.22}$

g. $\dfrac{200}{1} = \dfrac{P}{0.001}$ h. $\dfrac{1/2}{3} = \dfrac{e}{5}$ i. $\dfrac{\frac{3}{8}}{\frac{7}{8}} = \dfrac{\frac{1}{4}}{t}$

2. The proportion $\dfrac{R_1}{R_2} = \dfrac{S_1}{S_2}$ can be used to determine an unknown rise or span of the roofs shown in Fig. 5-38. Both roofs have the same pitch. Determine the unknown values for each of the following exercises.

   a. $R_1 = 10'$, $S_1 = 30'$, $S_2 = 6'$, $R_2 = ?$
   b. $R_2 = 8'$, $S_1 = 24'$, $S_2 = 20'$, $R_1 = ?$
   c. $R_1 = 12'$, $R_2 = 9'$, $S_1 = 30'$, $S_2 = ?$
   d. $R_1 = 16'$, $R_2 = 14'$, $S_2 = 42'$, $S_1 = ?$

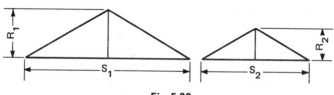

Fig. 5-38.

3. The proportion, $\dfrac{N_L}{N_C} = \dfrac{T_S}{T_L}$, is used for lathe thread cutting computations using simple gearing when the fixed stud gear and the spindle gear have the same number of teeth. N_L is the number of threads per inch on the lead screw, N_C is the number of threads per inch to be cut, T_S is the number of teeth on the stud gear, and T_L is the number of teeth on the lead screw gear. Determine the missing values in each of the following exercises.

   a. $N_L = 4$, $N_C = 8$, $T_S = 32$, $T_L = ?$
   b. $N_L = 7$, $T_S = 35$, $N_C = 15$, $T_L = ?$
   c. $N_C = 10$, $N_L = 6$, $T_L = 40$, $T_S = ?$
   d. $N_L = 8$, $T_L = 42$, $T_S = 28$, $N_C = ?$

Direct and inverse proportions

A tradesman may be required to convert word statements or other given data into proportions. When setting up a proportion it is important that the terms of the proportion be placed in their proper positions.

A problem which is to be set up and solved as a proportion must first be analyzed in order to determine where the terms are to be placed. Proportions are either direct or inverse.

Direct proportions

Two quantities are *directly proportional* if a change in one produces a change in the other in the same direction. If an increase in one produces an increase in the other, or if a decrease in one produces a decrease in the other, the two quantities are directly proportional. The proportions discussed will be those that change at the same rate. An increase or decrease in one quantity produces the same rate of increase or decrease in the other quantity.

When setting up a direct proportion in fractional form, the numerator of the first ratio must correspond to the numerator of the second ratio. The denominator of the first ratio must correspond to the denominator of the second ratio.

The following examples illustrate the method of setting up and solving direct proportions.

Example 1. If a sump pump discharges 450 gallons of water in 10 minutes, how long will it take to discharge 1,125 gallons?

Step 1. Analyze the problem: An increase in the number of gallons discharged (from 450 gal. to 1,125 gal.) requires an increase in time. Time increases as gallons discharged increase; therefore, the proportion is direct.

Step 2. Set up the proportion: Let x represent the number of minutes required to discharge 1,125 gallons.

$$\frac{450 \text{ gal.}}{1,125 \text{ gal.}} = \frac{10 \text{ min.}}{x \text{ min.}}$$

Notice that the numerator of the first ratio corresponds to the numerator of the second ratio: 450 gallons are discharged in 10 minutes. The denominator of the first ratio corresponds to the denominator of the second ratio: 1,125 gallons are discharged in x minutes.

Step 3. Solve for x: $\frac{450}{1,125} = \frac{10}{x}$; $450x = 10(1,125)$; $450x = 11,250$;

$x = \frac{11,250}{450}$; $x = 25$ min. *Ans.*

Step 4. Check: $\frac{450}{1,125} = \frac{10}{25}$; $450(25) = 10(1,125)$; $11,250 = 11,250$.

Example 2. A sheet metal cone is shown in Fig. 5-39. The cone is 16 inches high with an 18 inch diameter base. Determine the diameter 7 inches from the top of the cone.

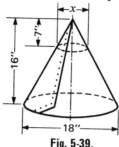

Fig. 5-39.

Step 1. Analyze the problem: As the height of the cone decreases from 16 inches to 7 inches, the diameter also decreases at the same rate. The proportion is direct.

Step 2. Set up the proportion: Let x represent the diameter 7 inches from the top.

$$\frac{7 \text{ inches in height}}{16 \text{ inches in height}} = \frac{x \text{ inches in diameter}}{18 \text{ inches in diameter}}$$

Notice that the numerator of the first ratio corresponds to the numerator of the second ratio: the 7 inch height corresponds to the x inch diameter. The denominator of the first ratio corresponds to the denominator of the second ratio; the 16 inch height corresponds to the 18 inch diameter.

Step 3. Solve for x: $\frac{7}{16} = \frac{x}{18}$; $16x = 7(18)$; $x = \frac{126}{16}$; $x = 7\frac{14}{16} = 7\frac{7}{8}$ in.

$Ans.$

Step 4. Check: $\frac{7}{16} = \frac{7\,7/8}{18}$; $16(7\,7/8) = 7(18)$; $126 = 126$

Inverse proportions

Two quantities are *inversely* or indirectly proportional if a change in one produces a change in the other in the opposite direction. If an increase in one produces a decrease in the other, or if a decrease in one produces an increase in the other, the two quantities are inversely proportional. For example, if one quantity increases by 3 times its original value, the other quantity decreases by 3 times or is 1/3 of its original value.

When setting up an inverse proportion in fractional form, the numerator of the first ratio must correspond to the denominator of the second ratio. The denominator of the first ratio must correspond to the numerator of the second ratio.

The following examples illustrate the method of setting up and solving inverse proportions.

Example 1. Two gears in mesh are shown in Fig. 5-40. The driver gear has 40 teeth and revolves at 360 rpm. Determine the rpm of a driven gear with 16 teeth.

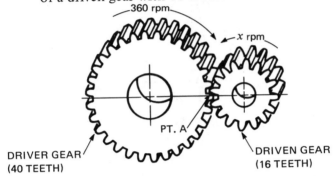

DRIVER GEAR
(40 TEETH)

DRIVEN GEAR
(16 TEETH)

Fig. 5-40.

Step 1.   Analyze the problem: When the driver gear turns one revolution, 40 teeth pass point A. The same number of teeth on the driven gear must pass point A. Therefore, the driven gear turns more than one revolution for each revolution of the driver gear. The gear with 16 teeth (driven gear) revolves at greater rpm than the gear with 40 teeth (driver gear). A decrease in the number of teeth produces an increase in rpm. The proportion is inverse.

Step 2.   Set up the proportion: Let x represent the rpm of the gear with 16 teeth.

$$\frac{40 \text{ teeth}}{16 \text{ teeth}} = \frac{x \text{ rpm}}{360 \text{ rpm}}$$

Notice that the numerator of the first ratio corresponds to the denominator of the second ratio; the gear with 40 teeth revolves 360 rpm. The denominator of the first ratio corresponds to the numerator of the second ratio; the gear with 16 teeth revolves x rpm.

Step 3.   Solve for x: $\dfrac{\overset{5}{\cancel{40}}}{\underset{2}{\cancel{16}}} = \dfrac{x}{360}$; $2x = 5(360)$; $2x = 1{,}800$; $x = \dfrac{1{,}800}{2}$;

$x = 900$ rpm. *Ans.*

Step 4.   Check: $\dfrac{40}{16} = \dfrac{900}{360}$; $40(360) = 16(900)$; $14{,}400 = 14{,}400$.

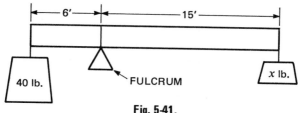

Fig. 5-41.

Example 2. A balanced lever is shown in Fig. 5-41. A 40-pound weight is placed 6 feet from the fulcrum. Determine the weight required 15 feet from the fulcrum in order to balance the lever.

Step 1. Analyze the problem: The 40-pound weight is closer to the fulcrum than the unknown weight. An increase in the distance from the fulcrum produces a decrease in weight required to balance the lever. The proportion is inverse.

Step 2. Set up the proportion: Let x represent the weight 15 feet from the fulcrum.

$$\frac{\overset{2}{\cancel{6}}\text{ feet}}{\underset{5}{\cancel{15}}\text{ feet}} = \frac{x \text{ pounds}}{40 \text{ pounds}}$$

Notice the numerator of the first ratio corresponds to the denominator of the second ratio: the 40-pound weight is 6 feet from the fulcrum. The denominator of the first ratio corresponds to the numerator of the second ratio; the unknown weight is 15 feet from the fulcrum.

Step 3. Solve for x: $\frac{6}{15} = \frac{x}{40}$; $5x = 2(40)$; $5x = 80$, $x = \frac{80}{5}$; $x = 16$ pounds. *Ans.*

Step 4. Check: $\frac{6}{15} = \frac{x}{40}$; $\frac{6}{15} = \frac{16}{40}$; $6(40) = 16(15)$;
$= 240 = 240$.

PROBLEMS

Analyze each of these problems to determine whether the problem is a direct or inverse proportion, set up the proportion, and solve.

27. Determine height x of the retaining wall shown in Fig. 5-42.

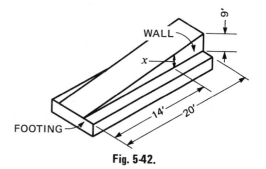

Fig. 5-42.

28. A machine produces 2,700 parts in 13 1/2 hours. How many parts are produced by the machine in 16 1/2 hours?
29. An engine uses 4.5 gallons of gasoline when it operates for 6.25 hours. How long can the engine operate on 11.5 gallons of gasoline?
30. Three machines produce at the same rate. The three machines produce 720 parts in 1.6 hours. How many hours would it take two machines to produce 720 parts?
31. The lever shown in Fig. 5-43 is in balance. Determine distance x.
32. If 1,500 cubic yards of soil are removed in excavating the foundation of a building to a 3 foot depth, how many cubic yards are removed when excavating to a 10 1/2 foot depth?
33. Forty-five pieces of lumber are piled to a height 6 3/4 feet. How many pieces of lumber are required for a pile 9 feet high?
34. If five machine operators complete a job in 2.8 hours, how many operators are required to complete the same job in 3.5 hours?
35. The tank shown in Fig. 5-44 contains 1,200 gallons of water when completely full. How many gallons does it contain when filled to a height of 3-3/4'?

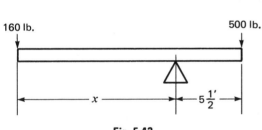

160 lb.   500 lb.

Fig. 5-43.

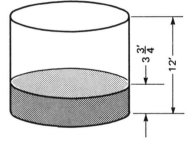

Fig. 5-44.

36. A sheet of steel 6 1/4 feet long weighs 400 pounds. A piece 2 3/8 feet long is sheared from the sheet. Determine the weight of the 2 3/8 foot piece.
37. A micrometer reading is made on the tapered piece shown in Fig. 5-45, page 259. Determine the micrometer reading (Dia. B) to 3 decimal places for the following:
    a. Length of piece = 10.000″, diameter A = 1.500″, and dimension C = 7.500″.
    b. Length of piece = 8.750″, diameter A = 1.250″, and dimension C = 3.875″.
    c. Length of piece = 5 1/2″, diameter A = 1 1/16″, and dimension C = 1 3/8″.
    d. Length of piece = 14 3/4″, diameter A = 2 1/4″, and dimension C = 11 1/4″.

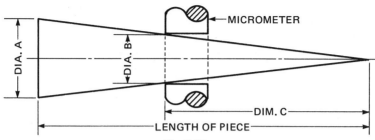

Fig. 5-45.

38. Of two gears that mesh as shown in Fig. 5-46, the one which has the greatest number of teeth is called the gear, and the one which has the fewer teeth is called the pinion. Refer to the table and determine x in each problem.

GEAR

PINION

	Number of Teeth on Gear	Number of Teeth on Pinion	Rpm of Gear	Rpm of Pinion
a.	48	20	120	x
b.	32	24	x	210
c.	35	x	160	200
d.	x	15	150	250
e.	54	28	80	x

Fig. 5-46.

REVIEW EXERCISES AND PROBLEMS

1. Raise each of the following problems to the indicated power.
   a. 2.3^3
   b. 0.07^2
   c. 0.2^5
   d. $(2/3)^3$
   e. $(6\ 1/4)^2$

2. Determine the roots of the following problems:
   a. $\sqrt{36}$
   b. $\sqrt[3]{64}$
   c. $\sqrt{16/81}$
   d. $\sqrt{86.49}$
   e. $\sqrt{21.50 - 3.01}$

3. Find A when $h = 8$, $b = 14$, and $b' = 10$ in $A = (h/2)(b + b')$.

4. Find r when $c = 10$, and $h = 2$ in $r = (c^2 + 4h^2)/8h$.

5. Add the following signed numbers as indicated:
   a. $-14 + (-8)$
   b. $(-7/8) + 5/8$
   c. $-2.3 + (-8.7) + 9.5 + (-1.3)$

6. Subtract the following signed numbers as indicated.

   a. $-15 - (-9)$ c. $(13.5 + 6.2) - (15.6 - 17.4)$

   b. $(11/32) - (-15/32)$

7. Multiply the following signed numbers as indicated:

   a. $8(-5)$ c. $(-1/3)(3/8)$

   b. $(-9.6)(-2.2)$ d. $(-3)(2.2)(-0.1)(-5)$

8. Divide the following signed numbers as indicated:

   a. $18/-3$ c. $-22.2 \div 5.55$

   b. $-4\,\overline{)\,-36}$ d. $(6\ 3/8) \div (-1/2)$

9. Raise each of the following signed numbers to the indicated powers.

   a. -3^2 d. -6^{-2}

   b. -3^3 e. $+25^{1/2}$

   c. $(-1/2)^3$

10. Determine the indicated root of each of the following signed numbers:

    a. $\sqrt[3]{-27}$ d. $\dfrac{\sqrt[5]{-32}}{3}$

    b. $\sqrt[5]{-32}$

    c. $\sqrt[3]{\dfrac{-8}{-27}}$ e. $\dfrac{-3}{\sqrt[3]{-64}}$

11. Solve each of the following combined operations exercises:

    a. $24 - (6)(-3) + 4^2$ c. $4^2 + \sqrt[3]{-27} - (3)(-1)(5)$

    b. $90 - (2)(3 - 5) - 12$ d. $\sqrt{(-20)(-2) + (2.5)(4^2)}$

12. Add the terms in each of the following exercises:

    a. $-7.3x^2,\ 4.7x^2$

    b. $(2a - 10b - 8c),\ (3a + 2b - 10c)$

13. Subtract the terms as indicated in each of the following exercises:

    a. $-0.52xy - 3.9xy$ b. $(3a^3 - 0.2a^2) - (-a^3 + 0.2a^2)$

14. Multiply the terms as indicated in each of the following exercises:

    a. $(-0.5m^2 p^3)(-2m^3 p)$ b. $-5a^3 b^2(6a^2 b^2 - b + 3)$

15. Divide the terms as indicated in each of the following exercises:

    a. $-4a^3 b^4 \div -1/2\ ab^3$

    b. $(18x^3 y^2 - 12x^4 y^3 + 6x^5 y^4) \div 6x^2 y^2$

16. Raise the terms to the indicated powers in each of the following exercises:

  a. $(7a^4 b^3)^3$ c. $(8x^2 - 4y)^2$
  b. $[-3m^2 (p^3)^2 t^3]^2$

17. Determine the roots of each of the following exercises:

  a. $\sqrt{25a^6 b^4 c^2}$ c. $\sqrt{4/25\ x^4 y^8}$
  b. $\sqrt[3]{-64r^3 t^6}$

18. Remove parentheses and combine like terms where possible for each of the following exercises:

  a. $8x + (3x + x^2)$ c. $(a^2 + y^2) - (2a^2 - y^2)$
  b. $-(x^3 + y - xy)$

19. Solve each of the following combined operations exercises. Simplify and combine like terms where possible.

  a. $18 - 3(-10x) - 3x$ c. $\dfrac{20xy^4}{4xy^3} - (y^3)^2 + 2y$

  b. $-4(d^2 + 3) - (d^2 e^3)^2$ d. $\sqrt{\dfrac{12y^6}{3}} \div y^3$

20. Solve each of the following equations:

  a. $y + 5 = 22$ j. $c/3 = 12$
  b. $P + 28 = -102$ k. $y/-5 = -0.6$
  c. $N + 1/2 = 4\ 3/8$ l. $S \div 2\ 1/8 = 5$
  d. $A - 7 = 35$ m. $L^2 = 49$
  e. $0.06 = x - 1.05$ n. $M^3 = -64$
  f. $H - 3/4 = 8$ o. $A^2 = 1/49$
  g. $8x = 72$ p. $\sqrt{M} = 12$

  h. $-0.2y = -18$ q. $\sqrt[3]{b} = -2$
  i. $3/16\ C = -8$ r. $\sqrt{x} = 3/8$

21. Solve each of the following equations which consist of combined operations:

  a. $7x - 24 = 3$ c. $(M + 4)(M - 4) = -20 + 2M^2$
  b. $10(A - 3) = -2(A + 6)$ d. $-5(x - 4) = \sqrt{x^2 - 3x}$

22. Solve each of the trade formulas for the quantity indicated by substituting numerical values for letter values. Calculate to three decimal places where necessary.

  a. $H.P. = 0.000016MN$. Solve for M when $H.P. = 22$ and $N = 50.8$.
  b. $M = E - 0.866P + 3W$. Solve for W when $M = 3.3700, E = 3.2200$, and $P = 0.125$.

c. $Do = \dfrac{Pc(N + 2)}{3.1416}$. Solve for Pc when $Do = 4.5000$ and $N = 16$.

d. $C = \dfrac{3.1416DN}{12}$. Solve for N when $C = 5{,}000$ and $D = 10$.

e. $C = \dfrac{0.7854D^2 L}{231}$. Solve for D when $C = 15$ and $L = 11$.

f. $I = \dfrac{E\,(ns)}{r\,(ns) + R}$. Solve for R when $I = 0.7$, $E = 1.5$, $(ns) = 5$, and $r = 1.9$.

23. Determine the unknown distances in:

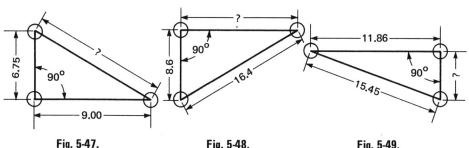

Fig. 5-47.   **Fig. 5-48.**   **Fig. 5-49.**

24. Determine distance x in Fig. 5-50.

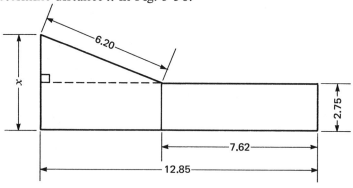

Fig. 5-50.

25. Rearrange each of the following trade formulas in terms of the designated letters:

a. $E = I(R + r)$, $r = ?$
b. $Ca = S(C - F)$, $F = ?$
c. $Do = 2C - d + 2a$, $a = ?$
d. $C = 0.7854D^2 L/231$, $D = ?$
e. $M = E - 0.866P + 3W$, $W = ?$
f. $Z = \sqrt{R^2 + X_L{}^2}$, $R = ?$

26. Express the following ratios in fractional form. Reduce where possible.

    a. 6:15

    b. $4x^3y:6xy$

    c. 78" to 100"

    d. 1/2 to 2/3

27. Write the following comparisons as ratios. Convert units and reduce where possible.

    a. 8" to 2'

    b. 2 yd. to 14 in.

    c. 3 cu. ft. to 0.5 cu. yd.

28. Solve for the unknown value in each of the following proportions:

    a. $\dfrac{4}{6} = \dfrac{y}{15}$ b. $\dfrac{3}{10} = \dfrac{x}{5.5}$ c. $\dfrac{1/4}{3/8} = \dfrac{1/2}{e}$

29. The proportion $\dfrac{h_1}{h_2} = \dfrac{\ell_1}{\ell_2}$ can be used to determine an unknown dimension if three dimensions are known in the triangle shown in Fig. 5-51. Determine the following missing values to three decimal places where necessary.

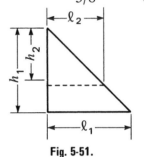

Fig. 5-51.

    a. $h_2 = 3$ ft., $\ell_2 = 2.75$ ft., $\ell_1 = 4.83$ ft., $h_1 = ?$

    b. $h_1 = 7.5''$, $\ell_1 = 6.7''$, $\ell_2 = 2.8''$, $h_2 = ?$

    c. $h_1 = 0.75$ yd., $h_2 = 1.5$ ft., $\ell_1 = 0.65$ yd., $\ell_2 = ?$

    d. $h_1 = 14.6''$, $h_2 = 6.6''$, $\ell_2 = 1.25$ ft., $\ell_1 = ?$

30. A pump discharges 1,500 gallons in 0.6 hr. How long does it take to discharge 250 gallons?

31. The metal bar shown in Fig. 5-52 weighs 512 lb. What is the remaining weight of the bar if 3.5 inches are machined from the top? Give answer to two decimal places.

32. Of two gears in mesh, the driver gear has 30 teeth and revolves at 320 rpm. Determine the rpm of a driven gear with 12 teeth.

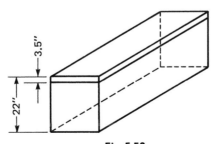

Fig. 5-52.

33. A gasoline tank which is 20 feet high contains 30,000 gallons of fuel measured at a depth of 17.5 feet. How many gallons does the tank hold when full? Give answer to the nearest whole gallon.

34. The crankshaft speed of a car is 3,200 rpm when the car is traveling 60 mph. What is the crankshaft speed when the car is traveling 47 mph.? Give answer to two decimal places.

35. Determine the diameter of the small pulley shown in Fig. 5-53.

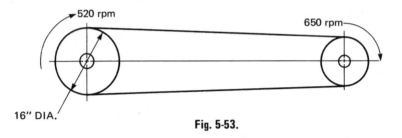

520 rpm

650 rpm

16″ DIA.

Fig. 5-53.

6 Computation, Accuracy, and the Use of the Slide Rule

In the preceding units much of what we learned involved making various computations and applying what was learned to our daily work. Computations will be of increasing importance. In this unit we shall learn to make computations systematically, to judge their accuracy, and to use the slide rule.

Arrangement and care in computations

A computation that is not correct is worthless. In fact, incorrect computations may cause loss of time, loss of assets, and even accidents. The following suggestions if followed thoughtfully will help you avoid errors. Study them and fix them in your mind.

1. Arrange the work systematically. Start at the top of the paper and place the successive operations either to the right of one another as in writing, or each below the other in a column. Follow a definite, consistent procedure. Make clear, distinct figures.

When working with a formula, write the formula first, substitute in place of the letters the numerical values that are known, and indicate clearly the operations to be performed.

When dealing with an equation, show clearly that the same operation, such as adding 7, or dividing by 3, is done on *both sides* of the equal sign.

2. Label each result of a preliminary step and the final result.

3. Check each operation before going on to the next.

In *addition*, if you first add each column from the bottom up, check by adding each column from the top down. In the example at the right, if you start with 6 and add 1, then 2, then 9, to check you start with 9, add 2, then 1, then 6.

$$
\begin{array}{r}
849 \\
382 \\
491 \\
\underline{576}
\end{array}
$$

To check a *subtraction*, add the result to the number that was subtracted. This sum should be the original number.

Example. From 7,242 subtract 3,829. Check the result.

THINK	DO THIS
1. Arrange the numbers in a column with the one to be subtracted below the other.	1. 7,242 3,829
2. Subtract.	2. 3,413 Result

3. Check by adding the result to the number that was subtracted to see if this sum is the same as the original number.	3. 3,829
	   3,413
	   7,242 Since this is the same as the original number 7,242, the subtraction is correct.

Ans. 3,413

Multiplication may be checked either by interchanging the factors and multiplying again, or by dividing the product by one of the factors. The result of this division should be the other factor.

Example. Multiply 429 and 52 and check.

THINK	DO THIS

1. Write one of the factors below the other, and multiply.	1. 429
	     52
	   858
	  2145
	22,308 Product

2. a. Check by multiplying after reversing the factors. If both products are the same, the multiplication is correct.	2. a. 52
	    429
	   468
	  1 04
	 20 8
	22,308 Product

2. b. Check by dividing the product 22,308 by one of the factors. If the result of this division is the same as the other factor, the multiplication is correct.	2. b. 52)22,308 429
	    20 8
	     1 50
	     1 04
	      468
	      468

Ans. 22,308

Division may be checked by multiplying the two factors to obtain the product, the original number.

Example 1. Divide 7,261 by 137.

THINK	DO THIS
Here 7,261 is the product of the two factors, one of which is 137. To find the other: 1. Divide 7,261 by 137.	1. $$\begin{array}{r} 53 \\ 137 \overline{)\,7{,}261} \\ \underline{6\,85} \\ 411 \\ \underline{411} \end{array}$$
2. To check, multiply the two factors 137 and 53. The product should be the original number.	2. $$\begin{array}{r} 137 \\ \underline{53} \\ 411 \\ \underline{685} \\ 7{,}261 \end{array}$$ Since the product here is the original number, 7,261, the division is correct. <div align="right">*Ans.* 53</div>

If the division is not exact, continue to two places beyond the decimal.

Example 2. Divide 7,296 by 137.

THINK	DO THIS

7,296 is the product of 137 and some other factor.
1. Divide 7,296 by 137.
  a. In this example, the other factor can be expressed exactly as 53 35/137; or
  b. The division may be continued to a decimal place and the other factor obtained as an approximation, here 53.3.
	1. a. $$\begin{array}{r} 53 \\ 137 \overline{)\,7296} \\ \underline{685} \\ 446 \\ \underline{411} \\ 35 \end{array}$$ b. $$\begin{array}{r} 53.25 \text{ or } 53.3, \\ 137 \overline{)\,7296.00} \text{ approx.} \\ \underline{685} \\ 446 \\ \underline{411} \\ 35\,0 \\ \underline{27\,4} \\ 7\,60 \\ \underline{6\,85} \end{array}$$

2. In either case, to check the division multiply the two factors. This product should be either exactly the original product or an approximate equivalent of it.	2. a. 137 x 53 35/137 =
	    (137 x 53) + 35 = 7,296
	  b. 137 x 53.3 = 7,302.1,
	    which is approximately 7,296
	
	<div align="right">*Ans.* 53 35/137 or 53.3 approx.</div>

Every operation can be checked by having some other person perform the operation. This other person should use a method different
from the one you used, if that is possible.

Estimate the result to detect large errors, like a misplaced decimal
point.

For example, suppose we write 72 x 3.5 = 25.2. To estimate the
product of 72 and 3.5, we would think that 70 x 3 = 210 and 70 x 4 =
280. Therefore, 72 x 3.5 should be between 210 and 280. Our result,
25.2, is incorrect. The correct product is 252.

Accuracy

The distance from the earth to the sun sometimes is given as
93,000,000 miles. Is that accurate? It may be or it may not be. If we
ask whether the distance is accurate to within 500,000 miles, the answer is
"Yes." When we say the distance is 93,000,000 miles, we mean that the
distance is somewhere between 92,500,000 and 93,500,000 miles. The
accuracy is to one-half million miles.

When we say that a line is 4.8″ long, we mean the line is somewhere
between 4.75″ and 4.85″; the accuracy is to one-half of 0.1″, or 0.05″.
The measurement might be written 4.8 ± 0.05″. This means the true
length is between 4.8″ + 0.05″ and 4.8″ – 0.05″. If a measured length
is given as 9.05, the true length is between 9.045 and 9.055.

If a measurement is given as 10 3/16″, it is understood that the line
is somewhere between 10 5/32″ and 10 7/32″. The measurement is
accurate to within one-half of 1/16″, or 1/32″. This measurement might
be written 10 3/16″ ± 1/32″.

When we ask about the accuracy of a number, we must indicate the
range within which the measurement is to be given. Unless this is done,
accuracy has little meaning.

Rounding off numbers

If we wish to express a value given in small units as an approximately
equal value given in larger units, we *round off* the first number. The process of "rounding off numbers" consists of changing the digits representing
the smaller units to zeros so that the number is given in terms of larger
units.

Only *one* digit in the original number needs to be considered to determine how to round off. If that *one* digit is 5 or more, you increase the
digit to its left by 1, replace that one digit with a 0, and replace all digits
to its right with 0's. If that *one* digit is 4 or less, you leave the digit to its
left as is and replace that one digit and all digits to its right with zeros.
If you wish to round off to the nearest ten, you inspect only the units

digit. When rounding off to the nearest thousand, you inspect only the hundreds digit. In rounding off to the nearest hundredth, consider only the digit in the thousandths place.

If we wish to express the measurement 8.276" to the nearest tenth of an inch, we note *the* digit in the hundredths place, 7, and we write 8.3", since 8.276" is nearer to 8.3" than to 8.2". We have rounded off the number 8.276 to tenths.

If we express 8,635 to the nearest hundred, we note *the* digit in the tens place, 3, and we write 8,600, since 8,635 is nearer 8,600 than it is to 8,700; that is, 8,635 rounded off to hundreds is 8,600. If we round 92,655,000 to millions we have 93,000,000.

Example 1. Round off the number 627,450 to thousands.

THINK	DO THIS
1. Since the number is to be rounded off to thousands, first determine the digit in the number that stands in the thousands place.	1. In 627,450 the digit 7 is in the thousands place.
2. Look at the digit which follows this digit.	2. The 4 follows 7.
3. If this following digit is less than 5, we replace that and all following digits by 0. If that following digit were 5 or more than 5, we replace that and all following digits by 0, but add 1 to the digit preceding 5.	3. 4 is less than 5, so we write 627,000. *Ans.* 627,450 rounded off to thousands is 627,000.

Example 2. Round off 6.8977 to hundredths.

THINK	DO THIS
1. The digit in the hundredths place is 9.	1. 6.8977
2. The digit following this one in the number is 7.	2. 6.8977
3. Since this following number, 7, is greater than 5, we replace 7 and all digits after it by 0, but add 1 to the digit preceding this 7.	3. 6.8977 becomes 6.9000 or 6.90. *Ans.* 6.8977 rounded off to hundredths is 6.90.

Summary

To round off any number to a specified place:

1. Determine the digit of the number that is in the specified place.

2. Look at the digit that follows the digit in the specified place.

3. If the following digit is less than 5, replace it and all the digits of the number that follow by zeros.

   If this digit is 5, 6, 7, 8, or 9, replace it and all following digits by zeros, but add 1 to the digit of the number that precedes this one.

Round off 86.594 to tenths.

1. Digit in the tenths place is 5.

2. The digit following this 5 is 9.

3. Since this is 9, we replace 9 and 4 by 0 and 0 but add 1 to the 5 that precedes the 9: 86.600.

Ans. 86.594 rounded off to tenths is 86.6.

EXERCISES

Round off the following numbers to the places specified:

1. 8,625 to hundreds
2. 7.92 to tenths
3. 0.94966 to thousandths
4. 286.455 to tenths
5. 0.0897 to thousandths
6. 92,846,728 to millions
7. 3,845,939 to hundred thousands
8. 12.809 to tenths
9. 371.195 to hundredths
10. 802,099 to thousands
11. 0.0395 to thousandths
12. 56,996 to hundreds
13. 37,392,000 to millions
14. 31.929 to hundredths
15. 8.0996 to hundredths
16. 1.72484 to ten thousandths

Significant digits

If 0.0375″ is obtained from careful measurement with proper instruments, the three digits 3, 7 and 5 are said to be *significant digits*. There is no question of the 3 and 7; there is some doubt about the 5, as this is an estimated digit. The number is said to have three significant digits.

As we saw above, the number of significant digits that is obtainable depends in large part upon the situation from which the number arises, upon the kind of scale used, and upon the care used by the worker.

Every nonzero digit is significant. The decimal point does not determine whether a digit is significant except where zeros occur in the numbers. For example, 485, 48.5, 0.485, etc. are all expressed to three significant digits. The 5 is the doubtful digit in each case.

Zeros to the left of nonzero digits are not significant. In 0.00485 the three zeros are not significant. Zeros to the right of nonzero digits and to the left of the decimal are not significant, as in 485,000.

Zeros between nonzero digits are significant. In 5,046 the zero is significant.

Zeros that follow nonzero digits to the right of the decimal point are significant. 32.0 has three significant digits. 3.200 has four significant digits.

In 6002.506 all zeros are significant.

Detecting large errors by estimation

By carefully estimating the result of a computation, we can often discover an error such as a decimal point in the wrong place. Can you estimate the product of 123 and 406? How many digits are there in the product? Study the following examples:

Example 1. Suppose you multiplied 2.8 and 3.5 and wrote the product as 98. Check this result by estimating the product of 2.8 and 3.5.

THINK	DO THIS
1. If we round off each number to the leftmost digit, we can multiply the resulting numbers mentally.	1. 2.8 is approx. 3.0. 3.5 is approx. 4.0.
2. Multiply these results.	2. 3.0 x 4.0 = 12.0
3. Compare this approximate product with the product obtained, 98.	3. Since the approximate product is 12.0, the product 98 must be in error.
4. If a serious discrepancy shows, then repeat the multiplication.	4. 2.8 x 3.5 = 9.8 *Ans.* The error was in placing the decimal point.

Example 2. The result of dividing 84.3 by 3.01 was written 2.8. Check this result by estimating the division of 84.3 by 3.01.

THINK	DO THIS
1. Round off both numbers to the leftmost digit.	1. 84.3 is approximately 80.0. 3.01 is approximately 3.00.

2. Divide these approximate numbers.

2. 80 ÷ 3 is approximately 27.

3. Compare the result, 27, with the written result, 2.8.

3. Given result 2.8, estimate 27. It appears an error has been made.

4. If there appears to be a serious discrepancy, then repeat the operation.

4.
$$\begin{array}{r} 28.06 \\ 3.01\,)\overline{84.30.0} \quad \text{or } 28.1 \\ \underline{60\ 2} \\ 24\ 10 \\ \underline{24\ 08} \\ 2000 \\ \underline{1806} \\ 194 \end{array}$$

Ans. Correct result is 28.1, approx.

Summary

To check a computation by estimating:

1. Round off each number to the first digit on the left.

2. Perform the operation on these rounded off numbers.

3. Compare this result with the result first obtained. If there appears to be a large difference in the two, repeat the original computation.

EXAMPLE

The square root of 486.5 was found to be 261.25. Check by estimating the square root.

1. 486.5 is approx. 500.

2. The square root of 500 is between 20 and 30 since 20 x 20 = 400 and 30 x 30 = 900.

3. 261.25 is not between 20 and 30. Repeat the operation.
$$\begin{array}{r} 22.05 \quad \text{or } 22.1 \\ \sqrt{486.50} \\ 4 \\ 42\,\overline{)86} \\ \underline{84} \\ 4405\,\overline{)25000} \\ \underline{22025} \end{array}$$

Ans. Square root is 22.1, approx.

EXERCISES

Check the given result by estimating. Is the result approximately correct? Do not actually solve these exercises.

1. 8.3 x 5.9 = 48.97

2. 12.7 x 4.1 = 520.7

3. 22.3 x 6.2 = 13.8

4. 98.6 ÷ 5.76 = 17.1

5. $21.4 \div 47.6 = 4.5$
6. $30.07 \times 21.2 = 63.75$
7. $32.4 \times 6.5 \times 3.4 = 71.61$
8. $42.7 \times 3.2 = 136.64$

9. $8.27 \div 3.54 = 23.3$
10. $2.05 \times 12.7 \times 1.6 = 41.65$
11. $6.72 \div 0.54 = 1.246$
12. $0.086 \div 1.73 = 0.95$

Accuracy in operations

In the operations of multiplication and division with measured or estimated numbers, keep in the result the number of significant digits that are in the *item with the least number of significant digits.*

Thus, 18.45×6.04 is, by computation, 111.4380. It should be rounded off to three significant digits, since 6.04 has three significant digits.

0.082×0.046 is by computation 0.003772. It should be rounded off to two significant digits, 0.0038, because both 0.082 and 0.046 have only two significant digits.

Principle of Accuracy

The result of multiplying or dividing measured or estimated numbers can be no more accurate than the least accurate of the given numbers.

EXERCISES

State the number of significant digits in each of the following:

1. 756
2. 0.0824
3. 17,500
4. 48.04
5. 0.00708
6. 34.0*

7. 7,600,000
8. 1.005
9. 92,866
10. 0.00012
11. 9.120
12. 0.06308

Do the following exercises, rounding off the result to the proper number of significant digits; check each result.

13. 4.7×9.26
14. 183×0.24
15. 5.67×0.03735
16. $7.62 \div 0.4855$
17. $18.62 \div 0.089$

18. $0.073 \div 0.00352$
19. $860 \div 3.72$
20. 52.9×60.7
21. $0.42 \div 1.005$

Some facts about the slide rule

We shall discuss only the A, B, C, and D scales on the slide rule illustrated in Fig. 6-1.

*The zero here is significant.

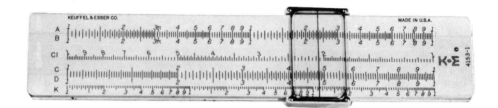

Fig. 6-1 Modern slide rule. *(Courtesy Keuffel & Esser Company.)*

The slide rule consists of three parts: a frame or body which we call the *rule*; a strip called the *slide* which fits into grooves in the rule and slides back and forth; and a transparent *indicator* or *runner* containing a hair line. The indicator may be moved back and forth along the rule as desired.

We shall study two scales on the body, A and D, and two on the slide, B and C. The figure 1 at the left end of any scale is the *left index* of that scale. The figure 1 at the right end of any scale is the *right index* of that scale.

Let us start by studying scale D on the rule (Fig. 6-2).

Fig. 6-2. The D scale of the slide rule.

The D scale

Note carefully the number of divisions. From large **1** at the left-hand end of scale D to large **2**, there are ten divisions, each of which is divided into ten smaller divisions numbered 1 through 9.

From large **2** to large **3**, and **3** to **4**, there are ten divisons, but each is divided into only five smaller divisions, and these are not numbered.

The remaining divisions **4** to **5**, **5** to **6**, etc., are each divided into ten divisions, each of these being divided into only two smaller divisions.

If we let large **1** represent 1,000; large **2**, 2,000; large **3**, 3,000, etc., then each of the ten divisions between **1** and **2** will represent 100, so that small 1 will be 1,100; small 2, 1,200; small 3, 1,300, etc. The smallest divisions between small 1 and small 2 will each represent 1/10 of 100, or 10.

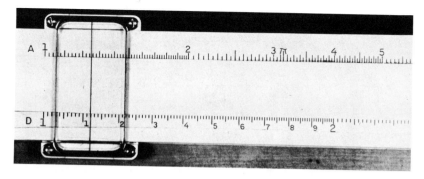

Fig. 6-3. Indicator set at 1,120.

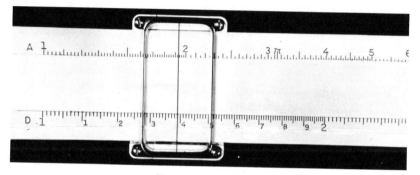

Fig. 6-4. Indicator set at 1,390.

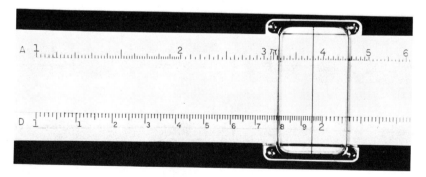

Fig. 6-5. Indicator set at 1,950.

(Note that in these and the following illustrations showing the slide rule, we have blanked out all scales other than the A, B, C, and D scales, and in some cases we have also removed the slide to make the illustration clearer.)

The spaces 2 to 3 and 3 to 4 are first divided into ten parts. Each of these represents 100. In Fig. 6-6 the reading indicated is 2,300 and the reading in Fig. 6-7 is 2,900. The smallest divisions here each represent 20 since there are 5 within the other divisions. For example, Fig. 6-8 indicates 2,040; Fig. 6-9, 2,240; Fig. 6-10, 3,230; and Fig. 6-11, 3,880.

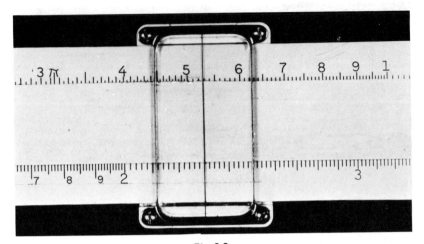

Fig. 6-6.

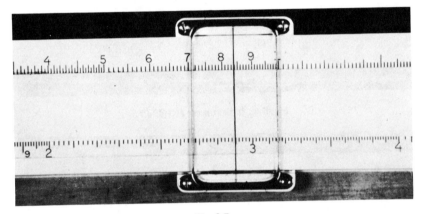

Fig. 6-7.

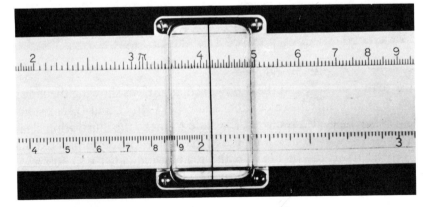

Fig. 6-8.

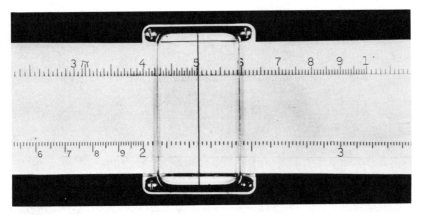

Fig. 6-9.

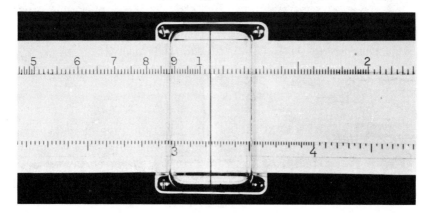

Fig. 6-10.

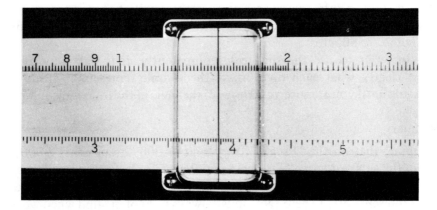

Fig. 6-11.

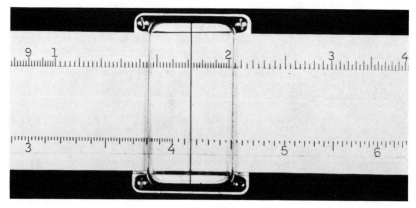

Fig. 6-12.

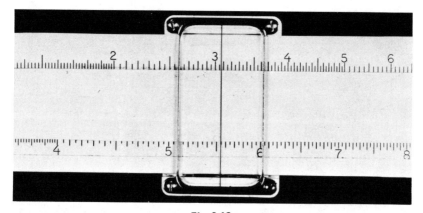

Fig. 6-13.

The spaces **4** to **5**, **5** to **6**, **6** to **7**, etc., are first divided into ten smaller spaces. Each therefore represents 1/10 of 1,000, or 100. Each of these is divided into two spaces so that the smallest space represents 50. Figure 6-12 indicates a reading of 4,150 and the rule in Fig. 6-13 shows a reading of 5,550, etc.

In order to use the slide rule competently, it is very important to know clearly what each division on the D scale represents. Study your slide rule until you are sure you know all the facts in the following Summary.

Summary

Assuming that on the D scale, **1** represents 1,000, **2** represents 2,000, and so on, the smallest space between **1** and **2** represents 10, the smallest space between **2** and **4** represents 20, and the smallest space between **4** and the end represents 50. (These facts must always be kept in mind when working with the slide rule.)

EXERCISES

Place the hair line of the indicator of your slide rule as indicated below and tell what the reading of the D scale indicates.

1.

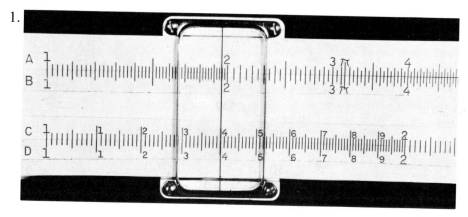

Fig. 6-14.

2.

3.

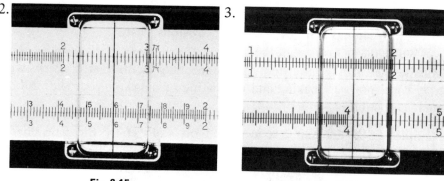

Fig. 6-15.

Fig. 6-16.

4.

5.

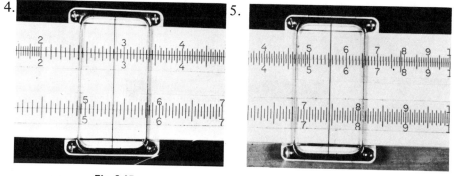

Fig. 6-17.

Fig. 6-18.

6.

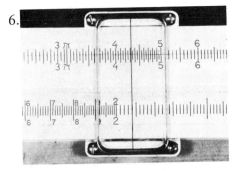

Fig. 6-19.

Set the hair line of the indicator so as to indicate each of the following on the D scale.

7. 1,750	13. 2,380	19. 4,300	25. 9,150
8. 1,290	14. 3,240	20. 4,450	26. 7,000
9. 1,460	15. 3,880	21. 5,250	27. 1,530
10. 1,830	16. 1,910	22. 6,850	28. 2,280
11. 1,990	17. 2,620	23. 7,200	29. 3,460
12. 2,160	18. 3,440	24. 4,650	30. 8,550

Estimating readings within the smallest divisions, D scale

When the hair line falls between two of the lines of the scale, we must estimate the reading indicated. Since accuracy in this matter comes with experience, practice making and estimating readings on your slide rule as much as possible.

Example 1. What reading is indicated on scale D by the hair line in Fig. 6-20?

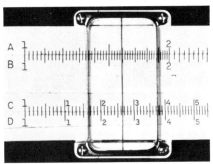

Fig. 6-20.

Since the hair line falls between 1,260 and 1,270 and appears to be about 1/3 the width of the space from 1,260, we estimate the value to be 1,263.

Ans. 1,263

Example 2. What reading is indicated in Fig. 6-21?

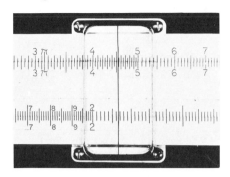

Fig. 6-21.

The line falls between 2,120 and 2,140. It appears to divide the space about 3/4 from 2,120 and 1/4 from 2,140. Since the smallest division here represents 20, 3/4 of 20 is about 15. Add this to the smaller value and call the value 2,135. *Ans.* 2,135

Example 3. What reading does the hair line indicate in Fig. 6-22?

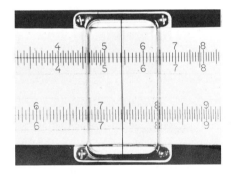

Fig. 6-22.

The hair line falls between 7,350 and 7,400. It appears to be about 1/3 of the space from 7,350 and 2/3 from 7,400. Since the smallest space represents 50, then 1/3 of 50 is about 16. Add this to the smallest number and call the reading 7,366 or 7,370. *Ans.* 7,370

Summary

When the hair line falls within the smallest subdivision, estimate that part of the subdivision beyond the left-hand end of the subdivision. Add this fractional part of the subdivision (10 or 20 or 50) to the reading indicated at the left-hand end of the subdivision. The sum is the reading required. Remember that the last figure is estimated, not exact.

EXERCISES

Estimate the readings indicated by the hair line in each of the following exercises:

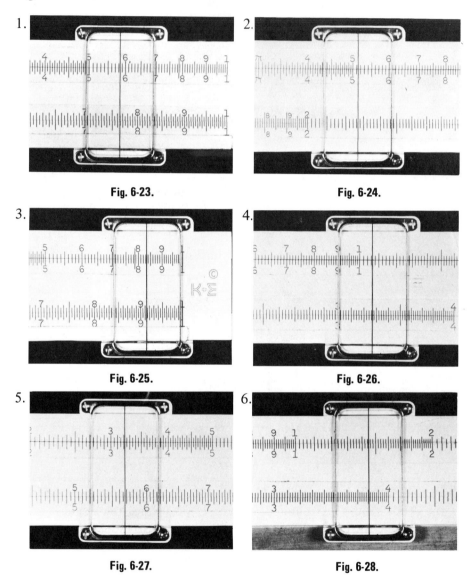

1.

Fig. 6-23.

2.

Fig. 6-24.

3.

Fig. 6-25.

4.

Fig. 6-26.

5.

Fig. 6-27.

6.

Fig. 6-28.

Set the indicator so that the hair line shows each of the following readings on scale D:

7. 1,572	10. 1,786	13. 3,107	16. 7,142
8. 1,293	11. 2,370	14. 3,633	17. 8,335
9. 1,845	12. 2,955	15. 5,290	18. 9,184

Other values for the large figures

In our study of scale D thus far, we have assumed that **1** represents 1,000, **2** represents 2,000, **3** represents 3,000, etc. The slide rule is so constructed that **1** may represent 10, 100, 1,000, 10,000, 1, 0.1, or 0.01. In

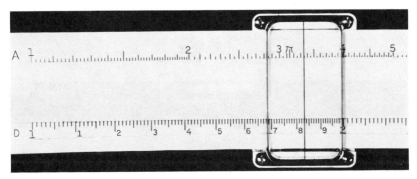

Fig. 6-29.

fact, the left index **1** may represent any number obtained by multiplying or dividing 10 by itself any number of times. We call these *powers* of 10. Hence **1** may represent any power of 10.

Thus, if **1** represents 1, then **2** represents 2; **3**, 3; etc. Each division between **1** and **2** would be 1/10 of 1, or 0.1. Hence, the reading in Fig. 6-29 would be 1.830.

In Fig. 6-30 the reading is 3.760, whereas the reading in Fig. 6-31 is 4.650, etc.

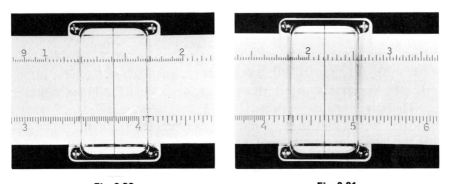

Fig. 6-30.   **Fig. 6-31.**

If the left index **1** represents 0.01, then **2** represents 0.02 and **3** represents 0.03, etc. Each of the large divisions between **1** and **2** would represent 1/10 of 0.01, or 0.001. Thus the reading in Fig. 6-32 would be 0.0136. In Fig. 6-33, the reading is 0.0465.

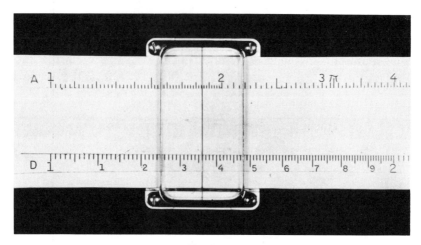

Fig. 6-32.

In Fig. 6-33, the reading is 0.0465.

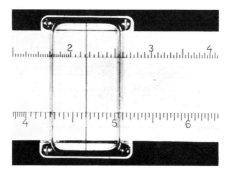

Fig. 6-33.

Summary

The left index of scale D may represent any power of 10, such as 0.0001, 0.001, 0.01, 0.1, 1, 10, 100, 1,000, etc. Then each of the large numbers 2, 3, etc., represents twice, three times, etc., what the index represents.

The larger spaces between 1, 2, 3, etc., represent 1/10 of what the index represents, and the smallest spaces represent 1/10, or 1/5, or 1/2 of what the larger spaces represent.

EXERCISES

Set the indicator to the reading indicated on scale D, and make the reading for the value of the index shown.

1. The index in Fig. 6-34 represents 1.
2. The index in Fig. 6-35 represents 100.

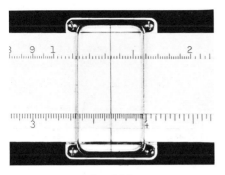

Fig. 6-34.

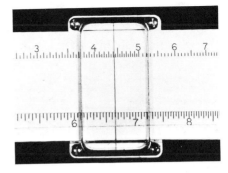

Fig. 6-35.

Set the indicator to make the following readings on scale D. Indicate the value you attach to 1 in each case.

3. 567.2	7. 17.6	11. 72,600	15. 84.60
4. 9.860	8. 57.3	12. 43,560	16. 1.926
5. 185.6	9. 0.156	13. 0.0268	17. 0.735
6. 2.560	10. 0.835	14. 0.0575	18. 36,590

The C scale

If you place the left index of the C scale directly over the left index of the D scale, you will see that the two scales are exactly alike. Whatever has been learned concerning the D scale also applies to the C scale.

Multiplying two numbers with the slide rule

When multiplying numbers with the slide rule, follow one of two methods for determining the decimal place in the result. First, the index may represent the power of 10 which will suit the numbers of the problem. Or, the first four significant digits of the numbers may be considered without regard to the position of the decimal point, the multiplication completed, and the position of the decimal point in the result determined by estimation. The latter method will be followed here.

Multiplication may be performed in several ways. The C and D scales are used in these examples. In general,

1. The first movement is made with the runner and the final movement is made with the runner.
2. Begin and end multiplication on the D scale.
   This means that multiplication is begun by setting the hair line of the runner on the D scale and is completed by setting the runner to show the answer on the D scale.

Example 1. Multiply 28 x 15.

Think of the numbers as 2,800 and 1,500.

1. Set the hair line of the runner on 2,800 on the D scale, as shown in Fig. 6-36.
2. Move the left index of the C scale under the hair line of the runner, as in Fig. 6-36.

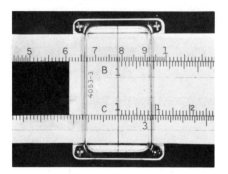
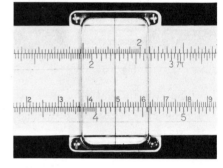

Fig. 6-36.	Fig. 6-37.

3. Set the indicator over 1,500 on the C scale, as in Fig. 6-37.
4. Make the reading that the indicator gives on the D scale, here 4,200.
5. Locate the decimal point by thinking 30 x 15 = 450; hence there must be three figures to the left of the decimal place.   *Ans.* 420

Example 2. Multiply 2.18 by 32.70.

Think of the numbers as 2,180 and 3,270.

1. Place the hair line of the indicator on 2,180 on the D scale, as shown in Fig. 6-38.
2. Move the left index of the C scale under the hair line, as in Fig. 6-38.

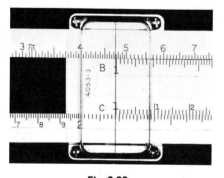

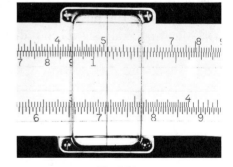

Fig. 6-38.	Fig. 6-39.

3. Set the indicator over 3,270 of the C scale, as in Fig. 6-39.
4. Make the reading that the indicator gives on the D scale, 7,140.
5. Locate the decimal point by estimating the product of 2 x 30 = 60.

This shows two places to the left of the decimal point. Give the answer to only three significant digits, since one of the factors has only three digits. *Ans.* 71.4

Example 3. Multiply 43 5/8 by 73 1/2.

1. When fractions are involved, we must first express them as decimals: 43.625 x 73.5.
2. Place the runner on 43,625 (approximately) of the D scale, as shown in Fig. 6-40.
3. Place the left index of the C scale under the hair line, as in Fig. 6-40.

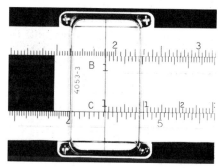

Fig. 6-40.

4. Move runner to 7,350 on the C scale. This however would place the indicator off the rule. We must stop now and change indexes. That is, we must place the hair line of the runner over the left index of the C scale, and then slide the right index of the C scale under the hair line (Fig. 6-41). Now we may proceed to move the runner to 7,350 on the C scale (Fig. 6-42).

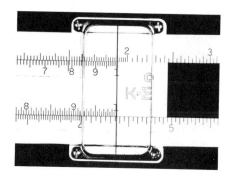

Fig. 6-41.

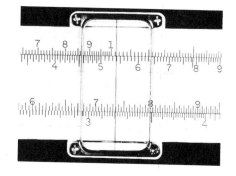

Fig. 6-42.

5. Make the reading that the indicator gives on the D scale, 3,206.
6. Estimate the product 40 x 70 = 2,800 and place the decimal point. This shows four places to the left of the decimal point. Since one of

the factors, 73.5, is given to only three significant digits, the answer should be expressed to no more than three significant digits.

Ans. 3,210

Summary

To multiply two numbers:
1. Set the hair line of the indicator over the first factor of the product on the D scale.
2. Place the left index of the C scale under the hair line.
3. Set the indicator over the second factor of the product on the C scale. If this would place the indicator off the rule, change the C indexes, then set the indicator on the second factor on the C scale.
4. Make reading given by indicator on the D scale.
5. Obtain number of digits to the left of the decimal point by estimating the product. Place the decimal point.
6. Determine number of significant digits to be retained in the answer.

EXERCISES

Multiply as indicated, using the slide rule.

1. 4.0 x 2.3
2. 1.7 x 46
3. 135 x 512
4. 176 x 32.0
5. 5.30 x 158
6. 384 x 210
7. 1.73 x 3.62
8. 306 x 25.8
9. 433 x 36.5
10. 81.5 x 6.64
11. 3.6 x 3.14
12. 48.3 x 48.3
13. 2.03 x 167.3
14. 746 x 5.56
15. 1.536 x 30.6
16. 7.15 x 3.05
17. 6.25 x 19
18. 743 x 0.0567

Multiplying more than two numbers

To multiply together several numbers, we start as for two numbers, but without reading that product, multiply by the third number, then that product by the fourth, etc.

Example 1. Multiply 17.49 x 3.14 x 5 1/2 x 0.093.
1. Place indicator over first factor (1,749) on D scale.
2. Move left index of C scale under indicator.
3. Move indicator over second factor (3,140) on C scale (Fig. 6-43).
4. Without bothering to read this product, place right index of slide under the indicator.
5. Move indicator to next factor (5,500) on C scale (Fig. 6-44).
6. Slide right index under hair line.
7. Set indicator over next factor (9,300) on C scale.
8. Read the product from D scale. This reading is 2,810 (Fig. 6-45).

9. Estimate the result: 20 x 3 x 5 x 1/10 = 30.
10. Place the decimal point and determine the number of significant digits in the answer. *Ans.* 28 (to two significant digits)

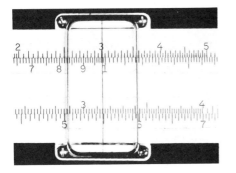

Fig. 6-43.

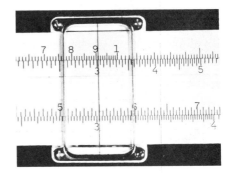

Fig. 6-44.

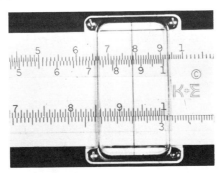

Fig. 6-45.

Summary

To multiply together more than two numbers:
1. Place indicator over first factor on D scale.
2. Move index of C scale under indicator.
3. Move indicator over second factor on C scale.
4. Move index of C scale under indicator.
5. Move indicator over next factor on C scale.
6. Repeat steps 4 and 5 until all factors have been used. Then make reading of product on D scale.
7. Estimate product and place decimal point.
8. Determine the number of significant digits to retain in the answer.

EXERCISES

Multiply as indicated, using the slide rule.

1. 27 x 96 x 45
2. 372 x 958 x 144
3. 83.6 x 17.7 x 72.55
4. 5.171 x 1.266 x 9.438

5. 8 1/2 x 7 3/4 x 15 5/8   9. 5,280 x 746 x 32.95
6. 7.2 x 7.2 x 3.14   10. 1,728 x 3.722 x 5,000 x 4.962
7. 0.7854 x 0.25 x 1.726
8. 2.75 x 2.75 x 3.1416 x 15.6 x 62.5

Dividing by means of the slide rule

Since division is the inverse of multiplication, we reverse the procedure on the slide rule.

Example 1. Divide 72 by 24.

1. Place the indicator hair line over the product, 7,200, on the D scale. Then move the slide so the known factor, 2,400, on the C scale, also is under the hair line (Fig. 6-46).

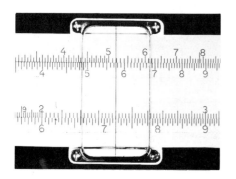

Fig. 6-46.   Fig. 6-47.

2. Read the other factor, 3,000, on the D scale, under the index of the C scale (Fig. 6-47).
3. Estimate the answer, 70 ÷ 20 = 3.5, to place the decimal point in the slide-rule answer, 3.   *Ans.* 3

Example 2. Divide 28.60 by 0.084.

1. Place the indicator hair line over the product, 2,860, on the D scale.
2. Move the slide so that the known factor, 8,400, on the C scale, is also under the hair line (Fig. 6-48).
3. Move the indicator to the index of the C scale and read the other factor, 3,410, on the D scale (Fig. 6-49).
4. Estimate the result (28 ÷ 0.08 is the same as 2,800 ÷ 8): 2,800 ÷ 8 = 350.
5. Place the decimal point in the answer and retain three significant digits.   *Ans.* 341

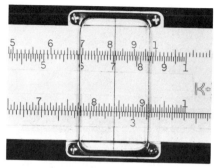

Fig. 6-48.

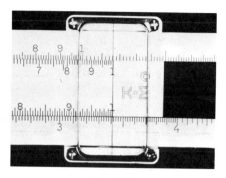

Fig. 6-49.

Summary

To divide with the slide rule:
1. Set the indicator hair line over the product, on the D scale.
2. Move the slide so that the known factor, on the C scale, is also under the hair line.
3. Move the indicator to the index of the C scale, and read the other factor on the D scale.
4. Estimate the result and place the decimal point in the answer.
5. Determine the number of significant digits to retain in the answer.

EXERCISES

Divide as indicated, using the slide rule.

1. 138 ÷ 46
2. 680 ÷ 136
3. 52 ÷ 12
4. 468 ÷ 3.5
5. 12 ÷ 25
6. 45 ÷ 83

7. 3.8 ÷ 2.4
8. 57.63 ÷ 3.14
9. 0.156 ÷ 96
10. 19 3/4 ÷ 12 1/2
11. 3.42 ÷ 2.72
12. 871 ÷ 0.468

13. 2,875 ÷ 37.1
14. 3.75 ÷ .0227
15. 3.14 ÷ 2.72
16. 3.42 ÷ 81.7
17. 0.00377 ÷ 529
18. 1,029 ÷ 9.70

Using the slide rule to perform both multiplication and division

Many calculations contain both multiplication and division. We can do any number of such operations on the slide rule without recording the intermediate results.

Example 1. $\dfrac{345 \times 1.75}{3.35} = ?$

This expression can be solved in three ways:
A. Multiply 345 x 1.75 and divide this product by 3.35;
B. Divide 345 by 3.35 and multiply this result by 1.75;
C. Divide 1.75 by 3.35 and multiply this result by 345.

A. Move runner to 345 on D scale.
Place left index under hair line.
Move runner to 1.75 on C scale (Fig. 6-50).
Slide 3.35 on C scale to hair line (Fig. 6-51).
Move runner to index.
Read 1800 on D scale.

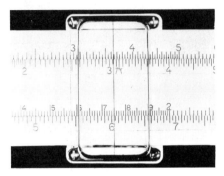

Fig. 6-50.   Fig. 6-51.

Estimate result. $\dfrac{300 \times 2}{3} = 200$.

Place decimal point. 180.   *Ans.* 180

B. Move runner to 345 on D scale.
Slide 3.35 on C scale to hair line (Fig. 6-52).
Move hair line to 1.75 on C scale.
Read 1800 on D scale.
Estimate result and place decimal point.

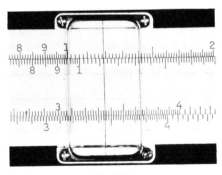

Fig. 6-52.

C. Move runner to 1.75 on D scale.
Slide 3.35 on C scale to hair line (Fig. 6-53).
Move hair line to 345 on C scale (Fig. 6-54).
Read 1800 on D scale.
Estimate result and place decimal point.

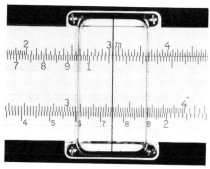

Fig. 6-53.

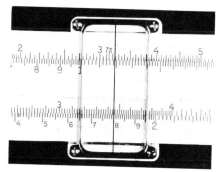

Fig. 6-54.

Example 2. $\dfrac{21.8 \times 49.7}{0.372 \times 0.643} = ?$

Two of the possible ways to find this result are:

A. Multiply 21.8 x 49.7, divide this product by 0.372 and then divide this result by 0.643;

B. Divide 21.8 by 0.372, multiply by 49.7, and divide by 0.643.

A. Move runner to 218 on D scale.
   Place right index under hair line.
   Move runner to 497 on C scale (Fig. 6-55).
   Slide 372 on C scale to hair line.
   Move runner to index.
   Slide 643 on C scale to hair line (Fig. 6-56).
   Move runner to index.
   Read 453 on D scale.

   Estimate result. $\dfrac{20 \times 50}{0.4 \times 0.6} = \dfrac{1000}{0.25} = \dfrac{100000}{25} = 4,000.$

   Place decimal point. 4,530.   *Ans.* 4,530

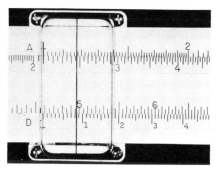

Fig. 6-55.

B. Move runner to 218 on D scale.
    Slide 372 on C scale to hair line (Fig. 6-57).
    Move runner to 497 on C scale.
    Slide 643 on C scale to hair line (Fig. 6-56).
    Move runner to index.
    Read 453 on D scale.
    Estimate result and place decimal point.

Method B is usually the best and quickest.

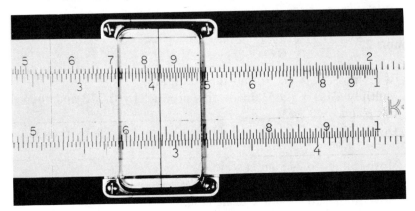

Fig. 6-56.

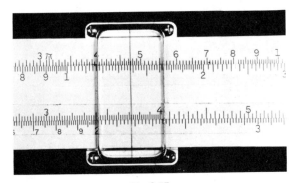

Fig. 6-57.

Example 3. $\dfrac{18.57 \times 3.91 \times 3\ 3/4}{9.113} = ?$

Move runner to 1,857 on D scale.
Slide 9,113 to hair line (Fig. 6-58).
Move runner to right index.
Slide left index to hair line.

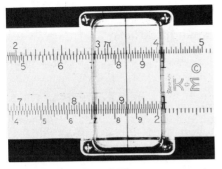

Fig. 6-58.

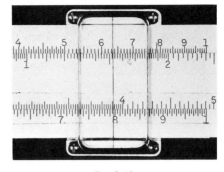

Fig. 6-59.

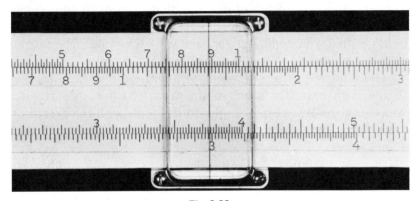

Fig. 6-60.

Move runner to 391 on C scale (Fig. 6-59).
Slide right index to hair line.
Move runner to 375 on C scale (Fig. 6-60).
Read result on D scale (2,985).

Estimate result. $\dfrac{20 \times 4 \times 4}{10} = 32$.

Place decimal point and retain three significant digits. 29.85 = 29.9, approx. *Ans.* 29.9

Summary

To perform both multiplication and division using the slide rule:
1. Divide the first factor of the denominator into the first factor of the numerator.
2. Multiply by the second factor of the numerator.
3. Divide by the second factor of the denominator.
4. Proceed in this manner until all the factors are used.
5. Estimate the result, place the decimal point, and determine the number of significant digits.

EXERCISES

Use a slide rule to perform the indicated operations:

1. $\dfrac{87 \times 3.26}{19.2}$ 3. $\dfrac{42.92 \times 26.77}{155.5}$ 5. $\dfrac{3.1416 \times 144 \times 18.5}{33.7}$

2. $\dfrac{5.22 \times 186.4}{73.3}$ 4. $\dfrac{0.0698 \times 93.29}{17.28}$ 6. $\dfrac{46.23 \times 68.85 \times 5\ 1/4}{231 \times 0.75}$

Squaring a number by means of the slide rule

Thus far we have worked only with scales C and D. If you examine scales A and B closely, you will discover that both of these scales are the same, and furthermore that each consists of two parts, one part beginning at the left with 1 and ending at the center of the scale. The second part begins at this point and continues to the end of the scale.

If you compare these two parts, you will see that each has the same length and each bears the same subdivisions. Also each part is exactly half as long as the D scale and similar to this scale with the exception that there are fewer subdivisions between the numbers **1** and **2**, **2** and **3**, **3** and **4**, and so on.

To square a number, or to find the square root of a number, we may use the D and A scales.

Example. What is the square of 16, or what is 16^2?

1. Set the indicator over the number 16 on the D scale.
2. Read the number under the indicator on the A scale (Fig. 6-61).
3. Estimate the result: 20 x 20 = 400. This indicates that the square of 16 is a 3-place number.   *Ans.* $16^2 = 256$

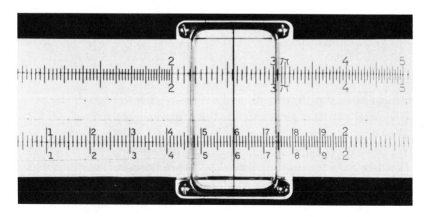

Fig. 6-61.

Summary

To square a number:
1. Place the indicator over the number on the D scale.
2. Read the number under the indicator on the A scale.
3. Estimate the result and place the decimal point.

EXERCISES

Square each of the following numbers using the slide rule.

1. 27
2. 115
3. 0.83

4. 1/8
5. 3.92
6. 0.0566

7. 5 3/4
8. 49.65
9. 3,948

10. 1.271

Finding the square root of a number with the slide rule

Finding the square root is the inverse of squaring a number. We therefore start with the A scales; remember there are two of them. In using these scales we must employ one of the following procedures:

A. 1. Use the *left-hand* section of the A scales to find the square root of a number whose *first significant digit* is either an *odd number* of places to the *left* of the decimal point, or an *even number* of places to the *right* of the decimal point.

2. Use the *right-hand* section of the A scales to find the square root of a number whose *first significant digit* is either an *even number* of places to the *left* of the decimal point, or an *odd number* of places to the *right* of the decimal point.

B. An alternative procedure for determining which section of the A scales to use is to mark off the given number into periods (see p. 190, Ex. 2) and, if the leftmost period containing significant digits is 10 or more, use the right-hand section and, if less than 10, use the left-hand section.

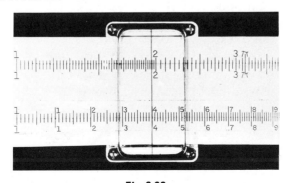

Fig. 6-62.

Example 1. Find the square root of 196.
1. Since the first significant digit, 1, of this number is an odd number of places (three) to the left of the decimal point or the leftmost period is less than 10, we select the left-hand section of the A scales.
2. Set the indicator over 196 in the section (Fig. 6-62).
3. Read the number under the indicator in scale D: 14.
4. Estimate the result: $10^2 = 100$ and $20^2 = 400$. Hence the square root of 196 must be between 10 and 20. *Ans.* 14.0

Example 2. What is the value of $\sqrt{0.00584}$?
1. Since the first significant digit is three places to the right of the decimal point or the leftmost period containing significant digits is more than 10, we use the right-hand section of the A scales.
2. Set the indicator over 5,840 in this section (Fig. 6-63).
3. Read the number under the indicator on scale D: 764.
4. Estimate the result: the square of 0.07 is 0.0049 and the square of 0.08 is 0.0064. Therefore the square root of 0.0058 lies between 0.07 and 0.08. *Ans.* 0.0764

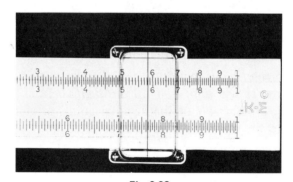

Fig. 6-63.

Summary

To find the square root of a number by using the slide rule:
1. Determine which section of the A scales is to be used. (See statement above).
2. Set the indicator over the number in this section.
3. Read the number under the indicator in scale D.
4. Estimate the result and place the decimal point.

EXERCISES

Using the slide rule, find the value of each of the following:

1. $\sqrt{730}$ 3. $\sqrt{1,890}$ 5. $\sqrt{7/8}$
2. $\sqrt{1,085}$ 4. $\sqrt{7.284}$ 6. $\sqrt{0.6154}$

7. $\sqrt{0.00234}$ 9. $\sqrt{923.6}$ 11. $\sqrt{5/16}$

8. $\sqrt{7.003}$ 10. $\sqrt{3.57}$ 12. $\sqrt{0.833}$

The B scale

If you place the left index of the B scale directly below the left index of the A scale, you will see that the two scales are exactly alike. Whatever has been learned concerning the A scale, therefore, also applies to the B scale.

Since the C scale is identical to the D scale, the B scales are identical to the A scales, and the C and B scales are both on one part of the slide rule (the slide), we can use the C and B scales also for finding squares and square roots of numbers.

REVIEW PROBLEMS

Use the slide rule to solve the following problems. Remember to estimate the result before placing the decimal point.

1. $\dfrac{3960}{0.85}$

2. $\dfrac{946}{\pi}$

3. 0.7854×25.3

4. $0.0625 \times 0.7854 \times 236$

5. $\sqrt{2} \times 110 \times 25$

6. $\dfrac{12 \times 850}{3.1416 \times 56}$

7. $\dfrac{2.54 \times (765)^2 \times 625}{0.55 \times 0.087}$

8. $\dfrac{2 \times 3,500,000 \times 280 \times 30}{10,000,000}$

9. $\dfrac{16.3 \times 1.34 \times 125}{3.12}$

10. $\dfrac{22 \times 45 \times 37}{27 \times 29}$

11. $\dfrac{1,268 \times 0.654}{33,000}$

12. $\dfrac{13.9^2 \times \pi}{15}$

13. $\sqrt{9^2 + \left(\dfrac{27-8}{2}\right)^2}$ *

14. $\dfrac{40 \times \sqrt{3}}{2 \times 3.1416}$

15. $45.3 \sqrt{0.625}$

16. $\dfrac{550}{\sqrt{3} \times 12}$

*First combine terms inside the parentheses, then square the expression. $\left(\dfrac{27-8}{2}\right)^2 = \left(\dfrac{19}{2}\right)^2 = \dfrac{361}{4}$. Next find 9^2, and add the result, 81, to $\dfrac{361}{4}$ before extracting the square root of the whole expression.

Proportions

One of the simplest uses of the slide rule is to solve proportions. Knowing that four numbers are related to each other in a certain ratio, and knowing three of these numbers, the other number may be found. The four numbers should be arranged in this, or an equivalent form:

$$\frac{a}{b} = \frac{c}{d}; \frac{a}{c} = \frac{b}{d}; \frac{b}{a} = \frac{d}{c}; \text{ etc.}$$

Example 1. $\frac{3}{6} = \frac{x}{8}$; $x = ?$

Use the C and D scales. Adjust the slide rule so that 3 on the C scale is directly above 6 on the D scale. The value of x is found on the C scale above the 8 on the D scale. *Ans.* $x = 4$

Example 2. $\frac{3}{x} = \frac{6}{8}$; $x = ?$

Adjust the slide rule so that 6 on the C scale is directly above 8 on the D scale. The value of x is found on the D scale below 3 on the C scale. *Ans.* $x = 4$.

Many fractions that are equal to a given fraction can be found by adjusting the slide rule to have the C and D scales lined up to represent the given fraction; then take readings at any location of the hairline of the indicator.

Example 3. $\frac{234}{312} = \frac{x}{y}$.

Place 234 on C scale over 312 on D scale.

Move runner to some number on C, say 6.
Read corresponding number on D, here 8.} then $\frac{234}{312} = \frac{6}{8}$

Move runner to some number on D, say 4.8.
Read corresponding number on C, here 3.6.} then $\frac{234}{312} = \frac{3.6}{4.8}$

Continue in the same manner for any other values of x or y.

Summary
To use the slide rule to solve proportions:
1. Adjust C and D scales to represent some fraction.
2. Read numerator and denominator for equivalent fractions, on C and D scales, by moving runner to new locations.

EXERCISES

Find the value of the variable in each exercise.

1. $\dfrac{x}{16} = \dfrac{16}{256}$

2. $\dfrac{152}{c} = \dfrac{645}{850}$

3. $\dfrac{m}{17.9} = \dfrac{9.45}{4.25}$

4. $\dfrac{80.8}{x} = \dfrac{142}{486}$

5. $\dfrac{7.80}{4.26} = \dfrac{21.6}{y}$

6. $\dfrac{54.3}{17.2} = \dfrac{98.0}{t}$

7. $\dfrac{7.32}{909} = \dfrac{x}{803}$

8. $\dfrac{1.56}{3.12} = \dfrac{s}{1,760}$

9. $\dfrac{0.00076}{385} = \dfrac{0.0040}{p}$

10. $\dfrac{262}{319} = \dfrac{800}{k}$

7 Pulleys, Belts and Gears

Power to run machines is transmitted from one machine to another, or from one part of a machine to another part, by means of belts on pulleys, chains on sprockets, or by gears.

Belts and pulleys

The intelligent worker knows how to calculate the diameters and speeds of pulleys and the lengths of belts.

Belts transmit power because of friction between belt and pulley. Belts are made of canvas, rubber, or leather. Flat belts are used with flat pulleys and V-belts are used with grooved pulleys.

Pulleys are used in pairs to increase speed, decrease speed, or to maintain the same speed as the source of power. Properly belted, pulleys are also used to change direction of rotation.

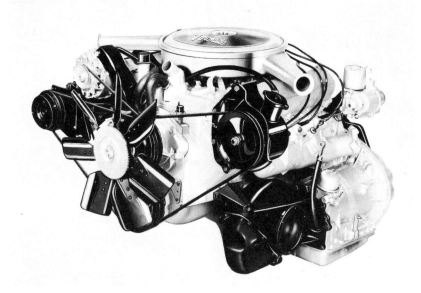

Fig. 7-1. V-belt drive in an automobile engine.
(Courtesy General Motors Corporation, Oldsmobile Division)

V-belt drives use V-belts and grooved pulleys called sheaves; they are especially good for very short centers.

V-belts have various cross-section sizes for different horsepower requirements. The S.A.E. standards vary from 5/8 x 3/8 to 1 x 9/16, with corresponding h.p. from 1/2 to 100. Sheaves for V-belts are made of cast iron, cast semi-steel, or pressed steel.

The surface (or peripheral) speed of a pulley

The surface speed, sometimes called run velocity, of a pulley is the number of feet a point on the rim of the pulley moves in one minute. If such a point moves 150 feet per minute as the pulley rotates, the surface speed of the pulley is said to be 150 feet per minute.

The surface speed of a pulley is equal to the circumference of the pulley multiplied by the number of revolutions the pulley makes in one minute.

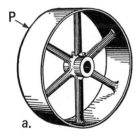 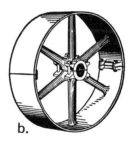

Fig. 7-2. Flat pulleys: (a) solid; (b) split.

If we represent the surface speed by V, the circumference (in inches) by C and the number of revolutions per minute by N, the surface speed, in inches per minute, is given by the formula

$$V = CN \text{ inches per minute}$$

Since the circumference is equal to π times the diameter, $C = \pi D$, we can write

$$V = \pi DN \text{ inches per minute}$$

When the diameter is given in inches, to obtain the number of feet per minute, we divide this result by 12.

$$V = (\pi DN)/12 \text{ feet per minute}$$

or, since $\pi = 3.1416$,

$$V = 0.2618DN \text{ feet per minute.}$$

Example 1. Find the peripheral speed of a 14″ pulley that makes 240 revolutions per minute (r.p.m.).

THINK	DO THIS
1. There is a formula for finding the speed. Write the formula.	1. $V = 0.2618DN$ $D = 14″$ $N = 240$ r.p.m.
2. Replace each variable by its known value.	2. $V = 0.2618 \times 14 \times 240$ The slide rule gives 880.
3. Complete the computations.	3. $V = 880$ *Ans.* Peripheral speed = 880 ft./min.

Example 2. What must be the r.p.m. of a 12-in. pulley if its surface speed is to be 2,300 ft. per minute?

THINK	DO THIS
1. Since the formula $V = 0.2168DN$ involves the r.p.m. (N), we may use this formula to find r.p.m. Write the formula.	1. $V = 0.2618DN$ $D = 12″$ $V = 2,300$ ft./min.
2. Replace the variables by their known values.	2. $2,300 = 0.2618 \times 12 \times N$
3. Solve this equation for N. Since 2,300 is the product of 0.2618 × 12 and N, divide the product by the factor 0.2618 × 12, to obtain the value of N.	3. $\dfrac{2,300}{0.2618 \times 12} = N$ $\dfrac{575 \times 4}{0.2618 \times 3 \times 4} =$ $\dfrac{575}{0.2618 \times 3} = N$ The slide rule gives 732. *Ans.* r.p.m. = 732

Note that when we have a formula that involves several variables, if all but one of these variables are known, we are able to obtain the value of the remaining letter by replacing the variables by their known values and solving the resulting equation for the required value.

Note also that, where applicable, the slide rule may be used to complete the computations.

Summary

To find the surface speed, the revolutions per minute, or the diameter of a pulley:

1. Write the formula $V = 0.2618DN$.

2. Replace the known variables by their values.

3. Solve the resulting equation for the remaining letter.

4. Complete the computations, using the slide rule if convenient.

EXAMPLE

What must be the diameter of a pulley that has a surface speed of 2,109 ft./min. and makes 560 r.p.m.?

1. $V = 0.2618DN$
   $V = 2,109$ $N = 560$

2. $2,109 = 0.2618 \times D \times 560$

3. $\dfrac{2,109}{0.2618 \times 560} = D$

4. $D = 14''$

Ans. Diameter must be $14''$

PROBLEMS

1. Find the speed of a belt running on a 36-in. pulley that is revolving at 450 r.p.m.

2. A grinding wheel having a diameter of 12 in. has a peripheral speed of 4,000 ft. per minute. Find the r.p.m.

3. How fast (r.p.m.) must a 12-in. pulley revolve in order to have a surface speed of 628 ft. per minute?

4. A belt having a surface speed of 6,689 ft. per minute drives a pulley at the rate of 350 r.p.m. What is the diameter of the pulley?

5. A driving pulley produces a peripheral speed of 3,456 ft. per minute. Find the diameter of the driven pulley if it revolves at the rate of 550 r.p.m.

6. Copy the following table and replace the question marks with the proper values.

	Diameter of pulley	N, or r.p.m.	Surface speed, ft. per min.
(a)	13″	?	1,021
(b)	?	350	7,056
(c)	54″	450	?
(d)	46″	?	5,412
(e)	6 3/4″	6,362	?
(f)	?	5,341	2,400

Two pulleys joined by a belt

When two pulleys are joined by a belt, the one in which the motion or power originates is called the *driver* and the one to which the motion and power are transmitted is called the *driven* pulley.

Fig. 7-2A. Jigsaw with V-belt drive.

Assuming that the belt does not slip, the surface speed of both pulleys is the same. This speed is also the same as the speed or number of feet per minute moved by a point on the belt. With the open-type belt both pulleys revolve in the same direction; with the cross-type belt the pulleys revolve in opposite directions.

The belt in Fig. 7-3 is the open type, and the one in Fig. 7-4 is the cross-type belt.

If we represent the diameter of the driver by D, its revolutions per minute by N, and its surface speed by V, and represent the diameter of

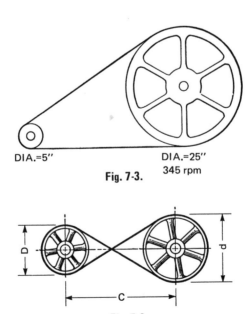

DIA.=5"   DIA.=25"

Fig. 7-3.   345 rpm

Fig. 7-4.

the driven pulley by d, its revolutions per minute by n, and its surface speed by v, we can write the following formulas:

Driver	Driven
$V = 0.2618DN$	$v = 0.2618dn$

Since for two pulleys connected by a belt the surface speeds are the same — that is, since $V = v$ — we can write $0.2618DN = 0.2618dn$, and dividing both parts of the equation by 0.2618,

$$DN = dn$$

Study this relation very carefully. It states that when two pulleys are connected by a belt, the product of the length of the diameter and the number of revolutions per minute of one equals the product of the length of the diameter and the number of revolutions per minute of the other (*assuming, of course, that there is no slip*).

Example. The $5''$ pulley pictured in Fig. 7-3 is the driver and the $25''$ one is the driven pulley. If the driven pulley is to make 345 r.p.m., what must be the r.p.m. of the driver?

THINK

1. Since the formula $DN = dn$ gives the relation between the diameters and the r.p.m.'s of the two pulleys that are connected by a belt, we use this to solve the problem. Write the formula.

2. Replace the known variables by their values.

3. Solve the resulting equation for the required quantity.

DO THIS

1. $DN = dn$
   $D = 5''$ $d = 25''$ $n = 345$

2. $5N = 25 \times 345$

3. $N = \dfrac{25 \times 345}{5}$

   $N = 5/5 \times 5 \times 345$
   $N = 5 \times 345 = 1{,}725$
   Ans. Driver makes 1,725 r.p.m.

Summary

If two pulleys are connected by a belt, the relation between their diameters and r.p.m.'s is given by the formula $DN = dn$, where D and N refer to the diameter and r.p.m. of the driver and d and n

EXAMPLE

A driver attached to a motor will make 350 r.p.m. This is to drive an $18''$ pulley making 290 r.p.m. What diameter pulley must be used on the motor?

to the diameter and r.p.m. of the driven. To find one of these quantities when the others are known:

1. Write this formula.

2. Replace the known variables by their values.

3. Solve the resulting equation for the remaining letter.

1. $DN = dn$

2. $D \times 350 = 18 \times 290$

3. $D = \dfrac{18 \times 290}{350} =$

$\dfrac{18 \times 29 \times 10}{35 \times 10} = \dfrac{18 \times 29}{35} = 15$

Ans. Driver must be 15″ in diameter

EXERCISES

Find the missing value in each of the following:

1. Driven 18″, driver 6″, r.p.m. of driver 1,500; r.p.m. of driven?
2. Driven 12″, driver 14″, r.p.m. of driver 1,800; r.p.m. of driven?
3. Driven 8″, driver 20″, r.p.m. of driven 2,250; r.p.m. of driver?
4. Driver 15″, driven 10″, r.p.m. of driver 972; r.p.m. of driven?
5. Driver 12″, driven 9″, r.p.m. of driven 1,460; r.p.m. of driver?

PROBLEMS

7. The diameter of a driving pulley is 6 in. and it revolves at the rate of 1,750 r.p.m. At what speed will the driven pulley revolve if it is 22 in. in diameter?
8. A 10-in. diameter pulley is to be operated at 525 r.p.m. when belted to a 1,750 r.p.m. motor. What diameter pulley is required for this motor?
9. A stepped-cone pulley driving a 13-in. engine lathe (Fig. 7-5) revolves at a speed of 150 r.p.m. The diameter of the largest step on the headstock pulley is 9 1/2 in., whereas that of the smallest step is 3 11/16 in. The corresponding steps of the countershaft cone are 3 11/16 and 9 1/2 in., respectively. What is the highest speed at which the spindle will revolve? What is the lowest speed?
10. The r.p.m. of a driven pulley is 400 and its diameter is 30 in. What should be the diameter of the drive pulley to be used, if it is to revolve at 1,200 r.p.m.?

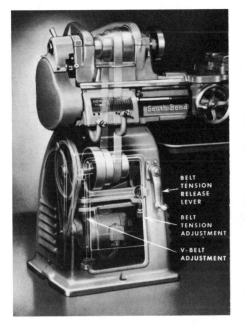

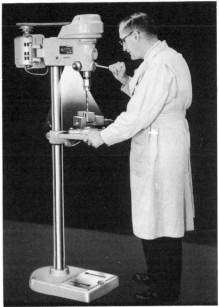

Fig. 7-5. Phantom view of a 13" lathe showing belt and pulley. *(Courtesy South Bend Lathe)*

Fig. 7-6. Drill press with V-belt cone pulleys. *(Courtesy South Bend Lathe)*

11. The diameters of the steps of the cone pulley of a 10" sensitive drill are 4 1/2", 3 1/2", and 2 1/2", respectively. The corresponding diameters of the motor cone pulley are 4 1/4", 5 1/4", and 6 1/4", respectively. Find the spindle speeds available when the motor turns at a rate of 400 r.p.m.

12. Copy the following table and replace the question marks with the proper values.

	D	N	d	n
a	6"	520	14"	?
b	8"	842	?	650
c	12"	720	18"	?
d	?	1,500	10"	2,250
e	15"	?	6"	1,800

Horsepower transmitted by belting

The horsepower transmitted by flat belting depends primarily upon the velocity or speed of the belting and the pull or force per inch of width to which it is subjected. For this purpose the following formula is used:

$$\text{h.p.} = \frac{SVW}{33,000}$$

where S equals the stress or pull per inch of width of belt in pounds, W the width of the belt in inches, and V the velocity or speed of the belt in feet per minute.

Since $V = 0.2618DN$, then if we replace V by $0.2618DN$ in the formula on page 309 we have

$$\text{h.p.} = \frac{0.2618SDNW}{33,000}$$

Since the effect of pull is an assumed quantity, it is not necessary to retain such an exact quantity as 0.2618 in the formula. A sensible approximation is 0.25 or 1/4. The formula then becomes

$$\text{h.p.} = \frac{SDNW}{4 \times 33,000} = \frac{SDNW}{132,000}$$

Note: If it is desired to find the width of the belt required to transmit a given horsepower, the same formula may be used if it is solved for W; this results in the formula

$$W = \frac{132,000 \times \text{h.p.}}{S \times D \times N}$$

The power-transmitting capacity of a given type or make of V-belt can be obtained from the manufacturer.

PROBLEMS

13. What horsepower will a light double leather belt which is 6" wide transmit if the driving pulley is 15" in diameter and revolves at 100 r.p.m.? Assume that the safe working pull is 90 lb. per inch of width.

14. What horsepower can be transmitted by a 2 1/2" belt if the driving pulley is 12" in diameter and the speed of the pulley 225 r.p.m.? Assume that the effective pull is 23 lb. per inch of width.

15. A 5-h.p. motor revolves at 1,750 r.p.m. at full load. The diameter of the pulley is 5". Find the width of the belt required for this job if the effective pull of the belt is 35 lb. per inch of width.

16. A single belt running over an 8" pulley transmits 25 h.p. when it is operating at 2,400 r.p.m. What is the width of the belt if the tension of the belt is 40 lb. per inch of width?

17. How much horsepower is transmitted by a 3" double belt in which the tension is 55 lb. per inch of width, if the driving motor is equipped with a 4 1/2" pulley revolving at 1,750 r.p.m.?

Chain drives

Chain drives consist of a combination of two sprockets and a chain and are usually operated at slower speeds than belts with pulleys. Chain

Fig. 7-7. Single and double roller chain with steel plate driver and driven sprockets. *(Courtesy Cullman Wheel Company)*

a.

b.

c.

d.

Fig. 7-8.   Common types of chain: (a) roller chain; (b) silent chain; (c) bushed roller chain; (d) detachable chain. *(Courtesy Link Belt Company)*

drives are used where the distance between driver and driven shafts is too great for gearing and too short for belting. Two common types of chain used are the link chain and roller chain. A sprocket may be a plain flat plate, may have a hub on one side only, or a hub on both sides.

Motion or power transmitted by gears

Gears are used to transmit motion from one shaft to another, without loss of power by slippage. When two gears mesh, as in Fig. 7-10, any movement of one will result in movement of the other. The gear that imparts the motion is the *driver* gear, and the one to which the motion is transmitted is the *driven* gear.

Fig. 7-9. **Cut-tooth sprocket.** *(Courtesy Link Belt Company)*

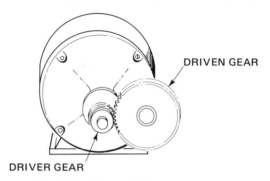

DRIVEN GEAR

DRIVER GEAR

Fig. 7-10. **Driver and driven gears.**

When one mating gear is quite small compared with the other, the smaller gear is sometimes referred to as a "pinion." The pinion is usually the driver.

Spur gears have straight teeth which are parallel to the axis of the shaft and are used when the shafts that are being connected are parallel to each other. It is not necessary that the shafts be parallel, as many types and sizes of gears are available.

Helical gears have curved teeth and will operate with their shafts at right angles to each other, as well as parallel. The *herringbone gear* is a double-helical gear used for high speeds and heavy loads. *Bevel gears* are used to connect shafts at angles with each other. If the angle happens to be a right angle, and the two gears have the same number of teeth, they are called *miter gears. Hypoid gears* are used when the centers of gears are set "in line" with each other. *Worm gears* are used in connection with a worm to transmit motion when two shafts are not parallel. A worm is like a screw-thread with the "threads" being the shape of gear teeth.

Fig. 7-11. Spur gears used in index head of a milling machine. *(Courtesy Browne & Sharpe Manufacturing Company)*

Fig. 7-12. Helical gears. *(Courtesy Charles C. Bond Company)*

Fig. 7-13. Herringbone or double-helical gears. *(Courtesy Charles C. Bond Company)*

Fig. 7-14. Bevel gears. *(Courtesy Charles C. Bond Company)*

A *rack and pinion* will change rotary motion to straight-line motion or straight-line motion into rotary motion. A rack is a straight bar with teeth cut in it to mate with a pinion. The rack may move to turn the pinion, or the pinion may rotate to move the rack.

Calculations concerning gears are similar to those concerning pulleys. The formula that shows how the number of teeth and number of revolutions of the driver and driven gears are related is the same, regardless of the type of gear:

$$T \times N = t \times n, \text{ or } \frac{T}{t} = \frac{n}{N}$$

Fig. 7-15. Worm gears. *(Courtesy Charles C. Bond Company)*

Fig. 7-16. Gears, pinions, and sprockets. How many different types can you pick out? *(Courtesy Charles C. Bond Company)*

Here T represents the number of teeth of the driver, N its number of revolutions, t the number of teeth of the driven, and n its revolutions. Notice that this formula has the same form as the formula for pulleys (p. 307). In calculations involving gears we use the number of teeth, while in calculations involving pulleys we use the diameters of the pulleys. Can you see the reason for the difference between the formulas?

Example 1. A driver gear has 40 teeth and makes 1 turn. How many turns will the driven gear with 15 teeth make?

THINK	DO THIS
Since the formula $TN = tn$ gives the relation between teeth and revolutions of two meshed gears, we	
1. Write the formula.	1. $TN = tn$ $T = 40 \quad N = 1 \quad t = 15$
2. Replace each known variable by its value.	2. $40 \times 1 = 15n$
3. Solve the resulting equation for the remaining variable.	3. $40/15 = n$ $n = 2\ 2/3$ *Ans.* The driven gear makes 2 2/3 turns for 1 turn of the driver

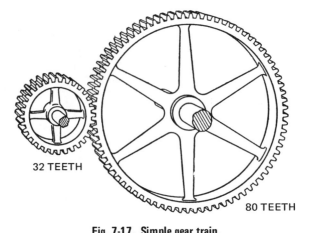

Fig. 7-17. Simple gear train.

Example 2. The driver of Fig. 7-17 has 32 teeth, the driven gear 80 teeth. If the r.p.m. of the driver is 120, what is the r.p.m. of the driven gear?

THINK	DO THIS
Since $TN = tn$ describes the relation between the number of teeth and the revolutions of the two gears we	
1. Write this formula.	1. $TN = tn$ $T = 32$ $N = 120$ $t = 80$
2. Replace each known variable by its value.	2. $32 \times 120 = 80n$
3. Solve the equation for the remaining letter.	3. $\dfrac{32 \times 120}{80} = n$ $48 = n$ *Ans.* Driven gear makes 48 r.p.m.

Example 3. A driver gear has 42 teeth; the driven gear has 70 teeth and turns at 120 r.p.m. How many r.p.m. does the driver gear make?

THINK	DO THIS
1. Write the formula.	1. $TN = tn$
2. Replace each variable by its given value.	2. $42N = 70 \times 120$

3. Solve the equation for N.

3. $N = \dfrac{70 \times 120}{42}$

$N = 200$

Ans. Driver makes 200 r.p.m.

Summary

When two gears mesh, the relation between their number of revolutions and number of teeth is given by the formula $TN = tn$, in which T is the number of teeth and N the revolutions of the driver gear, and t the number of teeth and n the revolutions of the driven. To solve a problem involving these quantities:

1. Write the formula.

2. Replace the known variables by their values.

3. Solve the resulting equation for the remaining letter.

EXAMPLE

A driver gear with 54 teeth makes 51 turns each minute. How many turns will the driven gear with 180 teeth make in one minute?

1. $TN = tn$
   $T = 54$, $N = 51$, $t = 180$

2. $54 \times 51 = 180n$
   $\dfrac{54 \times 51}{180} = n$

3. $15.3 = n$

Ans. Driven gear makes 15.3 turns each minute

EXERCISES

1. A driver gear with 28 teeth makes 75 turns a minute. How many turns will the driven gear with 52 teeth make in a minute?
2. A driven gear with 36 teeth must make 140 r.p.m. How many teeth must the driver have if it makes 210 r.p.m.?
3. Find the missing quantity in each line of this table.

	No. of teeth		Revolutions per minute	
	Driver	Driven	Driver	Driven
a	48	?	320	240
b	?	320	420	210
c	180	270	?	126
d	60	?	240	720
e	85	70	280	?

PROBLEMS

18. A motor running at 1,150 r.p.m. is fitted with a gear having 68 teeth. Find the speed of the driven gear if it has 288 teeth.
19. A gear running at 200 r.p.m. meshes with another revolving at 150 r.p.m. If the smaller gear has 24 teeth, how many teeth does the larger gear have?

Intermediate or idler gears

When two gears mesh, such as gears A and B in Fig. 7-18, they turn in opposite directions. When A turns clockwise, B turns counterclockwise. If a third, or *idler*, gear is inserted between the two, such as B between A and C, the effect is to cause the driven gear, C, to turn in the same direction as the driver, A. The idler, B, has no effect whatsoever on the number of revolutions of the driven gear. Such an arrangement of gears is called a *simple train.*

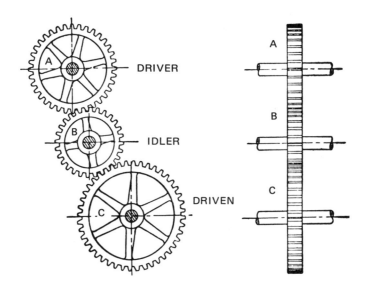

Fig. 7-18. Simple train of three gears.

Compound trains of gears

Study the gears in Fig. 7-19, page 319. Notice that C and B are on the same shaft. Here the motion goes from A to B through the teeth of the gears. Then the motion goes from B to C through their common shaft, and finally from C to D through the teeth of these gears. Such an arrangement of gears is called a *compound gear train*, or simply *compound gears.* Since B and C are keyed to the same shaft, they turn together in the same direction and make the same number of revolutions.

Fig. 7-18A. Simple train of three spur gears.

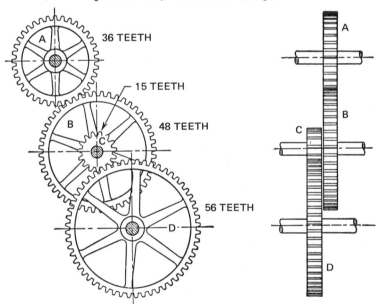

Fig. 7-19. Compound gear train.

The relative direction of rotation of the driving and driven gears can be determined by observing the number of shafts in the gear train. An odd number of shafts causes the first and last gears to rotate in the same direction and an even number of shafts causes them to turn in opposite directions.

This can readily be seen if you visualize the following: Two gears attached to shaft 1 turn in the same direction; gears on shaft 2 turn opposite to those on shaft 1; gears on shaft 3 turn opposite to those on shaft 2, and hence in the *same direction* as those on shaft 1; gears on shaft 4 turn opposite to those on shaft 3, and *opposite* to those on shaft 1; etc.

Example. In the gear train of Fig. 7-19, driver A has 36 teeth and makes 120 r.p.m., gear B has 48 teeth, gear C has 15 teeth, and driven gear D has 56 teeth. How many revolutions per minute does D make?

THINK	DO THIS
To solve this problem we must go from gear A to get B, then to gear C and finally to gear D.	
From gear A to gear B:	*A to B*
1. Write the formula.	1. $TN = tn$ $T = 36, N = 120, t = 48$
2. Replace the known variables by their values.	2. $36 \times 120 = 48n$
3. Solve this equation.	3. $\dfrac{36 \times 120}{48} = n$ $90 = n$
From gear B to gear C: Since these gears are keyed to the same shaft they make the same number of revolutions per minute.	Gear B makes 90 revolutions per minute. Thus, gear C makes 90 revolutions per minute.
From gear C to gear D:	*C to D*
1. Write the formula. T, N, t and n have new values here.	1. $TN = tn$ $T = 15, N = 90, t = 56$
2. Replace variables by their values.	2. $15 \times 90 = 56n$
3. Solve the resulting equation.	3. $24\ 3/28 = n$ *Ans.* Gear D makes 24 3/28 r.p.m.

In this train A and C both may be considered as drivers and B and D as driven gears. If T_A represents the number of teeth on A, T_C represents the number of teeth on C, t_B represents the number of teeth on B, and t_D represents the number of teeth on D, the relation between these quantities is shown in the formula:

$$NT_A T_C = nt_B t_D$$

where N is the r.p.m. of the first driver, and n, the r.p.m. of the final driven gear.

In words: the number of revolutions of the first driver times the product of the numbers of teeth on all the drivers equals the number of revolutions of the final driven times the product of the numbers of teeth on all the driven gears.

For any number of gears, drivers A_1, A_2, A_3...with teeth respectively T_1, T_2, T_3..., driven a_1, a_2, a_3...with teeth respectively t_1, t_2, t_3..., and using N to represent the number of revolutions of the *first* driver and n the number of revolutions of the *last* driven, the relation between the number of teeth and number of revolutions is $NT_1T_2T_3... = nt_1t_2t_3...$

Summary

In a compound gear chain the product of the revolutions of the first driver and the numbers of teeth of all the drivers equals the product of the number of revolutions of the last driven and the numbers of teeth of all driven gears. The formula for this relation is

$$NT_1T_2T_3... = nt_1t_2t_3...$$

1. Write the formula.

2. Replace the known variables by their values.

3. Solve the resulting equation for the remaining variable.

EXAMPLE

Solve the example on p. 320 using this extended formula.

1. $NT_A T_C = nt_B t_D$
   $N = 120$, r.p.m. of gear A.
   $T_A = 36, T_C = 15$, number of teeth in gears A and C, respectively. These are drivers.
   $t_B = 48, t_D = 56$. These are driven gears.

2. $120 \times 36 \times 15 = n \times 48 \times 56$

3. $\dfrac{120 \times 36 \times 15}{48 \times 56} = n$

   $n = 24\ 3/28$

   Ans. Gear D makes $24\ 3/28$ r.p.m.

EXERCISES

1. Find the revolutions per minute of gear D in the train pictured in Fig. 7-20. Will gears A and D turn in the same direction?

2. In a gear train, similar to Fig. 7-20, if gear A has 30 teeth, gear B has 24 teeth, gear C has 15 teeth, and gear D has 60 teeth, and gear A turns 1,750 r.p.m., how fast will gear D turn?

3. How many revolutions will D make while A makes 720 r.p.m. in the train in Fig. 7-21?

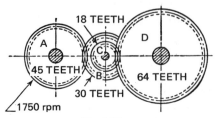

Fig. 7-20.

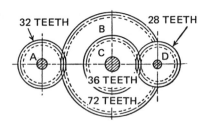

Fig. 7-21.

4. In a gear train, similar to Fig. 7-21, if gear *A* has 30 teeth, gear *B* has 65 teeth, gear *C* has 32 teeth, and gear *D* has 24 teeth, and gear *A* makes 720 r.p.m., how many r.p.m. will gear *D* make?

5. How many revolutions per minute must the driver *A* make in order that *D* will turn 180 times per minute in the train in Fig. 7-22?

6. In a gear train, similar to Fig. 7-22, if gear *A* has 36 teeth, gear *B* has 30 teeth, gear *C* has 76 teeth, and gear *D* has 32 teeth, and gear *D* turns 180 times per minute, how many revolutions per minute will gear *A* make?

7. In Fig. 7-23, what is the speed of the driven gear, *A*, if the speed of the driver is 250 r.p.m.? If gear *E* rotates clockwise, in which direction will gear *A* rotate?

8. In a gear train, similar to Fig. 7-23, gear *A* has 22 teeth, gear *B* has 38 teeth, *C* has 16 teeth, *D* has 30 teeth, and *E* has 18 teeth. What is the speed of gear *A* if gear *E* travels 250 r.p.m.?

9. In Fig. 7-24, page 323, driver *A* has 16 teeth, and gears *B*, *C*, and *D* have 32, 18, and 54 teeth, respectively. If *D* runs at 15 r.p.m., what is the speed of *A*?

10. In a gear train, similar to Fig. 7-24, gear *A* has 21 teeth, *B* has 42 teeth, *C* has 16 teeth, and *D* has 48 teeth. When gear *D* rotates at 15 r.p.m., how fast will gear *A* turn?

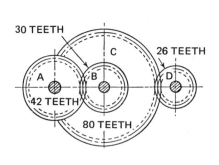

Fig. 7-22.

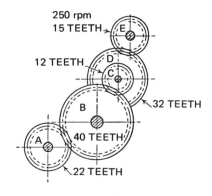

Fig. 7-23.

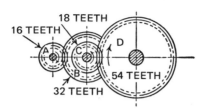

Fig. 7-24.

REVIEW PROBLEMS

1. In a simple train of gears, the driver has 72 teeth, the intermediate gear has 42 teeth, and the driven gear or follower has 36 teeth. If the driver makes 102 r.p.m., what will be the speed of the follower?

2. What horsepower can be transmitted by a belt 3″ wide? Assume that the driving pulley is 6″ in diameter, the speed of the pulley 1,180 revolutions per minute, and the effective pull 33 lb. per inch of width.

3. How wide a double belt is needed to transmit 6 horsepower if the effective working tension for a light double belt is 70 lb. per inch of width? The diameter of the pulley is 5″ and revolves 1080 r.p.m.

4. In the compound train of gears in Fig. 7-25, what is the speed of the driven gear if the driver runs at 60 r.p.m.?

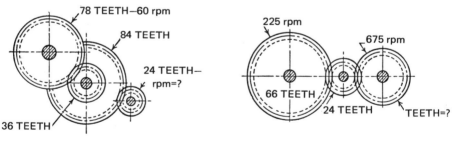

Fig. 7-25.   Fig. 7-26.

5. In Fig. 7-26 the speed of the driver is 225 r.p.m., whereas the speed of the driven gear is 675 r.p.m. How many teeth are there in the driven gear?

6. What is the speed of the driven gear in Fig. 7-27? If the speed of the driver is doubled, what will be the speed of the driven gear?

7. The 14″ pulley on a main shaft revolves at a rate of 175 r.p.m. and is belted to an 8″ pulley on the countershaft of the lathe. What is the speed of the countershaft?

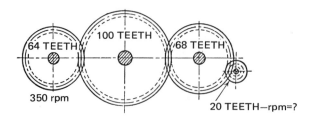

Fig. 7-27.

8. How many teeth should there be in a driven gear to provide a gear reduction of 3 1/2 to 1 if the driving gear has 84 teeth? (A gear reduction of 3 1/2 to 1 means that the ratio of the speed of the driving gear to the speed of the driven gear is 3 1/2 to 1.)

9. In a compound gear train, gear B meshes with gear A; gear C is on the same shaft as B; D meshes with C; E meshes with D; F is on the same shaft as E and meshes with G. If gear A rotates clockwise, in what direction does gear G rotate?

8 Some Ideas from Geometry

Many industrial and technical jobs can be done more intelligently with a knowledge of some of the facts from geometry. In this unit we shall study angles, triangles, and basic geometric constructions. We shall also get acquainted with some of the tools draftsmen and mechanics use to lay out shop work.

Angle, measure of an angle, right angle, perpendicular lines

When we speak of a *line*, we think of it extending infinitely far in both directions (Fig. 8-1).

Fig. 8-1.

By a *line segment* or simply a segment we mean the part of a line consisting of two points together with all the points of the line that are between the two points (Fig. 8-2). *AB* is a segment. *A* and *B* are the end points of the segment.

A   B

Fig. 8-2. Segment *AB*.

By a *ray* we mean that part of a line consisting of a point (called the end point) and those points of the line on one side of this end point. The point *B* and all points of the line on the *A* side of *B* form a ray. It is called the ray *BA*.

An *angle* is the plane figure formed by two rays, not in the same line, that have a common end point (Fig. 8-3). The rays *BA* and *BC* are the *sides* of the angle. The common end point *B* is the *vertex* of the angle.

We refer to the angle as ∠ *ABC*, read "angle *ABC*," or as ∠ *B* or ∠ *CBA*. When a single letter is used to indicate an angle it must be the letter at the vertex. When three letters are used to name the angle, the letter at the vertex must be placed between the other two.

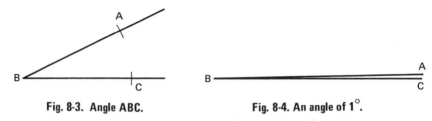

Fig. 8-3. Angle ABC.　　　　　Fig. 8-4. An angle of 1°.

Every angle has a measure. The unit of measure is a small angle called "one degree" (Fig. 8-4), written 1°.

Most of the angles with which we shall work have measures between 0° and 180°.

When two angles have the same measure the angles are *congruent*. The angle *ABC* of the bevel gage (Fig. 8-5) is congruent to angle *DEF* of the piece of oak. Fig. 8-6 shows that congruent angles can be made to fit perfectly when placed together.

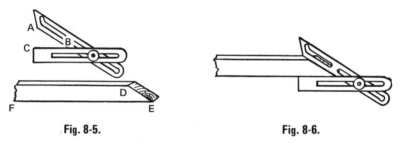

Fig. 8-5.　　　　　　　　　Fig. 8-6.

When two segments such as *AC* and *DB* in Fig. 8-7(a) have a point in common, point *O*, we say they *intersect* at *O*. *D'B'** and *A'C'* intersect at *O'*. *GH* and *FE* intersect at *F*. When two lines intersect and all four angles formed are congruent as in Fig. 8-7(a), the angles thus formed are *right angles*. The *measure of a right angle is 90°* and 1° is 1/90 of a right angle.

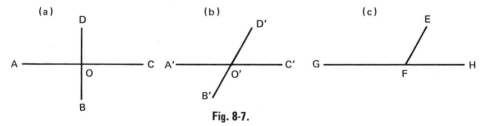

Fig. 8-7.

When two segments intersect and form a right angle the segments are *perpendicular* and each is said to be perpendicular to the other. In each of these figures *AB* is perpendicular to the other segment (Fig. 8-8).

D'B' is read "D prime B prime."

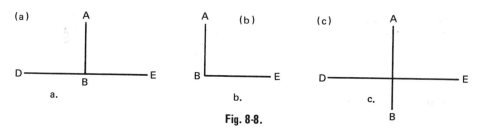

Fig. 8-8.

In a square, such as the cross section of the piece of steel shown in Fig. 8-9, all four angles are right angles, and every side is perpendicular to each of the two sides which it intersects.

An angle that is smaller than a right angle is called an *acute* angle; an angle larger than a right angle (larger than 90°) but less than 180° is called an *obtuse* angle (Fig. 8-10). An angle of exactly 180° is called a *straight* angle. Angle *DBE* of Fig. 8-8(a) is a straight angle.

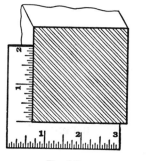

Fig. 8-9.

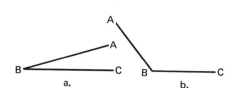

Fig. 8-10. (a) Acute angle; (b) obtuse angle.

Constructing an angle congruent to a given angle

Many of the advances in modern mass production depend upon man's ability to produce exact duplicates of objects. Let us learn an elementary technique in exact duplication, using the draftsman's simplest tools, the straightedge and compass. Constructing an angle congruent to a given angle means drawing with straightedge and compass an angle whose measure is exactly equal to the measure of a given angle. Study the instructions that follow; then make a similar construction on your paper.

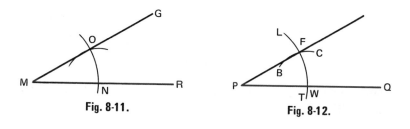

Fig. 8-11.   Fig. 8-12.

Example. Construct an angle that is congruent to ∠GMR (Fig. 8-11). (*Note:* The student should construct on paper as he studies this solution.)

THINK	DO THIS
1. Draw a ray. Congruent angles have the same measure. The following procedure, making use of a compass and straightedge, accomplishes this.	1. Make a segment of a ray *PQ* at least as long as *MR* (Fig. 8-12). Use the straightedge.
2. With compass point at *M* and opened to any convenient size less than *MR*, draw an arc intersecting the side *MG*. Do not change the compass opening.	2. Place compass point at *M* and open compass to some point *N* on *MR*. Draw arc *NO* intersecting side *MG* at *O*.
3. With same compass opening and with point of compass placed at point *P*, draw an arc intersecting *PQ*.	3. Without changing compass setting, place point of compass at *P* and draw arc *LT* intersecting *PQ* at *W*.
4. Place compass point at *N* on *MR* and open compass to point *O* on *MG*.	4. Get the distance *NO* on the compass.
5. With this compass opening and point of compass on *W* draw an arc.	5. Place compass point at *W* and draw an arc *BC* to intersect arc *LT* at *F*.
6. With straightedge draw line from *P* to complete the required angle.	6. With a straightedge placed on *P* and *F*, draw ray from *P* through *F*.
	Ans. The angle *FPQ* is the required angle congruent to ∠GMR.

EXERCISES

Using the compass and straightedge, construct angles congruent to each angle in Fig. 8-13.

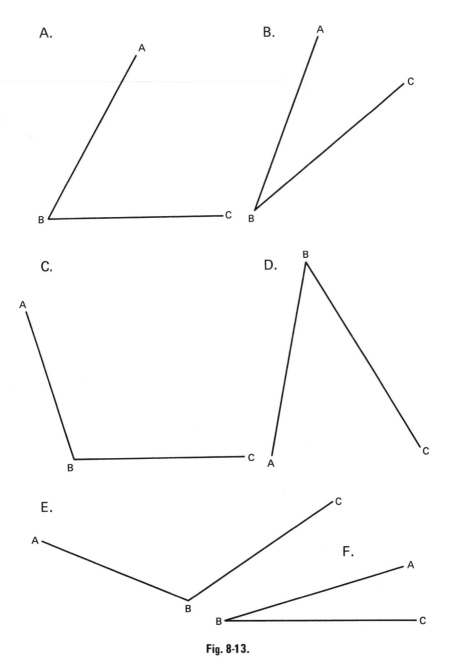

Fig. 8-13.

Measuring angles: the protractor

One instrument used for measuring an angle, that is, for finding the number of degrees in the angle, is the protractor (Fig. 8-14). This instrument is divided into 180 equal parts along its curved side. Each large

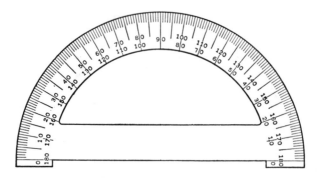

Fig. 8-14. Protractor.

division represents 10 degrees, and each small division, which is 1/10 of a large division, is the measure of one degree.

The machinist's bevel protractor, Fig. 8-15, for laying out angles in the shop, follows the same principle.

Fig. 8-15. Machinist's bevel protractor.

Example 1. Find the number of degrees in ∠BRV (Fig. 8-16).

THINK	DO THIS
1. The size of an angle is determined by the opening between the sides.	1. Place the center point, R, of the protractor on the vertex, R, of the angle, and the 0° line of the protractor along one side, RB, of the angle (Fig. 8-16).
2. The reading on the protractor must agree with the type of angle. An acute angle measures less than 90°; an abtuse angle measures more than 90° but less than 180°. Take reading from correct scale.	2. Note where the other side, RV, of the angle meets the protractor. Read the size of the angle at this point. The side RV crosses the protractor between 47° and 48°.

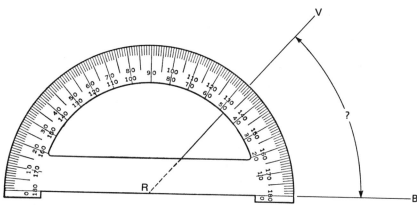

Fig. 8-16.

| 3. Accuracy of the reading cannot exceed accuracy of the instrument, here 0.5°. | 3. Since the reading is about midway between 47° and 48°, we estimate the angle to be 47.5°.
Ans. ∠*BRV* measures 47.5° |

Example 2. Draw an angle of 157° using a protractor.

THINK	DO THIS
1. An angle has two sides and its size is the measure of the opening between them.	1. Use a straightedge to draw one side of the angle. Mark the vertex *R* and another point along the side *S* (Fig. 8-17). This is the first side of the angle.
2. Position the protractor carefully.	2. Place the center point of the protractor on the point *R* of the line *RS*, and the 0° line of the protractor along side *RS*.

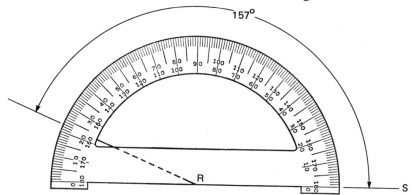

Fig. 8-17.

3. Be careful to use the correct scale on the protractor, so you get the proper type of angle – obtuse, in this case.

4. Complete the drawing.

3. Starting from $0°$, move along the scale to the $157°$ line. Place a mark on the paper at this point (point Q on Fig. 8-17).

4. Draw QR, the second side of angle QRS, which is an angle of $157°$ if you work accurately.

Summary

To measure an angle with a protractor:

1. Put the center point of the protractor at the vertex of the angle and the $0°$ line of the protractor along the one side of the angle.

2. Note where the other side of the angle meets the protractor. Estimate to a half of a degree.

3. Write the measurement; be careful to note the correct scale on the protractor.

EXAMPLE

Measure the angle FGH with a protractor.

1. Put center of protractor at vertex G with $0°$ line along GF.

2. GH meets protractor between 69.0 and 69.5 but nearer to 69.0.

Ans. $\angle FGH$ measures $69.0°$ approx.

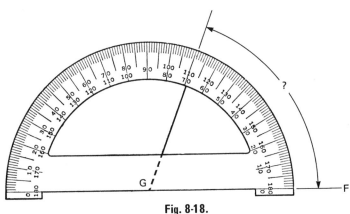

Fig. 8-18.

EXERCISES

1. Measure each of the angles in Fig. 8-19 to the nearest half of a degree. State which are acute and which are obtuse angles.

   In Exercises 2-5, use the protractor to draw the angle indicated.

2. An angle of $38°$

3. An angle of $75°$

4. An angle of $120°$

5. An angle of $158°$

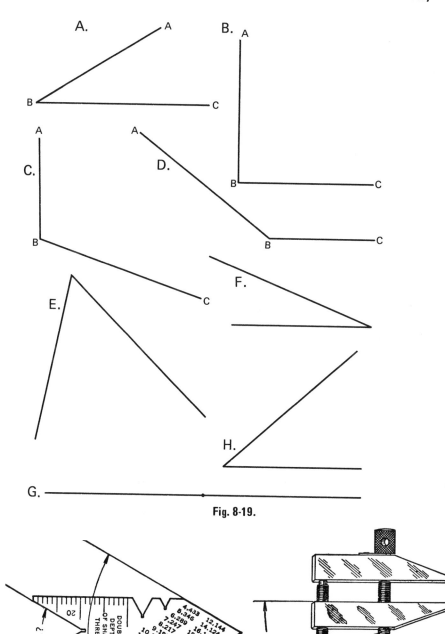

Fig. 8-19.

Fig. 8-20. Center gage.

Fig. 8-21. Toolmaker's parallel clamps.

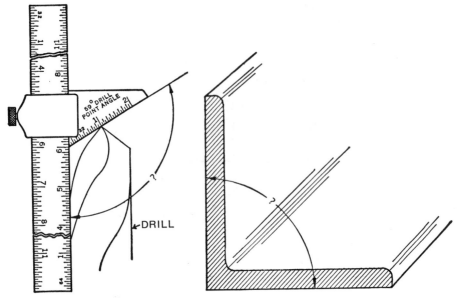

Fig. 8-22. Drill point gage.

Fig. 8-23. Section of angle iron.

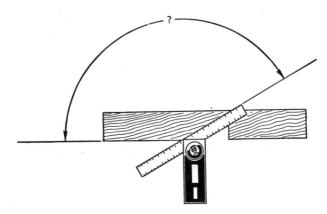

Fig. 8-24. Combination tool.

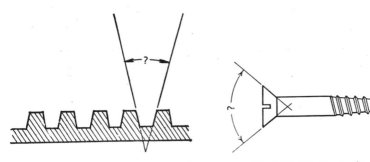

Fig. 8-25. Acme thread.

Fig. 8-26. Flathead wood screw.

PROBLEMS

1. Using the protractor, measure the angle or angles indicated in Figs. 8-20 through 8-26.
2. The drawings in Fig. 8-27 show the angle to which drills should be ground for drilling certain materials. Measure each of these angles.

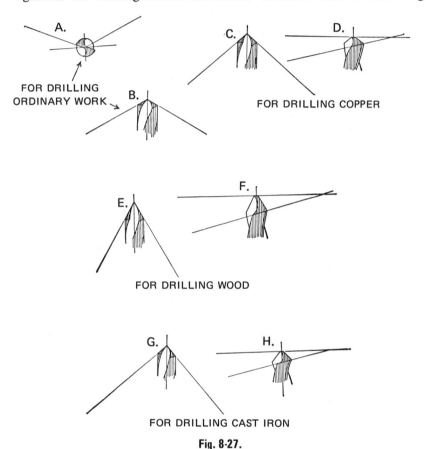

Fig. 8-27.

Triangles

A closed figure made up of three segments and three angles is a *triangle.*

Triangles are classified according to the relative lengths of their sides or the sizes of their angles. If all of the sides of a triangle are of different lengths, the triangle is called a *scalene* triangle (Figs. 8-28, 8-29, 8-30). If two sides of a triangle are the same length, the triangle is called an *isosceles* triangle (Figs. 8-31, 8-32, 8-33). If all three sides of a triangle are of equal length, the triangle is called an *equilateral* triangle (Fig. 8-34). If all the angles of a triangle are acute, the triangle is called an *acute*

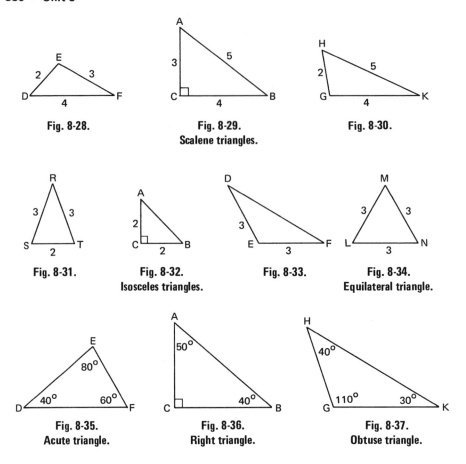

Fig. 8-28.

Fig. 8-29.
Scalene triangles.

Fig. 8-30.

Fig. 8-31.

Fig. 8-32.
Isosceles triangles.

Fig. 8-33.

Fig. 8-34.
Equilateral triangle.

Fig. 8-35.
Acute triangle.

Fig. 8-36.
Right triangle.

Fig. 8-37.
Obtuse triangle.

triangle (Fig. 8-35). If one of the angles of a triangle is a right angle, the triangle is called a *right* triangle (Fig. 8-36). If one of the angles of a triangle is an obtuse angle, the triangle is called an *obtuse* triangle (Fig. 8-37).

EXERCISES

Draw triangles illustrating each of the following:

1. An isosceles triangle that is (a) acute; (b) right; (c) obtuse.
2. An equilateral triangle that is acute. Can an equilateral triangle be obtuse? Explain.
3. A right triangle that is (a) scalene; (b) isosceles.
4. An acute triangle that is (a) scalene; (b) isosceles; (c) equilateral.
5. An obtuse triangle that is (a) scalene; (b) isosceles.
6. An equiangular triangle. Is this also equilateral?
7. A triangle with two angles equal. Is this also equilateral? Is it isosceles?

Drawing an equilateral triangle

You can make a scale drawing of any triangle using ruler and protractor, or straightedge and compass, if you know (a) the lengths of all three sides, (b) the lengths of two sides and the measure of the angle between the two known sides, or (c) the length of one side and the measures of the two adjacent angles. Drawing an equilateral triangle is particularly easy, since all sides are the same length.

Example. Make a layout for three 1/2″ holes with centers 3 5/8″ apart. Each center is to be a vertex of an equilateral triangle.

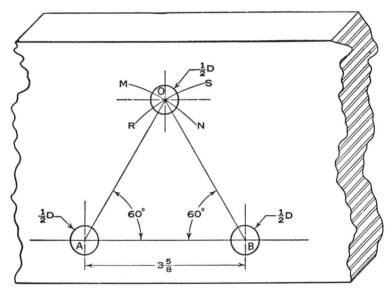

Fig. 8-38. Layout of three holes, each at an angle of 60° with the others.

THINK

1. Any triangle can be constructed if lengths of all three sides are known.

2. All sides of an equilateral triangle are the same length.

3. If you do not change the compass setting, you can mark off equal distances from any point.

DO THIS

1. Draw a line 3 5/8″ long (Fig. 8-38). Call this line *AB*.

2. Place the needle point of the compass at *A*, and using the distance *AB* as a radius, strike the arc *MN*.

3. Without changing the setting, place the compass point at *B* and strike the arc *RS*. These arcs meet at *O*.

4. *OA* and *OB* are each equal to *AB*. Triangle *AOB* is equilateral.

4. Join *OA* and *OB* by straight lines to form triangle *AOB*.

EXERCISES

1. Construct an equilateral triangle whose side is 2 1/2" long.
2. Construct an equilateral triangle whose side is 4.3" long.
3. One side of a building is a rectangle with an equilateral triangle on top. Construct a drawing of the equilateral triangle if the building is 24 ft. wide. Let 1/8 in. on your drawing represent 1 ft. of the building.

Expressing the altitude of an equilateral triangle in terms of the side

The *altitude* of any triangle is the perpendicular distance from the vertex of one angle to the opposite side or base, such as *a* in Fig. 8-39. The altitude of an equilateral triangle divides the opposite side into two equal parts, or it *bisects* the base: $VT = WT$. If we represent the length of the side by *s*, then $TW = \frac{s}{2}$. It is often necessary in industry to express the altitude of the equilateral triangle in terms of the length of the side of the triangle.

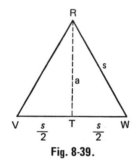

Fig. 8-39.

Example. Express the altitude of an equilateral triangle in terms of the lengths of its sides.

THINK

1. If we had a drawing of an equilateral triangle and its altitude, we could think better about it.

In order that what we do shall apply to any equilateral triangle we represent the length of its side by a letter, and its altitude by a different letter.

DO THIS

1. Draw an equilateral triangle. Let *s* represent the length of the sides and *a* the length of the altitude (Fig. 8-39).

2. Note that since RT is the altitude, the angle RTV is a right angle and $\triangle RTW$ is a right triangle. Use $\triangle RTW$ to write the equation concerning s, a and $\frac{s}{2}$.

2. $s^2 = a^2 + \left(\frac{s}{2}\right)^2$

3. Solve this equation for a. That is, use the properties of equality to have a^2 alone on one side. Subtract $\left(\frac{s}{2}\right)^2$ from both sides.

3.
$$s^2 = a^2 + \left(\frac{s}{2}\right)^2$$
$$\frac{-\left(\frac{s}{2}\right)^2 \qquad -\left(\frac{s}{2}\right)^2}{s^2 - \left(\frac{s}{2}\right)^2 = a^2}$$
$$s^2 - \frac{s^2}{4} = a^2$$
$$\frac{3}{4}s^2 = a^2$$
$$\frac{s^2}{4} \times 3 = a^2$$

4. To find a, find the square root of both sides of the resulting equation.

4.
$$\sqrt{\frac{s^2}{4} \times 3} = a$$
$$\text{or } \frac{s}{2}\sqrt{3} = a$$

Ans. Altitude $= \frac{s}{2}\sqrt{3}$

This result says that the altitude of an equilateral triangle is $1/2$ of the length of the side multiplied by $\sqrt{3}$, or 1.732. Briefly, $a = \frac{s}{2}\sqrt{3}$. This is a formula for finding the altitude of an equilateral triangle in terms of the length, s, of the side. It may be used to find the altitude in any of our work.

Formula for the altitude a of an equilateral triangle of side s is

$$a = \frac{s}{2}\sqrt{3}$$

Example. A gable roof in the form of an equilateral triangle is $15'$ wide. What is the altitude of the gable?

THINK	DO THIS
1. Write the formula.	1. $a = \frac{s}{2}\sqrt{3}$
2. Substitute 15 for s, and 1.732 for $\sqrt{3}$. (See table, in Appendix).	2. $a = \frac{15}{2} \times 1.732$
3. Perform the operations indicated.	3. $a = 12.99$ or approx. 13 feet

Ans. Altitude $= 13$ feet, approx.

Fig. 8-40. Gable roof.

EXERCISES

Find the altitudes of equilateral triangles with sides of the following lengths:

1. 12″	3. 1/18″	5. 1/32″	7. 16′	9. 30′
2. 1/8″	4. 1/20″	6. 18′	8. 24′	

Bisecting a line, an angle, an arc

Example 1. Bisect, that is, divide into two equal parts, the segment *AB* in Fig. 8-41.

THINK

On your own paper, make the construction as you study this solution.

1. Draw a segment *AB* that is equal to the length of segment *AB* in Fig. 8-41.

2. Place the point of the compass at *A*. Open the compass a distance nearly equal to the length of *AB*. Keep the point of the compass at *A* and strike two arcs, one above *AB* as *M* and one below as *N*.

3. Without changing the opening of the compass, place the point of the compass at *B* and strike two arcs *R* and *S*. These arcs will intersect arcs *M* and *N* at *C* and *O* respectively.

4. Draw the segment *CO*.

5. The point where *CO* intersects *AB* is the midpoint of *AB* and *AB* is bisected at *E*.

DO THIS

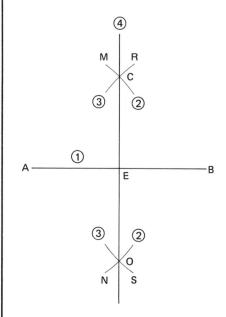

Fig. 8-41.

Ans. CO is the bisector of *AB*. The measure of *AE* equals the measure of *EB*.

Example 2. Bisect the angle *RST* of Fig. 8-42.

THINK	DO THIS
1. On your own paper, construct an angle equal to angle *RST* of Fig. 8-42. Letter the angle *RST*. To bisect this angle means to construct a ray that will divide the given angle into two congruent angles. With the compass point on *S*, open the compass any convenient distance (about *SF*) and strike an arc intersecting *SR* at *E* and *ST* at *F*.	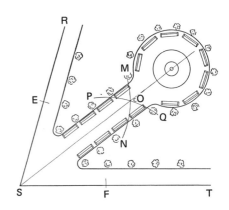

Fig. 8-42. Layout of landscape.

THINK	DO THIS
2. Place the compass point at *E*. Strike an arc *MN*.	
3. With the same radius and with compass point at *F*, strike the arc *PQ* intersecting arc *MN* at *O*.	*Ans.* The line that contains segment *SO* bisects the angle *ESF*. It makes ∠*ESO* congruent to ∠*OSF*.
4. Draw the line through *S* and *O*.	

Example 3. Bisect the arc *PQ* in Fig. 8-43.

THINK	DO THIS
1. Place the point of the compass at *P* and open the compass to *Q*.	
2. Keep the point of the compass at *P* and with *PQ* as the radius, strike the arc *MQN*.	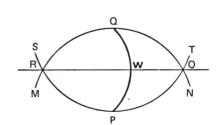
3. Using the same radius, place the point of the compass at *Q* and strike the arc *SPT*.	
4. These two arcs intersect at *R* and *O*. Draw the segment *RO*. *RO* intersects arc *PQ* at *W*. *W* is the midpoint of arc *PQ*.	

Fig. 8-43. Bisecting an arc.

Ans. Arc *PQ* is bisected at *W*.

EXERCISES

Bisect as indicated in each of the following exercises:

1. The base of an equilateral triangle that has dimensions twice as large as the dimensions shown in Fig. 8-38.

2. One angle of the triangle you drew for Exercise 1.
3. An arc *PQ* of a circle that has a radius of 2 1/2″. (The student will select the arc to be bisected.)
4. Each of the angles you made congruent to the angles of Fig. 8-13, Exercises (A) to (F).

The sum of the measures of all the angles of a triangle

There are many different kinds of geometry; the one we are studying is called Euclidean geometry. One of the facts that distinguishes it from other geometries is the idea that the sum of all the angles of any one triangle is exactly 180°. In each of the triangles shown on p. 336, the sum of the angles is 180°. This is true of every triangle, even though the measures of the individual angles of the triangle are not known. *The sum of the measures of the angles of any triangle is 180°.*

The 45° right triangle and the 30°-60° right triangle

Two triangles which draftsmen frequently use are the 45° and the 30°-60° right triangles.

(A) The 45° right triangle. The 45° right triangle is an isosceles triangle in which one of the angles, ∠*B* in Fig. 8-44, is a right angle and the two sides, *AB* and *BC*, are congruent.

In an isosceles right triangle the angles opposite to the congruent sides are congruent angles, and each measures 45°.

1. In △*ABC*, *AB* = *BC*.

2. The angle *A* is opposite to the side *BC*.

3. The angle *C* is opposite to the side *AB*.

4. Since *BC* = *AB*, ∠*A* = ∠*C*.

5. ∠*B* is a right angle.

6. The sum of the measures of all the angles of a triangle is 180°.

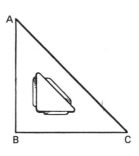

Fig. 8-44. The 45° right triangle.

7. Then ∠*A* + ∠*C* must equal 90° by the subtraction principle of equality.

$$7. \angle A + \angle B + \angle C = 180°$$
$$\underline{\qquad \angle B \qquad\qquad = \quad 90°}$$
$$\angle A \qquad\quad + \angle C = \quad 90°$$

8. But ∠*A* = ∠*C*, hence ∠*A* + ∠*C* = ∠*C* + ∠*C*.

$$8. \angle C + \angle C = 90°$$
$$2 \angle C = 90°$$
$$\angle C = 45°$$

9. ∠*A* also equals 45°.

$$9. \angle A = 45°$$

In the isosceles right triangle each of the acute angles measures 45°. Such a triangle is called the 45° right triangle, or also simply "the 45° triangle."

(B) The 30°-60° right triangle.

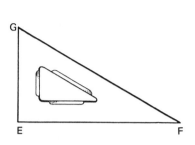

Fig. 8-45. The 30°-60° right triangle.

1. In triangle *GEF*, the measure of ∠ *F* is 30° and that of ∠ *G* is 60°. ∠*E* is 90°.

2. In such a triangle the side opposite to the 30° angle, the side *GE*, is always 1/2 of the hypotenuse *GF*, the side opposite to the right angle *E*.

1. ∠*G* + ∠*F* + ∠*E* = 180°
   ∠*G* = 60°; ∠*F* = 30°

2. *GF* = the hypotenuse
   GE = 1/2 *GF*

In a right triangle in which one acute angle is 30°, the other acute angle is 60°. Such a triangle is called the 30°-60° right triangle. In a 30°-60° right triangle the length of the hypotenuse is twice the length of the shortest side. Whenever the sum of two angles is 90° the angles are called *complementary angles*. The angles 30° and 60° are complementary since their sum is 90°.

EXERCISES

1. Measure the sides of each of the following right triangles and state the size of each angle. Name the kind of triangle.
2. A piece of a square bar (Fig. 8-47) is cut in half by making the cut through two opposite corners. What angles would the new pieces have? Why?
3. A carpenter placed his steel square as shown in Fig. 8-48. What are the measures of the angles *A* and *B*? Why?
   Hint: Find the angles by measuring the sides.
4. Determine the measure of each of the angles in the drawings shown in Figs. 8-49 through 8-53. Use your protractor. State the kind of triangle in each case where the triangle is complete.
5. Find the length of the rafter, and the run in Fig. 8-54, if the rise is 8 ft.

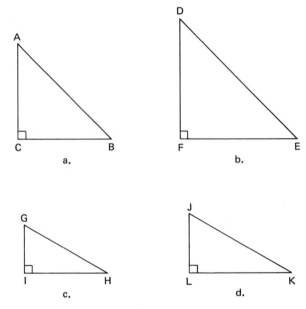

Fig. 8-46.

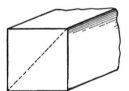

Fig. 8-47. Square steel bar.

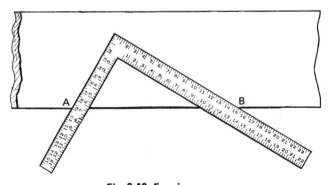

Fig. 8-48 Framing square.

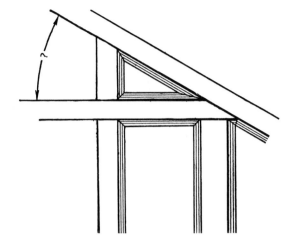

Fig. 8-49.

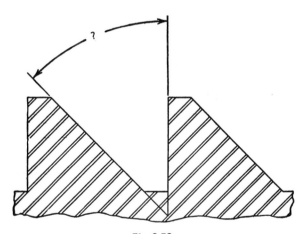

Fig. 8-50.

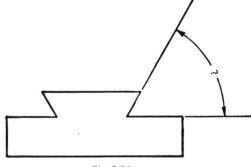

Fig. 8-51.

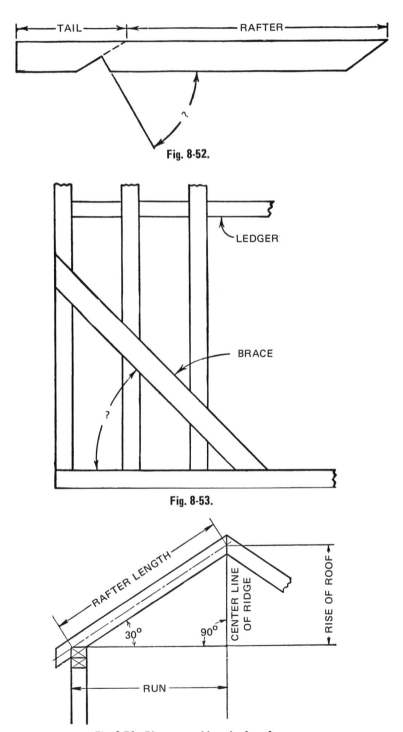

Fig. 8-52.

Fig. 8-53.

Fig. 8-54. Rise, run and length of a rafter.

Constructing a line perpendicular to a given line through a given point

There are several methods of procedure for making this construction. We shall illustrate four situations in each of which one line is drawn perpendicular to another.

In what follows we shall use "line" and "segment" interchangeably to mean a segment or an extended segment.

Method A. Use the *T*-square and triangle to construct a line perpendicular (⊥) to a given line through a point *not* on the line.

THINK	DO THIS
(We assume the paper containing the line and point is properly mounted on a drawing board.)	

1. Place the T-square in proper drawing position slightly below line *AB*.

2. Place a triangle (30°-60° or 45° triangle) so that one side of the right angle rests squarely against the T-square and the other side of the right angle is near the point *P*.

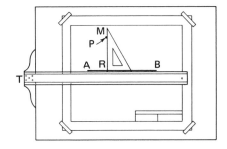

Fig. 8-55.

3. Place point of pencil at point *P*. Slide the triangle along the T-square until the triangle touches the pencil point, keeping the triangle squarely against the T-square.

4. Draw the line *RM* along the side of the triangle.

5. Since the triangle has a right angle at *R*, line *MR* will be at right angles to *AB*. Hence *MR* is perpendicular to *AB* and passes through point *P*.

Ans. MR is the required perpendicular.

Method B. Use a straightedge and triangle to construct a line through a given point and perpendicular to any line.

THINK	DO THIS
1. *P* is the given point.	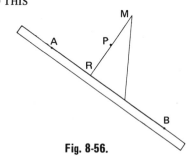
2. Draw a line such as *AB* in no particular position.	
3. Place a straightedge (the edge of a T-square may be used) along *AB*. Keep the straightedge firmly fixed in this position.	Fig. 8-56.
4. Follow step 2 of Method A.	*Ans.* *MR* is the required
5. Follow steps 3 and 4 of Method A.	perpendicular.

Method C. Use a straightedge and compass to construct a line through a point and perpendicular to a given line:

THINK	DO THIS
(C-1) If the point is *not* on the line: (If the point is *on* the line use Method C-2.)	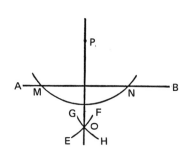
1. Place the point of the compass on *P*.	
2. Keep the compass point at *P* and open the compass to a radius that extends beyond line *AB*, as *PM*.	
3. Keeping point of compass at *P*, strike an arc that intersects *AB* in two points *M* and *N*.	Fig. 8-57.
4. Keeping the same radius, place the point of the compass first at *M*. Strike the arc *FE*; then at *N* strike the arc *GH* to intersect *FE* at *O*.	
5. With the straightedge draw the segment *PO*. Then *PO* is the required line.	*Ans.* *PO* is the line through *P* and perpendicular to *AB*.

(C-2) If the point *P* is on line *AB*:

1. Place the point of the compass on *P*. Open the compass to a radius of about 1 inch.

2. With point of compass on *P* and with this radius strike two arcs, one to intersect *AB* at *N*, the other to intersect *AB* at *M*. (*Note:* Extend line *AB* with straightedge if necessary.)

3. With point of compass at *M* and radius longer than *MP*, strike arc *CD*.

4. With point of compass at *N* and with the same radius, strike arc *EF* to intersect *CD* at *O*.

5. With straightedge draw *OP*.

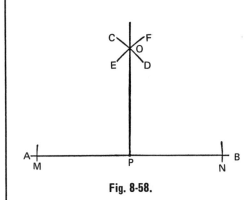

Fig. 8-58.

Ans. OP is the required line through *P* and perpendicular to *AB*.

EXERCISES

1. Draw a line segment *AB* 4″ long. Construct a perpendicular to this line through a point on *AB* that is 1″ from *A*.

2. Draw an equilateral triangle that is 4″ on a side and construct the altitude from one vertex to the opposite side.

3. A perpendicular is to be drawn from the point C on the circle (Fig. 8-59) to the diameter *AB*. Make this circle on paper and show how the perpendicular may be constructed. Use two methods.

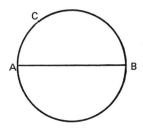

Fig. 8-59.

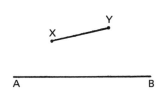

Fig. 8-60.

4. To project the line *XY* on the line *AB* (Fig. 8-60), perpendiculars are drawn from the ends of *XY* to *AB*. Show how this is done and find the projection of *XY* on *AB*.

Parallel lines

We stated (p. 120) that *parallel lines* are lines in the same plane which will not meet even though extended indefinitely.

Draw a line parallel to a given line through a given point.

Method A.   Using T-square and triangle. (This method can be used only when the given line, *CD*, is drawn at an angle of 30°, 60° or 45° to the T-square, when T-square and triangle are in proper position.)

THINK

1. Place T-square and triangle in same position as given line, but some distance from *P*.

2. Place pencil point on given point, *P*.

3. With T-square firm, slide triangle along T-square until *RT* of the triangle touches the pencil point.

4. Draw the line through *P* along the edge of the triangle.

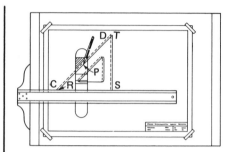

Fig. 8-61. Using T-square and triangle to draw parallel lines.

Ans. The line drawn along *RT* is parallel to *CD* and passes through point *P*.

Method B.   Using straightedge and triangle, draw a line through a given point, *P*, parallel to any given line, *AB*.

THINK

1. Place the triangle *MNO* with one of its edges *MN* along the given line *AB*.

2. Place the straightedge *ST* (it could be the edge of the T-square) against the edge *MO* of the triangle.

3. Holding the straightedge firm, slide the triangle along the straightedge away from the line *AB* and toward point *P*. Be sure the edge *MO* of the triangle is kept firm against the straightedge. Slide the triangle until the edge *MN* of the triangle is on point *P*.

4. Draw a line along the edge *MN* and through *P*.

DO THIS

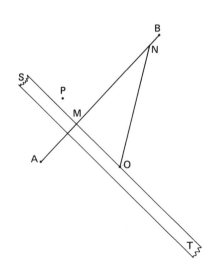

Fig. 8-62.

Ans. The line drawn along *MN* is parallel to *AB* and passes through point *P*.

Method C. Use the compass and straightedge to draw a line through a given point, *P*, and parallel to a given line, *AB*.

THINK

1. Construct a line through *P* that is perpendicular to *AB* (see p. 347). This is the line *MN*.

2. Construct a line through *P* perpendicular to *MN*. This is the line *RPS*.

3. Lines *AB* and *RS* are both perpendicular to line *MN*. When this is so, the two lines *AB* and *RS* are parallel to each other.

DO THIS

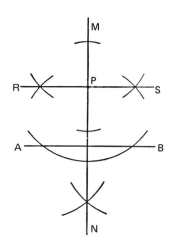

Fig. 8-63.

Ans. RS is the line through the point *P* parallel to line *AB*.

EXERCISES

1. Draw a line *TV*. Place a point *M* not on *TV*. Construct a line through *M* parallel to *TV*.
2. Draw an equilateral triangle 3″ on each side. Construct a line through one vertex parallel to the opposite base.
3. Given a triangle *ABC* with point *D* the midpoint of *AB*, through *D* draw a line parallel to side *AC*.

Dividing a given line into a certain number of equal parts

To divide a given segment, *AB*, into a specified number of congruent parts, for instance 9:

THINK

1. Through *A* draw a line *AC* that makes a convenient acute angle with the given segment *AB*.

DO THIS

2. Place the point of the compass at *A* and with a convenient radius strike an arc that intersects *AC* at *E*, making segment *AE*. Then using the compass point at *E* and with the same radius as before, strike an arc intersecting *AC* at *F*, making segment *EF*.

3. Repeat this procedure until nine such segments are made, ending at point *M*.

4. With a straightedge draw the line through *M* and *B*.

5. Then at each of the points *L*, *K*, *J*, etc., construct angles *ALY*, *AKX*, etc., all congruent to ∠*AMB* (p. 327).

6. Then the points *R*, *S*, *T*, *U*, V, *W*, *X*, and *Y* divide the segment *AB* into 9 equal parts.

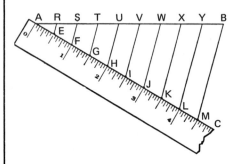

Fig. 8-64. Dividing line segment *AB* into nine equal parts.

Ans. AR is congruent to *RS*, *RS* is congruent to *ST*, etc.

EXERCISES

1. Draw a line 4 3/16″ long and divide it into three equal parts. Test the result with compass, or dividers.

2. Draw an equilateral triangle with sides 2 3/16″ long and trisect the base (divide into three equal parts). Test the result with compass, or dividers.

The hexagon

A *hexagon* is a closed plane figure of six sides. A regular hexagon has six equal sides and six equal angles (Fig. 8-65). Each angle of a regular hexagon contains 120 degrees (120°).

Fig. 8-65. A regular hexagon.

Example. Construct a regular hexagon with a given side, say 2″ long.

THINK	DO THIS
1. Construct a segment 2″ long, *AB*.	
2. With *AB* as a radius and with compass point first at *A*, then at *B*, construct arcs *MN* and *PQ* intersecting at *O*.	
3. With *O* as center and radius equal to *AB*, draw the circle.	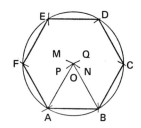
4. With this same radius and compass point at *B*, strike an arc that intersects the circle at point *C*.	
5. With *C* as center and same radius, strike an arc that intersects the circle at point *D*, and repeat this process for obtaining points *E* and *F*.	**Fig. 8-66. Construction of a regular hexagon.**
6. Draw the segments *BC*, *CD*, *DE*, *EF* and *FA*.	*Ans.* The figure *ABCDEF* is a regular hexagon.

Some interesting facts about regular hexagons

Notice that the hexagon is made up of three pairs of parallel lines (Fig. 8-67). *AD* is parallel to *CB*, *DF* to *CE*, and *FB* to *EA*. In connection with hexagon- and square-headed bolts, these parallel sides are called "flats."

The diagonals *AB*, *CD*, and *EF* all meet at *O*. This point is the center of the circle in which the hexagon is inscribed. An *inscribed hexagon* is one with all its vertices on the circle.

Each side of the hexagon is the same length as the radius of the circle (Fig. 8-67). Hence the six triangles *BOC*, *COE*, *EOA*, etc., are all equilateral triangles and are all congruent.

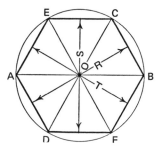

Fig. 8-67.

Fig. 8-68.

The distances "across the flats" R, S, and T (Fig. 8-67) are all equal, and one-half these distances equals the lengths of the altitudes of the equilateral triangles; hence these altitudes are likewise all equal.

A circle whose center is O and whose radius is equal to OM will pass through all the points M, R, T, N, S, and V (Fig. 8-68, page 353). The diameter of this inscribed circle is the distance "across the flats."

PROBLEMS

3. Draw a full-size end view of a hexagonal head bolt (Fig. 8-69) when the distance across its corners is 1 1/2".

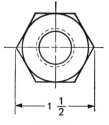

Fig. 8-69.

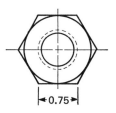

Fig. 8-70.

4. If the side of the hexagonal head is 0.75" (Fig. 8-70), what is the distance across the flats?
   Hint: Remember that half the distance across the flats is the altitude of the equilateral triangle shown.
5. Each side of a hexagon is 1" (Fig. 8-71). Reproduce this hexagon and draw the inscribed circle.
   Hint: Draw a perpendicular from the center to one side to determine the radius of the inscribed circle.

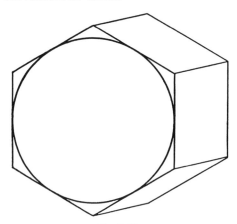

Fig. 8-71.

6. Construct a regular hexagon each side of which is 0.875″ (7/8″).
7. Compute the area of a regular hexagon that is 14″ on each side. *Hint:* Find the area of one of the equilateral triangles and multiply by 6.
8. Construct a regular hexagon inscribed in a circle whose radius is 1 1/2″. Join the alternate vertices of the hexagon. The figure formed is the inscribed equilateral triangle. This is the method used to inscribe an equilateral triangle in a circle. Find the area of this triangle.
9. (a) Inscribe an equilateral triangle in a circle of radius 2 1/2″.
   (b) Construct the inscribed circle of your equilateral triangle.

Constructing a tangent to a circle

A line is *tangent* to a circle when the line has only one point on the circle and can have no more even though extended. The line *AB* is tangent to circle *O* (Fig. 8-72). It has one point on the circle, the point *T*. This is the point of tangency or point of contact.

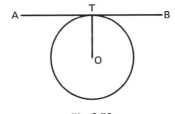

Fig. 8-72.

The sides of the hexagon are all tangent to the inscribed circle (p. 354).

A radius that is drawn to the point of contact is perpendicular to the tangent line. *OT* and *AB* are perpendicular to each other.

Example 1. To construct a tangent to a circle, *O*, at a point, *B*, on the circle.

THINK

1. In the circle *O* draw the radius *OB*.

2. Construct a line that is perpendicular to *OB* at point *B* (see p. 349). To do this extend *OB*.

   With *B* as center and with *BO* as radius, strike an arc at *C*. This makes *BC* congruent to *OB*.

DO THIS

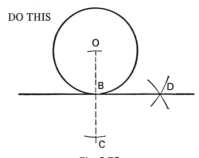

Fig. 8-73.

With O and C as centers and radius greater than OB, strike arcs that intersect at D. DB is the line perpendicular to OB at B.

3. Any line that is perpendicular to the radius at a point on the circle is tangent to the circle.

Ans. DB is tangent to the circle at point B.

Example 2. To construct a tangent to a circle, O, from a point, B, not on the circle. Point O is the center of the circle.

THINK

1. Draw the segment OB and bisect OB at C.

2. With C as center and with a radius equal to BC, strike the arc $BROT$ which intersects the circle at R and T.

3. Draw lines BR and BT.

DO THIS

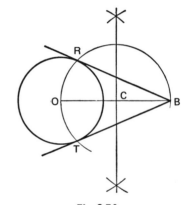

Fig. 8-74.

Ans. Both line BR and line BT are tangents to circle O through B.

EXERCISES

1. On your paper reproduce the circle O with point C (Fig. 8-75), then construct a tangent line at C to the circle O.

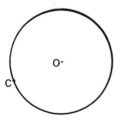

Fig. 8-75.

2. Construct a circle whose radius is 1 1/2″. Construct two diameters that are perpendicular to each other. Wherever each diameter meets the circle, construct a tangent line. Extend these tangent lines until they meet. What kind of figure is formed by them? Why?

3. Inscribe a square in a circle. (Use the perpendicular diameters mentioned in Exercise 2.)

Finding the center of a circle when only an arc of the circle is available

Example. Find the center of the circle containing arc *ABC* in Fig. 8-76.

THINK	DO THIS
1. Select any three points on the arc, such as *A*, *B* and *C*.	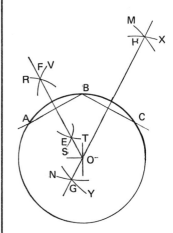
2. Draw the segments *AB* and *BC*.	
3. With compass point, first at *A*, then at *B*, and with a convenient radius, strike arcs *RS* and *TV*. These arcs intersect at *F* and *E*.	
4. Draw the line through *F* and *E*, *FE*.	
5. With *B* and *C* as centers and a convenient radius, strike arcs *MN* and *XY*, which intersect at *H* and *G*. Draw the line through *H* and *G*, *HG*.	Fig. 8-76.
6. Extend *FE* and *HG* until they meet at *O*.	*Ans.* Point *O* is the required center of the circle.

This same method enables one to draw the circle that passes through any 3 points, so long as the points do not all lie in one straight line.

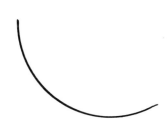

Fig. 8-77.

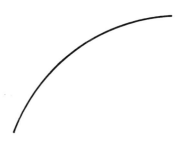

Fig. 8-78.

EXERCISES

1. The curve shown in Fig. 8-77, page 357, is the arc of a circle. Trace this on a piece of paper and find the center of the circle.
2. Trace the arc of the circle shown in Fig. 8-78, page 357, and find its center.
3. Figure 8-79 shows a piece of a flywheel. Find the diameter of this flywheel to the nearest inch. Trace the drawing before doing the construction.

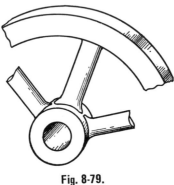

Fig. 8-79.

The ellipse

The *ellipse* is a closed, oval-shaped curve that is symmetrical to two lines or axes that are perpendicular to each other. The curve shown in Fig. 8-80 is an ellipse and is symmetrical to the axes *AB* and *CD*.

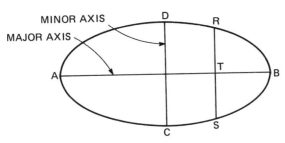

Fig. 8-80. Major and minor axes of an ellipse.

For a curve to be symmetrical to an axis means that if a perpendicular is dropped from a point on the curve, such as *R*, to the axis at *T* and extended the same length *RT* beyond the axis, such as *TS*, the other end of this extended line, *S*, is also on the curve.

The longer axis, *AB*, is the *major axis*, and the shorter one, *DC*, is the *minor axis* of the ellipse. *Note:* If the two axes were of the same length, we would have a circle; the area of the circle would be represented by the formula $A = \pi a^2$, where *a* is the radius.

If half the length of the major axis is represented by a, and half the length of the minor axis is represented by b, the area, A_E, of an ellipse can be obtained by using the formula

$$A_E = \pi ab$$

EXERCISES

Find the area of each of the following ellipses:

1. Major axis 12″, minor axis 8″
2. Major axis 4 1/2″, minor axis 2 1/2″
3. Minor axis 7.5″, major axis 12.25″
4. Minor axis 3 1/6″, major axis 5 1/6″
5. Minor axis 2 1/3″, major axis 4 1/9″

Constructing an ellipse when the major and minor axes are known

Method A. When the major axis is not more than 3/2 of the minor axis we proceed as follows:

THINK

1. Subtract the minor axis, *CD*, from the major axis, *AB*. *AB – DC = OF*, also *OE*.

2. Lay off *OF* on *OC* and *OE* on *OD*.

3. Find a segment *OG* equal to 3/4 of *OE*.

4. Lay off *OG* on *OA*, and *OH*, equal to *OG*, on *OB*.

5. From *E* draw lines of indefinite length through *G* and *H*. Do the same from *F*.

6. Using *H* as a center and *HB* as a radius, strike an arc *IBJ*.

7. Use the same radius, with *G* as center, and strike arc *LAK*.

8. With *F* as a center and *FL* as radius, strike arc *LDI*. And with *E* as center and the same radius, strike arc from *J* to *K*.

DO THIS

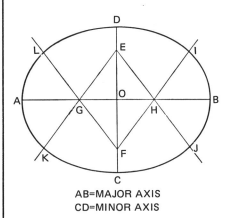

AB=MAJOR AXIS
CD=MINOR AXIS

Fig. 8-81. Ellipse drawn by approximate method.

Ans. The curve *ALDIBJCK* is approximately an ellipse.

Method B. A second approximate method for constructing an ellipse when the major and minor axes are known:

THINK	DO THIS
1. Construct two circles with the same center, one circle having diameter equal to the major axis, *AB*, the other circle with diameter equal to the minor axis, *DC*.	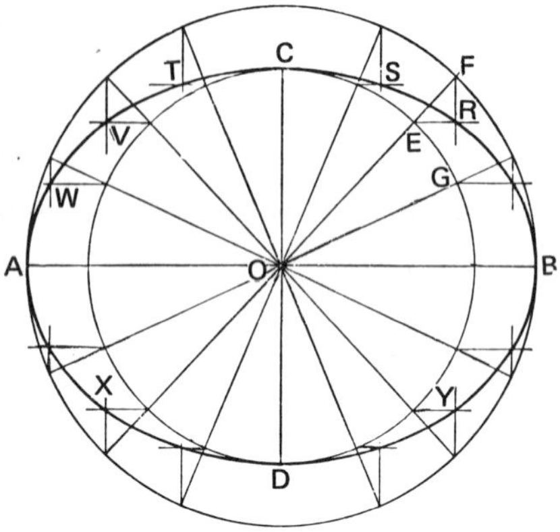 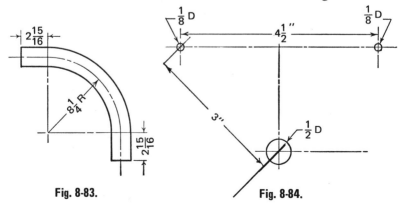

1. Construct two circles with the same center, one circle having diameter equal to the major axis, *AB*, the other circle with diameter equal to the minor axis, *DC*.

2. Draw radial lines through both circles, such as *OEF*, etc.

3. From *F* construct a line, *FR*, parallel to the minor axis, and from *E* one parallel to the major axis. Extend both of these lines until they intersect at *R*. *R* is one point on the ellipse.

4. Repeat Step 3 at each radial intersection obtaining points *R, S, T, V, W, X, Y*, etc.

5. Draw a smooth curve through these points and points *ABCD*.

AB=MAJOR AXIS
CD=MINOR AXIS

Fig. 8-82. Construction of an ellipse.

Ans. The resulting curve is approximately an ellipse.

REVIEW PROBLEMS

1. An elliptical tank for an oil truck is 12′ long. The major axis of the elliptical section is 4.75′. The minor axis is 3 1/4′.
   (a) Make a drawing of the ellipse by two methods. Use a scale of 1″ to represent 1′.
   (b) If the measurements given represent inside dimensions, find the capacity of the tank in gallons.
2. Find the length of the 90° elbow measured along the centerline.

Fig. 8-83.

Fig. 8-84.

3. Two 1/8″ holes are to be located 3″ from an existing 1/2″ hole and 4 1/2″ from each other (Fig. 8-84). These are all center-to-center distances. Make a drawing showing how you would lay out these holes.

4. A certain grade of iron used in the manufacture of pipe weighs 0.285 lb. per cubic inch. What is the weight per linear foot of pipe if the inside diameter is actually 0.622″ and the external diameter 0.840″?

5. Find the radius of the fillet in Fig. 8-85.

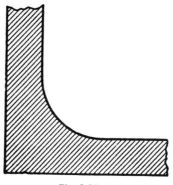

Fig. 8-85.

6. Construct a regular hexagon 1 3/8″ on a side. Measure the distance across the flats and compare the value with that found by multiplying the length of one side by the $\sqrt{3}$.

7. Bisect a line 3 7/8″ long using your compass.

8. The bolt circle of a flange is 7″ in diameter. Lay out the centers of 8 equally spaced 1/2″ holes on this circle.

9. Construct a triangle with sides of 1 3/8″, 1 3/4″, and 2 1/8″.

10. Divide a line 4 5/8″ long into seven equal parts.

11. The ellipse in Fig. 8-86 represents a cross section of an elliptical tank. If the tank is 15′ long, find its capacity in gallons.

12. Using a scale of 1 1/2″ to the foot, construct an ellipse similar to Fig. 8-87.

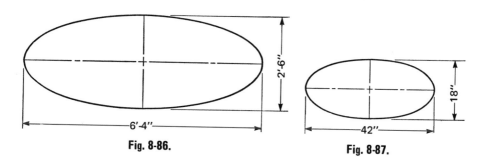

Fig. 8-86.   **Fig. 8-87.**

9 Screws, Bolts and Nuts

The very important industrial materials such as screws, bolts and nuts, and the facts associated with them, will be studied in this unit.

Before you begin this material, review the pitch and lead of screws on page 60.

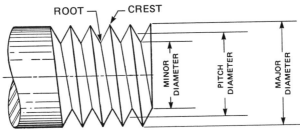

Fig. 9-1. Major, minor, and pitch diameter of a sharp V thread.

Thread terminology

Figure 9-1 illustrates a sharp V thread. The top, or outside, edge of the thread is the *crest*.

The *major diameter* is the diameter measured from the crest to the opposite crest. This is usually the same as the diameter of the material on which the threads are formed.

The bottom of the thread is called the *root* of the thread.

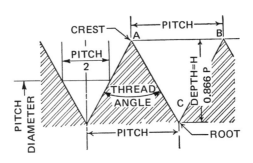

Fig. 9-2. The sharp V thread.

The *depth* of the thread is the perpendicular distance from *AB* (the crest) to *C* (the root) (Fig. 9-2). The *double depth* is twice this depth.

The *clearance* of the thread (square thread in particular) is an extra depth provided at the root (Fig. 9-6).

The *minor diameter* is the diameter of the material left uncut by the formation of the thread. It is equal to the major diameter minus the double depth (Fig. 9-1).

The *pitch diameter* is an imaginary diameter which would pass through the thread profile at points which would make the width of the groove equal to one-half the pitch (Fig. 9-2).

Types of threads

The *60° V thread* is the basic shape of several of the more common types of threads. If a line such as *AB* is drawn in the profile of this type thread (Fig. 9-2), an equilateral triangle *ABC* is formed. *AB* is the pitch of the screw, and the depth of the thread is the altitude of this equilateral triangle of which the pitch *AB* is one side.

The *American National Form thread* (Fig. 9-3) is a variation of the sharp V thread, and is to be preferred, as the sharp edges of the V are most easily broken off, and the bottoms are hard to keep clean. The sides of the A.N.F. thread make an angle of 60° with each other.

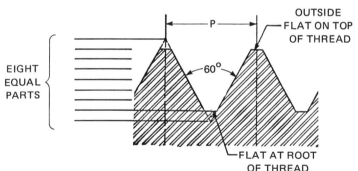

Fig. 9-3. American National form of thread.

If the depth of a sharp V thread is divided into eight equal parts and the top eighth is left off and the bottom eighth filled in, the effect will be an American National Form thread. The depth of the American National Form is therefore 3/4 of that of the sharp V thread. It should be noted that the crest of the thread is flat and the major diameter is the distance between opposite outside flats.

The *Metric Standard thread* is similar to the American National Form thread, but its diameter and pitch are given in terms of millimeters.

The *British Standard Whitworth thread* is another variation of the sharp V thread. Here 1/6 of the depth of the V thread is taken from the

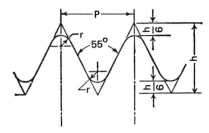

Fig. 9-4. British Standard Whitworth thread.

top and bottom. The sides of the thread if extended would make an angle of 55 degrees with each other. The top and bottom are rounded. Figure 9-4 illustrates such a thread.

The *Square thread* is used primarily for transmitting power and is illustrated in Fig. 9-5, which shows that the cross section is nearly a square, although an allowance for clearance of 0.005″ makes the depth slightly more than half the pitch (Fig. 9-6).

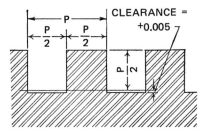

Fig. 9-5. Square thread.   Fig. 9-6. Detail of square thread.

The *Acme 29° thread* is also used primarily for transmitting power and is a variation of the square thread (Fig. 9-7). The sides of the Acme thread make an angle of 29° with each other, and each of these sides makes an angle of 14 1/2° with the vertical. The width of the flat top is 0.3707 times the pitch (Fig. 9-8).

The *Buttress thread* is especially adapted for transmitting power in one direction only.

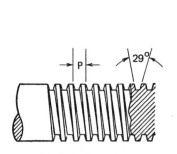

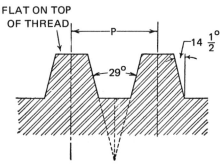

Fig. 9-7. Acme thread   Fig. 9-8. Detail of Acme thread.

The *Unified American Standard thread* is a modification of the American National Form thread, varying slightly in the root dimension, and is replacing the older forms of threads. This series was designed to permit interchangeability of parts with those of other nations.

The American National Form thread is provided in six series of sizes for heavy-duty, lightweight fasteners. These are known as National Coarse (N.C.), National Fine (N.F.), 8-Pitch, 12-Pitch, 16-Pitch, and Extra Fine (N.E.F.) Series. Some examples are: 1/4"-20 N.C., 1/4"-28 N.F., 1/2"-13 N.C., 1/2"-20 N.F., where the first part of the name designates the thickness or diametral distance, the second part indicates the number of threads per inch, and the third part tells the series. Tables 11 and 11A in the Appendix give the dimensions of many Unified and American Standard series threads.

The 8-Pitch series is used on diameters from 1" to 6", the 12-Pitch series on diameters from 9/16" to 6", and the 16-Pitch series on diameters from 3/8" to 6".

Screw threads are also classified as to the closeness of fit, one thread to another, and are listed as class 1 (loose), class 2 (mostly used), and class 3 (very percise).

Example. Find the minor diameter and the depth of a 5/8"-11 N.C. thread.

THINK

1. Use Table 11 in the Appendix to find the minor diameter.
2. Trace down the "Nominal Size" column to the given size, 5/8, then find the number of threads, 11, in the "NC" column.
3. Trace across this row to the column headed "Root Diameter." The minor diameter is 0.5069.
4. To get depth we must get the major diameter. In row 5/8-11 trace over to column headed "Major Diameter." Write the number found here.
5. Since the double depth is the difference between minor and major diameters, subtract the number found.

DO THIS

1, 2, 3. Minor Diameter = 0.5069

4. Major Diameter = 0.6250

5. 0.6250 − 0.5069 = 0.1181

6. The depth is 1/2 of that difference.

6. $\dfrac{0.1181}{2} = 0.05905$

Ans. Depth = 0.05905

Summary

To find the major and the minor or root diameters of an American National Form Thread use Table 11, in the Appendix.

1. Trace down the column headed "Nominal Size" to the dimension of the thread in question.

2. In this line trace over to column headed "Major Diameter." Read the major dimension of the given thread.

3. In the same line trace to column headed "Root Diameter." The number here is the required minor diameter.

4. The depth of the thread is half the difference of these two dimensions.

EXAMPLE

Find the major diameter and thread depth of a 5/16″-18 N.C. thread.

1, 2. Major diameter = 0.3125

3. Minor diameter = 0.2403

4. $\dfrac{(0.3125 - 0.2403)}{2} = 0.0361$

Ans. Depth = 0.0361

EXERCISES

Find the minor diameter and depth of thread of each of the following American National Form threads:

1. 1/4″-20 N.C.
2. 3/8″-24 N.F.
3. 1/2″-13 N.C.
4. 1″-8 N.C.
5. 7/8″-14 N.F.

6. 3/4″-10 N.C.
7. 3/4″-16 N.F.
8. 1/2″-20 N.F.
9. 2 1/2″-4 N.C.
10. 9/16″-18 N.F.

11. 7/16″-14 N.C.
12. 5/16″-18 N.C.
13. 8-32 N.C.
14. 7/16″-20 N.F.
15. 1 3/4″-5 N.C.

The tap drill

The inside of a nut, or the inside of the hole in a piece of metal into which a threaded screw or bolt is to fit, must itself have a thread. To prepare such a thread, a hole is first drilled into the metal with a tap drill. Then a tap or set of taps is used to cut the thread (Fig. 9-9).

Fig. 9-9. Tap.

Determining the proper size for the hole to accommodate a given thread

A generally accepted method of determining the size of the tap drill for a given thread is to deduct 75 percent of the double depth from the major diameter of the thread. The drill whose decimal size is nearest to this result is the size of commercial tap drill necessary to drill the hole.

However, the quickest and most satisfactory way to determine the proper size drill to use for the tap drill is to consult a tap drill chart. Table 11 in the Appendix includes the information for American National Form threads.

Holes for square threads cannot be tapped in one operation, and require a set of four taps, each tap removing only a portion of the metal from the hole.

Example. What size tap drill is needed to drill a hole of proper size for an American Standard 3/4″-10 N.C. thread?

THINK	DO THIS
1. Since Table 11 gives the sizes of tap drills needed to drill holes for American National Form threads, trace down column headed "Nominal Size" to the line marked 3/4-10.	1, 2. 21/32
2. Trace across this line to the column headed "Commercial Tap Drill." The number here is the size of the tap drill needed.	*Ans.* Tap drill for a 3/4″-10 N.C. thread is 21/32″

Summary	EXAMPLE
Table 11 gives the size of tap drills needed for various types and sizes of threads. To find the tap drill size for an American National Form thread:	What size tap drill is needed to drill the hole to accommodate a 1 1/8″-7 N.C. thread?
1. Trace down Table 11 column headed "Nominal Size" to the dimension and type of the thread.	1. Table 11 column one to 1 1/8″-7.

2. Trace across this line to column headed "Commercial Tap Drill." The number found here is the required size tap drill.	2. 1 1/8"-7 to tap drill gives 63/64. *Ans.* 63/64" tap drill needed for 1 1/8"-7 N.C. thread

EXERCISES

Find the commercial tap-drill size to allow an engagement of 3/4 of the full depth of each of the following screw threads:

1. 2 1/4"-8 N.F.
2. 1/2"-13 N.C.
3. 1/4"-20 N.C.
4. 9/16"-18 N.F.
5. 1/4"-28 N.F.
6. 1"-8 N.C.

7. 3/4"-16 N.F.
8. 1 1/2"-6 N.C.
9. 2 1/2"-4 N.C.
10. 5/8"-18 N.F.
11. 7/8"-9 N.C.
12. 1 3/4"-10 N.F.

13. 3/4"-10 N.C.
14. 5/16"-18 N.C.
15. 1 1/4"-12 N.F.
16. 3"-8 N.F.
17. 2 3/4"-4 N.C.

American standard bolt heads and nuts (Heavy Series)

Figure 9-10 shows a hexagonal head and a square head bolt with the corresponding nut.

The dimensions of the heads of the bolts, and of the nuts to go with the bolts, are expressed in terms of the diameter, D, of the bolt.

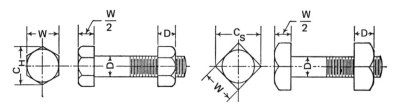

Fig. 9-10. American standard bolt and nut heads; (a) hexagon head; (b) square head.

For both kinds of bolts the distance across the parallel sides or "flats," W [Fig. 9-10(b)], is 1 1/2 times the diameter of the bolt plus 1/8 inch. Or, as a formula,

$$W = 1.5D + 0.125$$

For the hexagon head bolt, the distance across corners, C_H [Fig. 9-10(a)], is 1.15 times the distance across the flats, W. The formula is

$$C_H = 1.15W$$

For the square head, the distance across corners, C_S [Fig. 9-10(b)], is 1.414 times the distance across the flats, and the formula is

$$C_S = 1.414W$$

The thickness of the head in both cases is 1/2 the distance across the flats.

The thickness of the nut in both cases is the same as the diameter of the bolt [Fig. 9-10(a) and (b).]

These various dimensions can best be found by reference to a table giving this information. Manufacturers often supply such tables.

Example. Find the dimensions of the head and nut of a hexagonal bolt that has a shank that is 5/8″ in diameter.

THINK	DO THIS
1. For a hexagonal-head bolt the distance across flats is given by the formula $W = 1.5D + 0.125$. Write this formula.	1. $W = 1.5D + 0.125$
2. Replace the variable by the value given in the problem.	2. $D = 5/8$ $W = 1.5 \times 5/8 + 0.125$
3. Perform the computations.	3. $W = 1.0625$
4. Distance across corners of hexagonal bolt $C_H = 1.15W$.	4. $C_H = 1.15W$ $C_H = 1.15 \times 1.0625$ $C_H = 1.221875$ or 1.222

Ans. Distance across flats = 1.0625″
Distance across corners = 1.222″

Summary	EXAMPLE
To find the sizes of heads and nuts to go with hexagonal-head and square-head bolts:	Find the dimensions of the head and nut of a hexagonal-head bolt with shank 3/4″ in diameter.
1. Write the formula for the required dimension. (a) Across flats of both square- and hexagonal-head bolts, $W = 1.5D + 0.125$ (b) Across corners of hexagonal-head bolts, $C_H = 1.15W$. (c) Across corners of square-head bolts, $C_S = 1.414W$.	1, 2, 3. $W = 1.5D + 0.125$ $W = 1.5 \times 3/4 + 0.125$ $= 1.250$ $C_H = 1.15W$ $C_H = 1.15 \times 1.250 = 1.4375$ Thickness of head = $1/2W$ $= 1/2 \times 1.250$ Thickness of head = 0.625 Thickness of nut = D = $3/4″ = 0.75″$

(d) In both cases thickness of head = 1/2W. Thickness of nut = D, the diameter of the bolt.

2. Replace the variables by their known or computed values.

3. Perform the computations.

Ans. Distance across flats = 1.250″
Distance across corners = 1.4375″
Thickness of head = 0.625″
Thickness of nut = 0.75″

EXERCISES

Find all dimensions of heads and nuts for bolts as indicated:

1. Hexagonal, 3/4″ shank.
2. Square, 1″ shank.
3. Square, 5/8″ shank.
4. Hexagonal, 1 1/8″ shank.

5. Hexagonal, 7/8″ shank.
6. Square, 3/4″ shank.
7. Square, 1 1/4″ shank.
8. Hexagonal, 5/8″ shank.

REVIEW PROBLEMS

1. Find the double depth of an 8-32 N.C. thread.
2. What is the pitch of a 1 1/8″-12 N.F. thread?
3. The pitch of a sharp V thread is 1/24″. What is the single depth of this thread?
4. Find the minor diameter of a 5/16″-18 N.C. thread.
5. Find the proper size tap drill for a 5/8″-18 N.F. thread.
6. A 7/8″-9 N.C. thread has a major diameter of 0.8750″ and a minor diameter of 0.7307″. What is the depth of the thread?
7. The tap drill for a 3″-8 N.F. threaded bolt is 2.875″. What will be the actual depth of the thread produced?
8. A 2 1/2″ pipe thread has eight threads per inch. What is the pitch of the thread?
9. Find the pitch and the minor diameter of the special screw in Fig. 9-11.
10. What size tap drill should be used to drill the holes in Fig. 9-12?

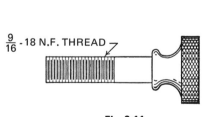

Fig. 9-11.

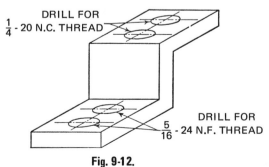

Fig. 9-12.

11. What is the area at the minor diameter of the stud in Fig. 9-13?
12. What size hexagon stock is required to make 1 1/2″ hexagonal-head bolts, heavy series?
13. A shop order calls for one gross of 7/8″ square-head bolts 10 3/4″ long.
    (a) What size square stock is required for this job?
    (b) If 1/8″ is allowed for cutting off and facing each bolt, how many inches of steel will be used for each dozen bolts?
14. Find the minor diameter, and the dimension of the top flat, of a 1 3/4″-5 N.C. thread.
15. How deep should the hole shown in Fig. 9-14 be bored in order that the square head of the bolt come just even with the surface?

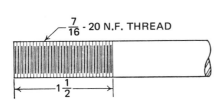

Fig. 9-13. Stud.

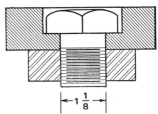

Fig. 9-14.

10 Work, Power, Energy and Stresses

Work

To be an intelligent industrial worker one must understand those ideas which have to do with force and power and energy. In this unit you will become acquainted with these ideas, and will also learn to make some important calculations concerning them.

Every time anything is moved, something or somebody does *work*. When a load of brick is raised by an elevator to a section of a building that is under construction, work is done. When an electrician pulls a cable through a conduit, work is done. When a casting is lifted from the floor to the bed of a boring mill, work is done. If the casting weighs 10 pounds and is lifted 3 feet, we say that 30 or 10 x 3 foot-pounds of work have been done. The unit generally used to measure the work done is the foot-pound. This is the amount of work equivalent to that done in raising one pound avoirdupois against the force of gravity a height of one foot. We speak of it briefly by saying the *foot-pound* is the amount of work done when a force of one pound has been exerted through a distance of one foot.

Example 1. How much work is done when a 100-lb. bag of cement is raised from the ground to an 8-ft. high platform in front of a concrete mixer?

THINK	DO THIS
1. The work done is the number of pounds raised (100) multiplied by the distance (8 ft.) through which the weight is moved.	1. 100 x 8 = 800
	Ans. 800 ft.-lb. of work

Example 2. How much work is done when a steel girder weighing 2.5 tons is raised to the top of a 150 ft. building?

THINK	DO THIS
1. Express the weight in pounds. Each ton weighs 2000 lb.	1. 2.5 tons = 2.5 x 2,000 = 5,000 lb.

2. The work done expressed in foot-pounds is the product of the weight (5,000) in pounds and the distance (150) in feet.

2. 5,000 x 150 = 750,000

Ans. Work done = 750,000 ft.-lb.

Summary

To find the amount of work done, expressed in foot-pounds:

1. Express the force exerted in pounds and the distance in feet.

2. Find the product of the force, in pounds, and the distance, in feet, through which the force moves. This is the amount of work done.

If *f* represents the force in pounds, *s* the distance in feet through which the force is exerted, and *w* the number of foot pounds exerted, the amount of work done, expressed as a formula, is:

$$w = f \times s$$

EXAMPLE

How much work is required to lift a 3 ton missile 10 ft. off its cradle?

1. 3 tons = 3 x 2,000
   = 6,000 lb.

2. 6,000 x 10 = 60,000 ft.-lb.

Ans. 60,000 ft.-lb. of work

f = 3 tons = 3 x 2,000
  = 6,000 lb.
s = 10 ft.
w = 6,000 x 10 = 60,000

Ans. w = 60,000 ft.-lb.

EXERCISES

Find the work in foot-pounds done in each of the following:

1. 40 lb. through 20 ft.
2. 500 lb. through 52 ft.
3. A ton through 52 ft.
4. A 20-lb. casting through 30".
5. 500 gallons of water through 74 ft. (A gallon of water weighs about 8 1/3 lb.)
6. A 56.5-lb. casting through 28".
7. A coil of wire, 74.3 lb., through 22 ft.
8. A 16-lb. sledgehammer through 7'.
9. When a 152-lb. man climbs a stairway 8'6" high.

PROBLEMS

1. Find the work done in raising a 250-lb. casting 2 1/2 ft. from the floor to the bed of a horizontal boring mill.
2. How much work is expended by an electric crane in raising one ton of steel to a height of 18'?
3. How much work is done in unloading 25 bars of steel from a truck to the ground, a distance of 4 1/4 ft.? Each bar of steel is 1 1/2" sq. and 12 ft. long, and weighs 7.650 lb. per lineal foot.

4. In punching a hole through a 1 1/4" steel plate, the punch press exerts a total pressure of 40 tons. How much work is done in punching each hole?

5. In turning a gear blank in a lathe, a piece of stock 2.75" in diameter is used. The tool making the cut is subjected to a constant pressure of 3.75 lb. If the stock revolves at 250 r.p.m., how much work is done in one minute?

Power

If you watch two men each lift a weight of 150 pounds, one will pick it up slowly and, after struggling for several seconds, will raise it to bench height. The other man will pick up this weight and raise it to bench height in a much shorter time. Both men have accomplished the same work, but the second is said to be "more powerful" than the first.

Power is the rate of doing work

Power is the work done in a certain unit of time, as one minute or one second. Power is usually expressed in *foot-pounds per minute, or per second.* In the case of the two men above, if the first man took 10 seconds to raise 150 pounds 3 feet, he did 450 foot-pounds of work in 10 seconds. Then 450 equals 10 times the number of foot-pounds per second. Therefore, 450 is the product of 10 and his power. Then his power is 450 ÷ 10 or 45 foot-pounds per second.

If the second man took 4 seconds to raise the 150 pounds 3 feet, he did 450 foot-pounds of work in 4 seconds. In his case, 450 equals 4 times his power, or 4 times the number of foot-pounds per second. Then his power is 450 ÷ 4 or 112 1/2 foot-pounds per second. This man would be said to have more power than the first.

Notice that work is expressed in foot-pounds, but *power* is expressed in *foot-pounds per minute, or per second.* It is the *rate at which work is done.*

Example 1. A 20-lb. force is required to pull a cable through a conduit. What power is exerted in pulling 50 ft. of this cable through this conduit in 4 minutes?

THINK	DO THIS
Since power is the number of foot-pounds per minute.	
1. Find the total work done in 4 minutes.	1. $w = f \times s$ $w = 20 \times 50$ $w = 1,000$ ft.-lb.

2. Work is the product of power (foot-pounds per minute) and time (number of minutes). Write the formula.

2. w = power x time

$w/4$ = power

3. Divide work by time to obtain power (the work per minute).

3. $1000/4$ = 250 ft.-lb. per min.

Ans. Power = 250 ft.-lb. per min.

Example 2. A force of 45 lb. is required to push a loaded dolly along a corridor. What power is exerted in pushing the dolly 150 feet along this corridor in 3 1/2 minutes?

THINK

1. Find the work done.

2. Divide the work (w) by the time (t) to obtain power.

DO THIS

1. $w = f$ x s

$w = 45$ x $150 = 6,750$ ft.-lb.

2. $\frac{w}{t} = \frac{6,750}{3.5} = 1,928.57$

$= 1,929$ (approx.)

Ans. 1,929 ft.-lb. per min.

Summary

To find power, that is foot-pounds per minute (or per second),

1. Find the work — the total number of foot-pounds.

2. Divide the work (w) by the time (t).

If P represnts power, w the work, t the time in minutes (seconds), the formula for power is:

$P = \frac{w}{t}$ or $P = \frac{fs}{t}$

expressed in ft.-lb. per min. (or ft.-lb. per sec.).

EXAMPLE

What power is exerted by a crane lifting a 1,500-lb. bucket of concrete 28 ft. in 8.4 seconds?

1. $w = f$ x s

$w = 1,500$ x $28 = 42,000$ ft.-lb.

2. $P = \frac{w}{t} = \frac{42,000}{8.4}$

$P = 5,000$

Ans. 5,000 ft.-lb. per sec.

EXERCISES

Find the power exerted in each of the following:

1. Raising a 46-lb. casting 3 ft. in 5 sec.
2. Raising 1,500 lb. of brick 38 ft. in 2 1/2 min.
3. Lifting a 2-ton steel I beam 54 ft. in 3 1/4 min.

PROBLEMS

6. A man exerts a force of 75 lb. in pushing a loaded dolly 85 ft. in 1 min. At what rate is he doing work?

7. A lathe weighing 1,475 lb. was raised from a truck to the third story of a factory building 43 ft. above the ground in 12 1/2 min. The bed of the truck was 42 in. above the ground. How many foot-pounds of work per minute were done?

8. A carpenter uses a force of 12 lb. in pushing a crosscut saw across a piece of lumber. It requires 85 strokes, each 20 in. long, to saw through the board. If 1 1/4 min. were consumed in this operation, at what rate was work being done?

9. A freight elevator lifts 2 1/2 tons to a height of 85 ft. in 3 min. How much work is done? At what rate is the work done? If 6 min. were required to lift the load, how much work would be done? At what rate would the work be done?

10. A man weighing 195 lb. carries a pipe fitting weighing 7 lb. up a 10-ft. ladder in 8 sec. Find the work done and the power utilized.

Horsepower

In the previous section power was expressed as the rate per minute (second) of doing foot-pounds of work. A familiar unit of power is the *horsepower* (h.p.). It is *the power that is equal to 33,000 foot-pounds of work in one minute.*

Example. Compute the horsepower of a motor that does 330,000 ft.-lb. of work in 4 minutes.

THINK	DO THIS
1. Since horsepower is the power equal to 33,000 ft.-lb. of work in one minute, we first find the total power. $$P = \frac{w}{t}$$	1. $P = \dfrac{330,000}{4}$ $= 82,500$ ft.-lb. per min.
2. Since each horsepower consists of 33,000 ft.-lb. per min., divide the total power (P) by 33,000 to obtain horsepower (h.p.).	2. $\dfrac{P}{33,000} = \dfrac{82,500}{33,000} = 2.5$ *Ans.* 2.5 h.p.

Summary

To find horsepower (h.p.):

1. Find the total power (*P*) utilized in foot-pounds per minute.

2. Divide the total power in foot-pounds per minute by 33,000.

$$\text{h.p.} = \frac{\text{ft.-lb. per min.}}{33,000}$$

$$\text{h.p.} = \frac{P}{33,000}$$

EXAMPLE

A motor is required to raise a maximum load of 7,500 lb. 20 ft. in 3/4 min. What must be its horsepower?

1. $P = \dfrac{w}{t} = \dfrac{7{,}500 \times 20}{3/4}$

$P = 7{,}500 \times 20 \times 4/3$
  $= 3/3 \times 2{,}500 \times 20 \times 4$
$P = 2{,}500 \times 80$
  $= 200{,}000$ ft.-lb. per min.

2. $\text{h.p.} = \dfrac{P}{33{,}000}$

$\text{h.p.} = \dfrac{200{,}000}{33{,}000} = 6.06$

$= 6.1$ (approx.)
Ans. 6.1 h.p. (approx.)

EXERCISES

Find the horsepower to the nearest tenth for each of the following:

1. 198,000 ft.-lb. are exerted in 6 min.
2. It takes 5.5 min. to do 375,000 ft.-lb. of work.
3. Three tons of coal are raised 25 ft. in 1.5 min.
4. A 2.6-ton steel girder is raised 32 ft. in 2 3/4 min.
5. A force of 3,500 lb. is exerted for 1/2 mile in 5.2 min.

PROBLEMS

11. What must be the horsepower of an electric motor to raise a maximum load of 2.5 tons 15 ft. in 30 sec.?
12. Find the horsepower of a pump that supplies water to a factory water tank 275 ft. above the ground. The maximum rate of pumping is 50 cu. ft. per minute. Water weighs 62.5 lb. per cubic foot.
13. An electric motor operating a derrick must be able to raise a maximum load of 12 tons a height of 100 ft. in 8 min. Compute the horsepower of the motor.
14. In how much time will a 15-h.p. motor raise a load of 5 tons through a distance of 100 ft.?
15. How many tons of iron ore can a 6-h.p. hoisting engine raise in 30 sec. from an open pit 18 ft. below the freight car to be loaded?

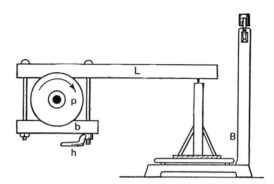

Fig. 10-1. Prony brake used to measure the brake horsepower of an electrical motor.

Brake horsepower

It is often necessary and convenient to know the actual power that is transmitted to a pulley or a shaft. In order to measure this power a device known as a *Prony brake* may be used (Fig. 10-1). If you study this figure carefully, you will notice that a brake shoe (*b*) has been clamped around a pulley (*p*). The handle (*h*) permits the clamp to be adjusted so that more or less friction is exerted between the brake lining and the surface of the drum of the pulley. An arm (*L*) is bolted to the brake and extends to a platform balance (*B*).

As the pulley turns in the direction of the arrow, it tends to turn the brake (*b*) and arm (*L*). In so doing, the arm, *L*, presses down on the platform balance, which records the pull created by the friction between pulley and brake. The force produced by the weight of the brake arm, *L*, must be deducted in order to obtain the true value of the pull.

If *L* represents length (in feet) of brake arm as shown, r.p.m. revolutions per minute of the pulley, and *F* the actual force exerted, the formula for the brake horsepower, b.h.p., is:

$$\text{b.h.p.} = \frac{2\pi L \text{ x r.p.m. x } F}{33,000}$$

Remember that *F* is the force recorded on the scales minus the weight of the brake arm.

Example. A Prony brake attached to an induction motor has an 18″ arm; the platform balance reads 12 lb. net (after the weight of the arm has been deducted) when the drum rotates at 1725 r.p.m. Find the brake horsepower developed in the motor at this load.

THINK	DO THIS
1. Write the formula for the brake horsepower.	1. b.h.p. $= \dfrac{2\pi L \text{ x r.p.m. x } F}{33,000}$

2. Write the value of each variable given in the problem.	2. $L = 18'' = 1.5'$, $F = 12$ lb., r.p.m. = 1,725, π = 3.1416
3. Replace, in the formula, each letter by its given value.	3. b.h.p. $= \dfrac{2 \times 3.1416 \times 1.5 \times 1,725 \times 12}{33,000}$
4. Make the indicated calculations. Use the principle of one where convenient.	4. b.h.p. = 5.9119 = 5.91 (approx.) *Ans.* b.h.p. = 5.91 h.p. (approx.)

A Prony brake similar to that of Fig. 10-2 is sometimes used to obtain the pull. Here T_1 and T_2 are spring balances. The actual pull is obtained by subtracting the reading on T_2 from that on T_1. (Note the direction of motion of the pulley.)

Example. In the case of the Prony brake of Fig. 10-2, if the scale T_1 reads 80 lb., the scale T_2 reads 30 lb., the diameter of the drum is 18'', and the drum makes 1,800 r.p.m., what is the brake horsepower developed?

THINK	DO THIS
1. Write the formula for the brake horsepower.	1. b.h.p. $= \dfrac{2\pi L \times \text{r.p.m.} \times F}{33,000}$
2. Write the value of each variable.	2. $L = 9'' = 0.75'$, r.p.m. = 1,800, $F = 80 - 30 = 50$, π = 3.1416
3. Substitute these values in place of the variables in the formula.	3. b.h.p. $= \dfrac{2 \times 3.1416 \times 0.75 \times 1,800 \times 50}{33,000}$
4. Make the indicated calculations.	4. b.h.p. = 12.85 = 12.9 (approx.) *Ans.* b.h.p. = 12.9 h.p. (approx.)

Summary	EXAMPLE
To find brake horsepower:	A pulley makes 2,500 r.p.m., the brake arm is 15'' long and the net pull is 60 lb. Find the brake horsepower.

1. Write the formula b.h.p. = $\dfrac{2\pi L \times \text{r.p.m.} \times F}{33,000}$	1. b.h.p. $= \dfrac{2\pi L \times \text{r.p.m.} \times F}{33,000}$
2. Write the numerical value of each of the variables in the formula.	2. $L = 15'' = 1.25'$, r.p.m. $= 2,500$, $F = 60$ lb., $\pi = 3.1416$
3. Replace each variable in the formula by its designated value.	3. b.h.p. $= \dfrac{2 \times 3.1416 \times 1.25 \times 2,500 \times 60}{33,000}$
4. Make the indicated calculations.	4. b.h.p. $= 35.7000 = 35.7$ (approx.) *Ans.* b.h.p. $= 35.7$ h.p. (approx.)

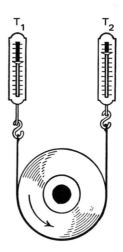

Fig. 10-2. Prony brake in which two spring balances are used in place of platform balance.

EXERCISES

Find the brake horsepower under each of the following conditions.

1. A pulley makes 1,800 r.p.m., the brake arm is 18″ long, and the net pull is 24 lb.
2. The r.p.m. of a motor is 2,250; the brake arm, 18″; and the net pull, 22.5 lb.
3. The r.p.m. is 1,475; the brake arm, 12″; and the net pull, 16.3 lb.
4. The r.p.m. is 1,680; the brake arm, 36″; and the net pull, 152 lb.

PROBLEMS

16. A Prony brake of the type shown in Fig. 10-1 attached to a shunt motor to be tested has a brake arm 18″ long. If the motor runs at 1,125 r.p.m. and scale reads 32 lb. net, what horsepower is delivered?

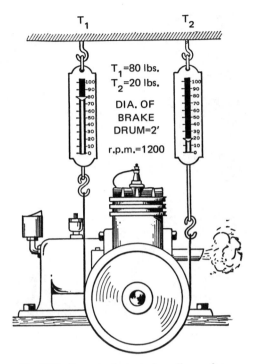

Fig. 10-3. Prony brake test of gasoline engine.

17. A direct-current motor is to be tested by means of a Prony brake when the motor is loaded and operating at a speed of 1,750 r.p.m. The following test data were obtained at that speed: weight of brake arm, 4 lb.; balance reading, 18 lb.; length of brake arm, 12". How many b.h.p. were developed at this load?

18. Figure 10-3 shows the data obtained in testing a gasoline engine by means of a brake band and two spring balances. Compute the b.h.p. developed.

19. The reading of the scale of a Prony brake when an engine was making 325 r.p.m. was 75 lb. net. The length of the brake arm was 42". Find the horsepower developed by the engine at this load.

20. The following readings were taken during a test with a Prony brake: length of arm, 4 ft. 6 in.; reading of balance, 295 lb.; r.p.m., 875. Find the horsepower of the motor. The balance reading is adjusted to include the weight of the arm.

S.A.E. horsepower

The horsepower of automobiles is computed according to a formula developed by the Society of Automotive Engineers (S.A.E.) and is called the *S.A.E. horsepower*. This formula is used by some agencies to determine the license fee or tax to be paid on automobiles.

Fig. 10-4. Power plant of a modern automobile.
(Courtesy General Motors Corporation, Oldsmobile Division)

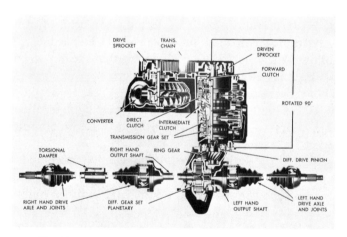

Fig. 10-5. Cross section of a power train in an automobile.
(Courtesy General Motors Corporation, Oldsmobile Division)

If D represents the bore of the cylinder (Fig. 10-6) in inches, and N the number of cylinders in the engine, then the S.A.E. horsepower is found by the following formula:

$$\text{h.p.} = \frac{D^2 N}{2.5}, \text{ or h.p.} = 0.4 D^2 N$$

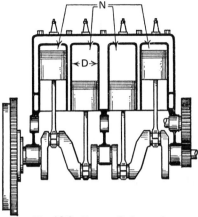

Fig. 10-6. Four-cylinder engine.

Example. Find the S.A.E. horsepower of a six-cylinder engine with a bore of 3 13/16″.

THINK	DO THIS
1. Write the formula for S.A.E. horsepower.	1. h.p. $= 0.4D^2N$
2. Write the value given for each of the variables.	2. D = 3 13/16″, N = 6
3. Replace the letters of the formula by these given values.	3. h.p. = 0.4 x (3 13/16)^2 x 6
4. Make the indicated calculations.	4. h.p. = 0.4 x (61/16)^2 x 6 = 34.88 = 34.9 (approx.) *Ans.* S.A.E. horsepower = 34.9 (approx.)

Summary	EXAMPLE
To find the S.A.E. horsepower of an engine,	What is the S.A.E. rated horsepower of an 8-cylinder engine with a bore of 3 5/16″?
1. Write the formula for S.A.E. horsepower.	1. S.A.E. h.p. $= 0.4D^2N$
2. Write the value of each of the variables.	2. D = 3 5/16″, N = 8
3. Replace each variable of the formula by the value given in the problem.	3. h.p. = 0.4 x (3 5/16)^2 x 8

4. Make the indicated computations.	4. h.p. $= 35.11 = 35.1$ (approx.)
	Ans. S.A.E. horsepower $=$ 35.1 (approx.)

PROBLEMS

21. Find the rated horsepower of an eight-cylinder V-8 engine having a bore of 3 3/16".
22. Find the horsepower of a six-cylinder engine having a bore of 3".
23. What is the rated horsepower of a V-8 engine which has a bore of 3 1/16" and a stroke of 3 3/4"?
24. A six-cylinder gasoline engine has a bore of 3 5/16" and a stroke of 3 1/2". What is its rated horsepower?
25. If the bore of the cylinder of the engine in Problem 24 is decreased by 3/16", what will be the loss in horsepower?

Indicated horsepower

Still another designation of horsepower is the measure of the power *developed in the cylinder* of the engine. This is called *indicated horsepower* (i.h.p.). It is obtained by means of the formula:

$$\text{i.h.p.} = \frac{P\,L\,A\,N}{33{,}000}$$

where P represents the mean effective pressure on the piston in pounds per square inch; L, the length of the stroke in feet; A, the area of the piston in square inches ($A = 0.7854d^2$ or, $A = \pi r^2$); and N, the number of strokes per minute.

Example. Find the indicated horsepower of a 16" x 30" engine that makes 150 r.p.m. and has a mean effective pressure of 50 lb. per square inch in the cylinder.

THINK	DO THIS
1. Write the formula for indicated horsepower.	1. Indicated horsepower $\text{i.h.p.} = \dfrac{P\,L\,A\,N}{33{,}000}$
2. Write the value of each variable that is given in the problem.	2. $P = 50$, $L = 30'' = 2.5'$, $A = 0.7854 \times (16)^2$, $N = 150$
3. Replace each variable in the formula by its designated value.	3. i.h.p. $=$ $\dfrac{50 \times 2.5 \times 0.7854 \times (16)^2 \times 150}{33{,}000}$

4. Make the indicated computations with the slide rule. Estimate the answer before placing the decimal point.

4. i.h.p. = 114.24

Ans. Indicated horsepower = 114.2 (approx.)

Summary

EXAMPLE

To find the indicated horsepower of an engine:

Find the indicated horsepower of an engine in which the pressure is 40 lb. per square inch, the stroke is 30 inches, the diameter of the piston is 15 inches, and the r.p.m. is 200.

1. Write the formula for indicated horsepower

$$i.h.p. = \frac{P\,L\,A\,N}{33,000}$$

1. $i.h.p. = \dfrac{P\,L\,A\,N}{33,000}$

2. Write the value of each variable in the formula: P = pressure, L = length of stroke in feet, A = area of the piston in square inches, N = number of strokes per minute.

2. $P = 40$, $L = 2.5'$,
$A = 3.1416 \times (7.5)^2$,
$N = 200$

3. Replace each variable in the formula by its indicated value.

3. $i.h.p. =$

$$\frac{40 \times 2.5 \times 3.1416 \times (7.5)^2 \times 200}{33,000}$$

4. Make indicated computations.

4. i.h.p. = 107.1 (approx.)
Ans. Indicated horsepower = 107.1 (approx.)

EXERCISES

Find the indicated horsepower in each of the following:

	Pressure	Stroke	Diameter of piston	r.p.m.
1.	50 lb.	28″	15″	190
2.	40 lb.	30″	18″	175
3.	60 lb.	32″	20″	200
4.	35 lb.	20″	10″	250

PROBLEMS

Find the indicated horsepower of each of the engines described in the following problems:

26. The pressure is 35 lb. per square inch; the stroke, 24"; the diameter of piston, 14"; and the r.p.m., 185.
27. The pressure is 75 lb. per square inch; the stroke, 48"; the diameter of piston, 18"; and the r.p.m., 170.
28. An engine making 580 r.p.m. develops a mean effective pressure of 30 lb. per square inch; its stroke is 10.2" and the diameter of its cylinder is 5 3/4".
29. The average pressure in the cylinder of an engine is 40 lb. per sq. in.; the area of the piston is 125 sq. in.; the length of the stroke is 12 in. What is the horsepower developed in the cylinder of the engine when it is turning at the rate of 275 r.p.m.?
30. An eight-cylinder gasoline engine turning at the rate of 1,800 r.p.m. develops 55 h.p. If the bore of the cylinder is 3 3/16" and the length of the stroke 5.6 in., find the mean effective pressure which is developed during the working stroke.

Watt and kilowatt

The unit of electrical power is the *watt*. It is the amount of power equal to 10^7 x 1 centimeter-gram per second. This is equivalent to 1/746 of a horsepower of mechanical power. This means that 746 watts is equivalent to one horsepower of mechanical power.

The watt is sometimes stated to be the rate of work represented by an electric current of one ampere under a pressure of one volt.

The *volt* is the unit of electrical pressure. It is the force of the electrical current through a circuit. A house is said to have sufficient "house power," or to be "adequately wired," if the electrical system has been designed to provide sufficient power to operate all the appliances and lights that might be installed in the house. Some appliances are built to operate on 220 volts, while others may need only 110 volts. It is important to know the voltage at which an appliance operates.

The number of watts (W) of power being used at any time is equal to the number of volts (E) times the number of amperes (I). Expressed as a formula, this is:

$$W = E \times I$$

The unit that is more frequently used to express electrical power is the *kilowatt* (k.w.). This unit is equivalent to 1,000 watts. The formula for the number of kilowatts of power is:

$$k.w. = \frac{W}{1,000} = \frac{E \times I}{1,000}$$

Example. Compute the power in watts and also in kilowatts supplied by a generator that delivers 250 amperes at 200 volts.

THINK	DO THIS
1. Write the formula for the number of watts.	1. $W = E \times I$
2. Replace the variables by their values as given in the problem.	2. $W = 200 \times 250$
3. Make the computations.	3. $W = 50,000$
4. Write the formula for expressing the power in kilowatts.	4. k.w. $= \dfrac{W}{1,000}$
5. Replace W by its value and make the computation.	5. k.w. $= \dfrac{50,000}{1,000} = 50$
	Ans. Power = 50,000 watts or 50 k.w.

Summary

	EXAMPLE
To obtain the number of watts (W) of electrical power that a circuit furnishes, multiply the number of volts (E) by the number of amperes (I).	Express in kilowatts the power rating of a generator that delivers 320 amperes at 220 volts.
1. Use the formula $W = E \times I$.	1. $W = E \times I = 220 \times 320$
2. Replace the variables by the values given, and make the computations.	2. $W = 70,400$ watts
3. To express the power in terms of kilowatts (k.w.) (1,000 watts), use the formula k.w. $= \dfrac{W}{1,000}$	3. k.w. $= \dfrac{W}{1,000} = \dfrac{70,400}{1,000}$ $= 70.4$ k.w. *Ans.* Power = 70.4 k.w.

EXERCISES

Find the required quantity in each of the following exercises:

1. The number of watts and kilowatts when a motor draws 40 amperes (amps.) at 220 volts.
2. The number of kilowatts when a motor draws 75 amps. at 110 volts.
3. The number of watts when an electric soldering iron draws 8 amps. at 115 volts.

Electrical power as horsepower

It is sometimes necessary to express electrical units of power, watts and kilowatts, in terms of the mechanical unit of power, horsepower. *One horsepower of mechanical power is equivalent to 746 watts of electrical power.* This is roughly three quarters of a kilowatt of electrical power.

To find the horsepower in an electrical circuit, divide the number of watts by 746.

$$\text{h.p.} = \frac{W}{746}$$

Example. An electric motor draws 25 amperes at 220 volts at full speed. What is the rating of this motor in horsepower?

THINK	DO THIS
1. Since we must first find the number of watts employed, write the formula for watts.	1. $W = E \times I$
2. Substitute the given values for the variables in the formula and make the computation.	2. $W = 220 \times 25$ $W = 5{,}500$
3. Write the formula for finding horsepower. Replace W by its value and make the computation.	3. $\text{h.p.} = \dfrac{W}{746}$ $\text{h.p.} = \dfrac{5{,}500}{746} = 7.37$ (approx.) *Ans.* Motor rating = 7.37 h.p. (approx.)

Summary	EXAMPLE
To find electric power, expressed in terms of horsepower:	Find the horsepower rating of a motor that draws 100 amperes at 220 volts.
1. Use the formula $W = E \times I$ to find the number of watts of electrical power the circuit furnishes.	1. $W = E \times I$ $W = 220 \times 100 = 22{,}000$
2. In the formula $\text{h.p.} = \dfrac{W}{746}$ replace W by the value found in Step 1. Make the computation.	2. $\text{h.p.} = \dfrac{W}{746} = \dfrac{22{,}000}{746}$ $= 29.49 = 29.5$ (approx.) *Ans.* Motor rating = 29.5 h.p. (approx.)

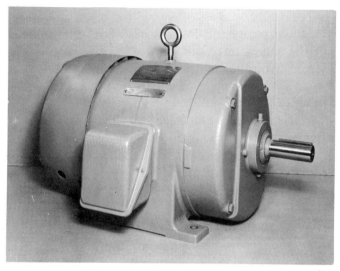

Fig. 10-7. Electric motor. *(Courtesy General Electric Company)*

EXERCISES

1. Find the horsepower for each of the exercises on page 387.
2. Find the horsepower equivalent to 120 amps. at 520 volts.
3. Find the horsepower of a motor that draws 125 amps. at 220 volts.

PROBLEMS

31. A motor (Fig. 10-7) takes 30 k.w. How many horsepower is it taking?
32. A heating element of a soldering iron takes 3.5 amps. at 110 volts. What power does it use?
33. How much power is drawn at full load from the line by a 5-h.p. motor operating at 220 volts?
34. An electric toaster carries a current of 6 amps. at 110 volts. Compute the power used in kilowatts.
35. A small electric furnace requires 2,100 watts for its operation, and is supplied from 120-volt mains. What current does the furnace draw from the line?
36. What horsepower will 24 lamps require if each takes 0.25 amps. at 100 volts?
37. A steam engine is rated as a 7.5-h.p. engine. What would be its kilowatt rating?
38. A generator delivers a current of 80 amps. at a pressure of 112 volts. What power does it supply in kilowatts? How many horsepower does it develop?

Efficiency

Energy is the capacity for doing work. The *efficiency of a machine* is the *ratio* of the energy given out, *output*, to the energy taken in, *input*, by the machine. Efficiency is found by dividing output by input. If a machine takes in 20 horsepower of energy and gives out 15 horsepower, its efficiency is 15/20 or 3/4.

Percent efficiency is the efficiency ratio expressed as a percent. Thus the efficiency of the machine just mentioned is 3/4 expressed as a percent. Remember that to express 3/4 as a percent we write 3/4 of 100 percent, or 75 percent (see percent, p. 102).

Example. Find the efficiency and the percent of efficiency of a motor that takes in 100 amperes at 210 volts and has an output of 16.5 k.w.

THINK	DO THIS
1. Since the efficiency is output divided by input, first find the input in k.w.	1. k.w. $= \dfrac{E \times I}{1,000}$ k.w. $= \dfrac{210 \times 100}{1,000} = 21$
2. Efficiency equals output divided by input ($e = o/i$). Divide output by input and express the result in decimal form.	2. $e = \dfrac{o}{i} = \dfrac{16.5}{21}$ $= 0.7857 = 0.786$ (approx.)
3. Express this result in percent by multiplying the result by 100% (percent efficiency, *e.p.* $= o/i$ x 100%).	3. *e.p.* $= 0.786 \times 100\% = 78.6\%$ *Ans.* Efficiency $= 0.786$ Percent efficiency $= 78.6\%$

Summary	EXAMPLE
To find the efficiency (*e*) of a machine and the percent of efficiency:	A motor at full speed draws 75 amps. at 300 volts. Its output is 18.5 k.w. Find its efficiency and the percent of efficiency.
1. Find the input and the output of the machine.	1. Input $= \dfrac{300 \times 75}{1,000} = 22.5$ Output $= 18.5$
2. Divide the output by the input ($e = o/i$).	2. $e = \dfrac{o}{i} = \dfrac{18.5}{22.5} = 0.822$

3. To express the efficiency in percent, multiply the efficiency by
100% (*e.p.* = *o*/*i* x 100%).

| 3. *e.p.* = 0.822 x 100% = 82.2%
Ans. Efficiency = 0.822
Percent efficiency = 82.2%

EXERCISES

Find the efficiency and the percent efficiency of each of the following:

1. Input is 30 k.w., output 22.8 k.w.
2. Input is 17.2 k.w., output 15.9 k.w.
3. Output is 12.6 k.w., input 14.8 k.w.
4. A machine that takes 55 amps. at 440 volts delivers 28 h.p.
5. A machine at full load operates ten lathes of 3/4 h.p. each. The machine draws 30 amps. at 220 volts.

PROBLEMS

39. A motor running at full load takes 75 amps. at 208 volts. The output is 13.9 k.w. What is the efficiency of the motor?
40. A drill press is operated by a 1 1/2 h.p. motor. The output of the drill press is 1.32 h.p. What is the efficiency of the machine in percent?
41. If the efficiency of a milling machine is 62% when taking a certain cut, what is the input at that load if the output is 3.295 h.p.?
42. A certain motor at full load supplies 12 h.p. to a Prony brake and takes 46 amps. at a pressure of 240 volts. What is the efficiency of the motor?
43. A certain shop motor has an input at full load of 6 1/2 h.p. It supplies 1/2 h.p. to a lathe, 3/4 h.p. to a second lathe, 1/2 h.p. to a sensitive drill, and 4 h.p. to a planer. Find the efficiency of the motor.

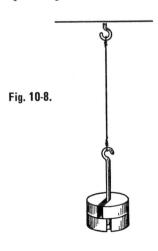

Fig. 10-8.

Direct stresses

If you attach a piece of wire to a hook as shown in Fig. 10-8, page 391, and hang weights on the hook, the wire will hold a number of these weights. As more and more weights are added, the wire will stretch, and eventually there will come a time when the wire will no longer be able to support the load, and it will fail or break. (A special name is applied to the situation where the wire breaks. This is explained later.)

The internal resistance in this wire that allows it to hold the weights or load is called *stress.* Stresses are measured in pounds, or tons, or kilograms.

There are three principle kinds of stresses:

(*a*) *Tension* is stress produced when a force, applied externally, tends to pull the materal apart (Fig. 10-8).

Fig. 10-9. Cables supporting the upper movable conveyors in this storage yard are under tension.
(*Courtesy Link Belt Company*)

Fig. 10-10. This column is under compression.

Fig. 10-11. The rivet in this single-riveted lap joint is under shear.

SHEARING FORCE →

SHEARING FORCE

(b) *Compression* is stress produced when a force, applied externally, tends to crush or push the material together (see Fig. 10-10).

(c) *Shear* is stress produced when a force, applied externally, acts in such a way as to cut through the material (see Fig. 10-11).

Unit stress is the stress in one unit of cross-sectional area. The unit stress is expressed in pounds per square inch, or in metric measure in kilograms per square centimeter and is found by dividing the stress by the cross-sectional area.

Example. A steel rod two inches in diameter is subjected to a load of 50,000 pounds. Find the unit stress within this rod.

THINK	DO THIS
1. Since unit stress is reckoned on one square unit of cross-sectional area, first find the cross-sectional area of the rod.	1. $A = \pi r^2$ $r = 1, \pi = 3.1416$ $A = 3.1416 \times 1^2 = 3.1416$ sq. in.
2. Since 50,000 is the product (L) of the area (3.1416) and the unit stress (U), to find unit stress divide the product (L) by the known factor (A). Formula for unit stress: $U = L/A$.	2. $U = L/A$ $50,000/3.1416 = 15,960$ or 16,000 (approx.) *Ans.* Unit stress = 16,000 lb./sq. in. (approx.)

Summary

To find unit stress:

1. Write the formula $U = L/A$, where U is the unit stress, L is the load and A the cross-sectional area in square inches.

2. Find the cross-sectional area.

3. Replace the variables in the formula by their known values and make the indicated calculations.

EXAMPLE

What is the unit stress on a rectangular bar 2" x 6" carrying a load of 100,000 lb.?

1. $U = L/A$

2. $A = 2 \times 6 = 12$ sq. in.

3. $U = 100,000/12$
$U = 8,334$ (approx.)
Ans. Unit stress = 8,334 lb./sq. in. (approx.)

EXERCISES

Find the unit stress in each of the following:

1. On a 3/4″ wire cable carrying a load of 8,500 pounds.
2. On a 3″ steel post carrying a load of 38,000 pounds.
3. If the ultimate strength of cast iron is 22,000 pounds per square inch, what is the greatest load that a cast iron post 2 1/4″ in diameter could carry? (But not safely. See the following section. Also note that the figure given, 22,000 pounds per square inch, is the ultimate strength of cast iron under tension. Check the table, page 396, for the ultimate strength of cast iron under compression and under shear.)

PROBLEMS

44. A steel tie rod 1 1/2″ in diameter is subjected to a load of 32,000 lb. Find the unit stress within the tie rod.
45. A cast-iron column which is to be subjected to a compression of 45,000 lb. is to be designed so that the unit stress shall not exceed 2,600 lb. per square inch. What should be the cross-sectional area?
46. A steel rod 6 ft. long and 1.125 in. in diameter is subjected to a tensile load of 12,500 lb. Find the unit stress.
47. The piston of a compressor 4 1/2″ in diameter is subjected to a pressure of 150 pounds per square inch. What is the total pressure on the piston?
48. A truss member is subject to a total load of 55,000 lb. tensile stress. If the safe working stress for yellow pine is 1,200 lb. per square inch, what size timber should be used? If only spruce were available, what size timber should be used? The safe working stress for spruce is 800 lb. per square inch.
49. A water tank 12′ in diameter and 11′ high is to be supported by four rectangular posts. Find the size of the hemlock posts to be used to support the tank when it is filled with water. Assume a working stress for hemlock of 800 lb. per square inch. Water weighs 62.5 lb. per cubic foot.

The elastic limit

Consider a bar of wrought iron in a tensile testing machine (Fig. 10-12). Assume this bar to be one square inch in cross-sectional area and 8 inches long. If a 5,000-pound load is applied, the bar will become longer by 0.0016 inch. If 10,000 pounds of pull is applied, the bar will be elongated 0.0032 inch.

Fig. 10-12. 60,000-pound testing machine likely to be found in modern technical school testing laboratories. *(Courtesy Tinius Olsen Testing Machine Company)*

If 15,000 pounds, the bar is elongated 0.0048 inch.
If 20,000 pounds, the bar is elongated 0.0064 inch.
If 25,000 pounds, the bar is elongated 0.0080 inch.

If you study these facts, you will see that *the bar is elongated exactly in proportion to the load applied.* Each time the load is increased 5,000 pounds, the bar is elongated 0.0016 inch.

This, however, continues true only as long as what is known as the *elastic limit* of the bar has not been reached. When that point is reached, a different proportion is found.

The elastic limit is the unit stress at which the deformation of the work begins to increase at a greater rate than the increase in load.

In the illustration above, the bar was elongated exactly in proportion to the load that was applied only because the elastic limit of the bar had not been reached by a load of 25,000 lb.

When the elastic limit is reached and exceeded, the length of the bar will increase at a greater rate than the increase in load.

When unit stress is less than the elastic limit, the object will return to its original form; when the unit stress is greater than the elastic limit, the object does not return to its original form. The point at which this change takes place is called the *yield point*.

Ultimate strength

As the load on a bar increases beyond its elastic limit, the bar extends rapidly until finally there comes a point where the load breaks or separates the bar into two parts. The *ultimate strength* is the highest unit stress attained before rupturing.

The ultimate strength of materials is from two to four times their elastic limits, and in some materials is much greater in compression than in tension.

The following tables give information on the ultimate strengths of a number of metals and other materials under tension, compression and shear stresses. Compare the figures in each column for aluminum, copper, iron and steel.

Average Ultimate Strengths of Some Metals, in Pounds per Square Inch

Metal	Tension	Compression	Shear
Aluminum, cast.	15,000	12,000	12,000
Aluminum, half hard sheet	19,000	60,000	19,000
Brass, cast	30,000	30,000	36,000
Copper, cast	24,000	40,000	25,000
Copper, rolled.	37,000	60,000	28,000
Duralumin, soft sheet.	35,000	50,000	30,000
Duralumin, treated and cold rolled . .	75,000	75,000	40,000
Iron, cast	22,000	90,000	25,000
Iron, cast (2 percent nickel).	50,000	150,000	50,000
Lead	3,000	. . . .	4,000
Monel Metal, rolled	95,000	90,000	65,000
Steel, casting	70,000	65,000	60,000
Steel, boiler plate	70,000	70,000	60,000
Steel, 0.10 carbon (soft)	60,000	60,000	45,000
Steel, 0.25 carbon (mild)	70,000	65,000	55,000
Tin, sheet.	5,000	6,500	5,000

Average Ultimate Strengths of Some Materials Other than Metals, in Pounds per Square Inch

Materials	Tension	Compression	Shear	
			With Grain	Across Grain
Granite	800	25,000	3,300	
Leather	4,000		7,000	
Sandstone	340	12,000	1,800	
Paper, flat punch.			8,500	
Paper, Bristol board, flat punch . .			4,800	
Wood:			*With Grain*	*Across Grain*
Oak			200	1,000
Yellow pine, long and short leaf . .			150	1,000
N.C. pine, Douglas fir.			100	1,000
White pine, spruce, fir			100	500
Hemlock			100	600

Example. If a 73,690 pound load crushes a 2 1/2-in. diameter cast aluminum rod 6 in. long, what is the ultimate strength of the cast aluminum? What is its probable elastic limit?

THINK

1. Since the ultimate strength is the crushing load per square inch of cross section, 73,690 is the product of the ultimate strength and the cross-sectional area. Find the cross-sectional area.

2. To find the ultimate strength (U), divide the crushing load by the cross-sectional area in square inches.

3. Since the ultimate strength is two to four times the elastic limit, divide the ultimate strength by two and then by four to find the probable elastic limit.

DO THIS

1. $A = (1\ 1/4)^2 \times 3.1416$
   $= 25/16 \times 3.1416$
   $A = 4.9125$ sq. in.

2. $U = \dfrac{73,690}{4.9125}$
   $= 15,000$ lb./sq. in. (approx.)

3. $\dfrac{15,000}{2} = 7,500$; $\dfrac{15,000}{4} = 3,750$

Ans. Ultimate strength = 15,000 lb./sq. in.
Elastic limit somewhere between 3,750 lb./sq. in. and 7,500 lb./sq. in.

Note that in the example we have found the ultimate strength and probable elastic limit of the cast aluminum rod under compression. Would you expect the figures to be different under tension?

Summary	EXAMPLE
When the crushing or breaking load of a bar is known:	A uniform steel bar 2" in diameter ruptures under a tensile stress of 220,000 pounds. What is its ultimate strength and its probable elastic limit under tension?
1. To find the ultimate strength, divide the crushing or breaking load by the cross-sectional area of the bar.	1. $A = 1^2 \times 3.1416 = 3.1416$ $U = \dfrac{220,000}{3.1416}$ $= 70,000$ lb./sq. in. (approx.)
2. To find the elastic limit of the material, divide the ultimate strength by 2 and by 4.	2. $\dfrac{70,000}{2} = 35,000;$ $\dfrac{70,000}{4} = 17,500$ *Ans.* Ultimate strength $= 70,000$ lb./sq. in. (approx.) Elastic limit between 17,500 and 35,000 lb./sq. in.

PROBLEMS

50. It requires 40,000 lb. to punch a 3/4" hole in a steel plate 1/2" thick. What is the ultimate strength of the steel in the plate?
51. A uniform steel bar 1.49" in diameter ruptures under a tensile stress of 200,000 lb. Find the ultimate strength.
52. The head of a 1" semifinished bolt is 7/8" thick. There is a tendency to pull off the head from the bolt when a force of 14,000 lb. is applied. What is the ultimate tensile stress?
53. A steel strut ruptures when a load of 85,000 lb. is applied. What is the cross-sectional area of the strut?
54. If the ultimate strength of cast iron under compression is 90,000 lb. per square inch, at what load did a hollow cast iron column rupture? The outside diameter of the column was 10" and the wall was 1" thick.

Factor of safety

The *factor of safety* (F) is a number obtained by dividing the ultimate strength (U) by the actual unit stress (S) that exists in a bar. The formula for the factor of safety is:

$$F = \frac{U}{S}, \text{ or } U = F \times S$$

This is an important item in all construction work.

Example. A timber 6″ x 6″ in cross section, ultimate strength 10,000 pounds per square inch, is under a tensile stress of 32,400 pounds. What is the factor of safety?

THINK	DO THIS
1. Since the factor of safety is the ultimate strength (U) divided by the unit stress (S), we first find the unit stress by dividing the total stress by the section area.	1. Unit stress = $\dfrac{\text{Total stress}}{\text{Section area}}$ Section area = 6 x 6 = 36 sq. in. $S = \dfrac{32{,}400}{36} = 900$ (approx.)
2. Write the formula for the factor of safety.	2. $F = \dfrac{U}{S}$
3. Replace the variables U and S by their values.	3. $F = \dfrac{10{,}000}{900} = 11.11$ $= 11$ (approx.)
4. Make the computations.	*Ans.* Factor of safety = 11 (approx.)

Summary	EXAMPLE
To find the factor of safety (F), divide the ultimate strength (U) by the unit stress (S) that actually exists.	A copper casting under compression was designed for a unit stress of 8,500 pounds. What factor of safety was used?
1. Write the formula.	1. $F = \dfrac{U}{S}$
2. Replace the variables by their values. Obtain the value for ultimate strength (U) from the table, page 396.	2. $F = \dfrac{40{,}000}{8{,}500} = 4.76$ (approx.)
3. Make the computations.	*Ans.* Factor of safety = 4.76 (approx.)

PROBLEMS

55. A steel casting under compression was designed for a unit stress of 12,500 lb. What factor of safety was used?
56. A leather belt is subjected to a unit tensile stress of 500 lb. Find the factor of safety.
57. The mortised oak girt shown in Fig. 10-13 is to be designed in such a way as to withstand a load of 1,500 lb. If a factor of safety of 5 is allowed, what should be the area of the section under shear?

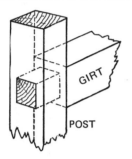

Fig. 10-13. Mortised oak girt.

REVIEW PROBLEMS

1. An elevator lifts a load of castings weighing 2 1/2 tons through a distance of 75 ft. in 15 sec. At what rate in horsepower is work being done?
2. A pulley 12" in diameter runs at the rate of 150 r.p.m. What is the efficiency if the tension in the belt averages 78 lb. and there is a loss of 1/4 h.p.?
3. The horsepower required to take a certain cut in an engine lathe is 0.567 while the corresponding input is 0.675 h.p. What is the percent efficiency?
4. The input of an electrical generator is 250 h.p. The corresponding output is 175 k.w. What is the efficiency of the generator at this load?
5. Find the efficiency of a drill press if the output is 32,500 ft.-lb. per minute for an input of 1 1/4 h.p.
6. The cylinders of a V-8 engine are 3 3/16" x 3 7/8". Find the rated horsepower of the engine.
7. The Prony brake lever used to test a gasoline engine is 5'6" long. What brake horsepower is developed if the engine under test makes 875 r.p.m. and the net scale reading is 22 lb.?
8. What is the horsepower of the motor of a pump that removes 500 gallons of water per minute from a sump in a tunnel 125 ft. below the surface of the ground? Assume the efficiency of the unit to be 62%.

9. A hoist motor does 66,000 foot-pounds of work in lifting a load 75 feet. How heavy is the load?

10. A six-cylinder gasoline engine has a bore of 3 5/16" and a stroke of 3 5/8". Find the rated horsepower.

11. A 7.5 h.p. 220-volt direct-current motor that drives a milling machine has an efficiency of 85% at full load. What current does this motor draw from the line at full load?

12. The indicated horsepower of a gasoline engine is 400 horsepower. The corresponding brake horsepower is 350 horsepower. What is the efficiency of the engine at this load?

13. An electric generator delivers 250 k.w. to the line. The power lost in the line is 1,100 watts. Find the efficiency of transmission.

14. The motor driving a radial drill receives 8.5 amperes, 440 volts. Find the h.p. supplied to the motor.

15. Find the horsepower of a twelve-cylinder 3 1/16" x 4 1/4" engine.

16. The windage, friction, and copper losses of an electric motor represent 12% of the output of the motor. What horsepower will this motor deliver if the input is 18.7 amperes at 112 volts?

11 Some Ideas from Trigonometry

In order to do some of the finer, more important jobs of industry, it is necessary to measure completely any triangle. For example, in order to be able to select the proper cutter to make a bevel gear, it is necessary to find out certain facts about a triangle.

The special study of triangles is called *trigonometry*. In this unit we shall learn some useful ideas about the subject.

Ratios associated with angles

A *ratio* is the quotient of two quantities. $\frac{5}{7}$ is a ratio. It is spoken of as the "ratio of 5 to 7." $\frac{AB}{CD}$ is the ratio of the length of AB to the length of CD.

In the right triangle, Fig. 11-1, we can write several ratios that include the sides. Thus $\frac{BC}{AC}$ or $\frac{1}{3/4}$ is one such ratio. $\frac{BC}{AB}$ or $\frac{1}{1\,1/4}$ is another. The student should write other ratios; there are four others.

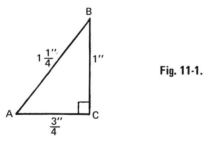

Fig. 11-1.

An interesting thing about these ratios is that if the angles in several right triangles are the same, the ratios of the corresponding sides will be the same regardless of how long the sides may be.

Thus, in the 45° right triangle ABC, Fig. 11-2, the ratio $\frac{AC}{AB}$ is $\frac{1}{1} = 1$. In the 45° right triangle BDE the corresponding ratio is $\frac{DE}{BD}$ or $\frac{2}{2}$, which also equals 1; and in the 45° right triangle FBG the ratio of

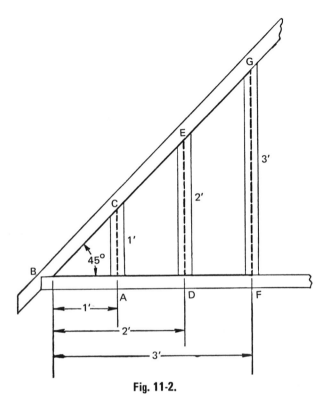

Fig. 11-2.

the corresponding pair of sides is $\frac{GF}{BF}$ or $\frac{3}{3}$, and this also equals 1. The student should try this on other ratios.

It is only by changing the size of the angles that the values of the ratios will change.

EXERCISES

In the following figures, measure each of the lines in the triangle to the nearest 16th of an inch. Find the value of the ratio BC/AB, the ratio BC/AC, and the ratio AC/AB.

1.   2.

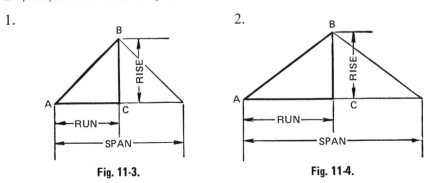

Fig. 11-3.   Fig. 11-4.

3.

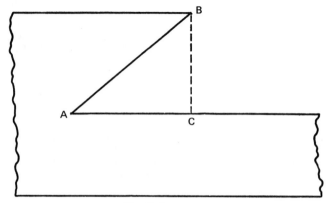

Fig. 11-5.

4.

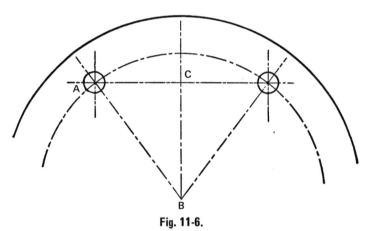

Fig. 11-6.

The sine of an angle

The ratios which we have been finding are of such great importance that they have been given names.

In the right triangle, *ABC* (Fig. 11-7), the side opposite angle *A* has been named side *a*; the side opposite angle *B* has been named *b*; and the side opposite angle *C* has been named *c*.

Fig. 11-7.

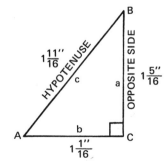

The side AC, or b, is said to be *adjacent* to angle A; it is one of the segments which are the sides of angle A. The side BC, or a, is said to be the side *opposite* to angle A. This side is *not* a part of angle A. The side AB, or c, which is opposite the right angle, is the *hypotenuse* of the right triangle.

Consider first the angle A. The ratio formed by dividing the length of the side opposite the angle A, the side a, by the length of the hypotenuse, c, is called the *sine* of angle A. It is written briefly sin A and is read "sine of A."

$$\sin A = \frac{\text{measure of side opposite } A}{\text{measure of hypotenuse}} \text{, or for Fig. 11-7, } \sin A = \frac{a}{c}$$

By measuring the sides, we find that in Fig. 11-7, $a = 1\ 5/16''$ and $c = 1\ 11/16''$.

If we substitute these numerical values for a and c, we obtain

$$\sin A = \frac{1\ 5/16}{1\ 11/16} \text{ or } \frac{21/16}{27/16} \text{ or } \frac{21}{16} \times \frac{16}{27} = \frac{21}{27} \text{ or } \frac{7}{9} \text{; } \sin A = \frac{7}{9}$$

This means that the sine of angle A, which is the ratio of the measure of the side opposite A to the measure of the hypotenuse of the triangle, is equal to 7/9.

If we express the fraction as a decimal, we obtain

$$\sin A = 0.7778$$

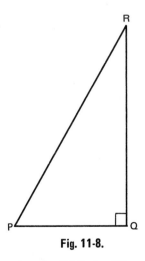

Fig. 11-8.

Example. In the right triangle PQR in Fig. 11-8, Q is the right angle. Find the sine of angle P, sin P.

THINK	DO THIS
1. Since sin *P* equals the ratio of the length of the side opposite *P*, the side *RQ*, to the length of the hypotenuse *PR*, measure *RQ*.	1. RQ = 2 1/16"
2. Measure the hypotenuse *PR*.	2. PR = 2 3/8"
3. Divide the measure of side *RQ* by the measure of hypotenuse *PR*.	3. $\sin P = \dfrac{2\,1/16}{2\,3/8}$ *Ans.* sin *P* = 33/38 = 0.8684

Summary

The sine of an angle *R*, written sin *R*, in a right triangle is equal to the ratio of the measure of the side of the triangle opposite angle *R* to the measure of the hypotenuse. To find the sine of angle *R*:

	EXAMPLE
	In triangle *PQR*, find sin *R*.
1. Measure the side opposite angle *R*.	1. Side opposite angle $R = 1\,1/16$"
2. Measure the hypotenuse, *PR*.	2. Hypotenuse = 2 3/8"
3. Divide the first by the second.	3. $\sin R = \dfrac{1\,1/16}{2\,3/8} = \dfrac{17}{38} = 0.4474$ *Ans.* sin *R* = 0.4474

EXERCISES

In the following exercises express the numerical value of the sine in both the fractional and the decimal forms.

1. From Fig. 11-9, find the sine of *A*.
2. Compute the numerical value of sin *T* in Fig. 11-10.
3. In Fig. 11-11, compute the numerical value of sin *M* and also compute the numerical value of sin *N*.
4. Compute the numerical value of sin *T* and also of sin *S* (Fig. 11-12). Each small division represents 1/10 in.

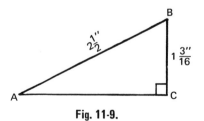

Fig. 11-9.

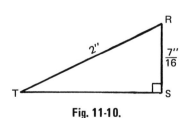

Fig. 11-10.

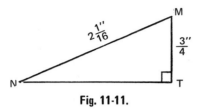

Fig. 11-11.

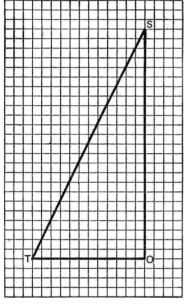

Fig. 11-12.

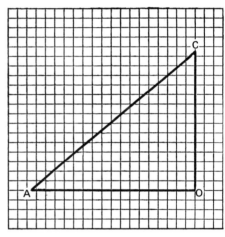

Fig. 11-13.

5. Compute the numerical value of sin C and also of sin A (Fig. 11-13). Each small division represents 1/10 in. Can you explain why sin C is not the same value as sin A?

The reciprocal of a number

The reciprocal of a number is the result obtained by dividing 1 by that number.

Example 1. Find the reciprocal of 8.

THINK	DO THIS
1. The reciprocal is 1 divided by the number.	1. Divide 1 by 8.
	Ans. 1/8

Example 2. Find the reciprocal of 5/16.

THINK	DO THIS
1. The reciprocal is 1 divided by the given number.	1. Divide 1 by 5/16. $$1 \div \frac{5}{16} = 1 \times \frac{16}{5} = \frac{16}{5}$$ *Ans.* 16/5

Note: In Fig. 11-14 sin R is 3/5. The reciprocal of sin R, that is, 1/sin R or $\frac{1}{3/5}$, is equal to $1 \div \frac{3}{5}$ or $\frac{1}{1} \times \frac{5}{3}$ or $\frac{5}{3}$. If you look at Fig. 11-14 you will notice that 5/3 is the ratio of the measure of the hypotenuse to the measure of the side opposite angle R, that is side ST.

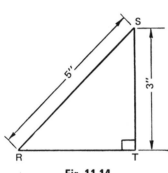

Fig. 11-14.

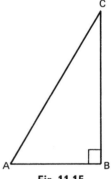

Fig. 11-15.

This ratio also has a name, the *cosecant* of angle R, written csc R. If you ever meet cosecant R in your work, always replace it by the reciprocal of sin R, 1/sin R.

Summary

Summary	EXAMPLE
1. The reciprocal of any number is 1 divided by the number.	1. In Fig. 11-15 find the reciprocal of sin A. Find cosecant A. (a) Measure AC = 2 9/16″ (b) Measure BC = 2 3/16″
2. The reciprocal of sin A equals $\frac{1}{\sin A}$.	2. $\sin A = \frac{2\,3/16}{2\,9/16} = \frac{35}{41}$ Reciprocal of $\frac{35}{41} = \frac{1}{35/41} = \frac{41}{35}$
3. Cosecant $A = \frac{1}{\sin A}$, written $\csc A = \frac{1}{\sin A}$.	3. Cosecant A, or csc A = 41/35 = 1.1714 *Ans.* 41/35 = 1.1714 = cosecant A

Note. You may wish to learn that the cosecant is the ratio of the measure of the hypotenuse to the measure of the side opposite the angle.

EXERCISES

1-5. Find sin A and sin B in each of the following right triangles.

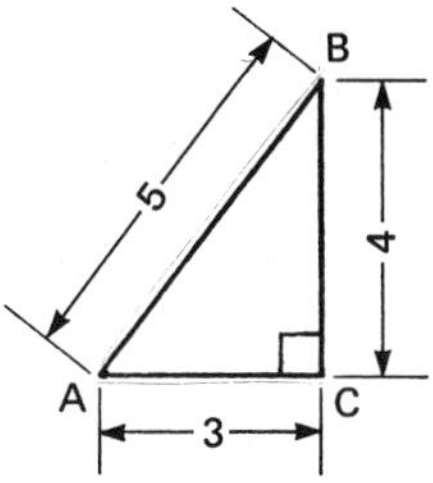

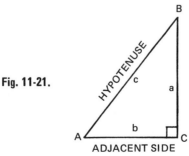

Fig. 11-16.

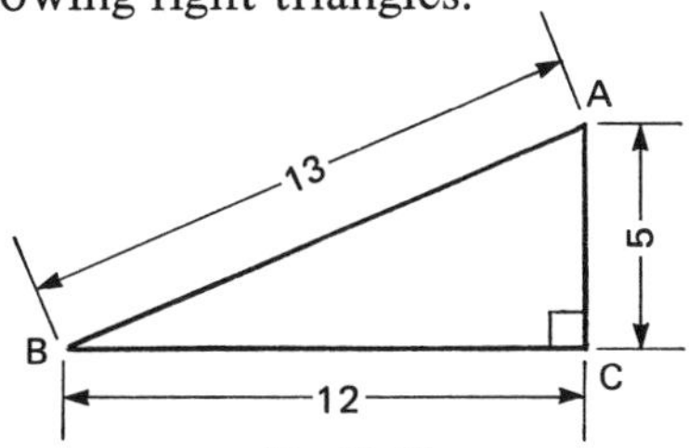

Fig. 11-17.

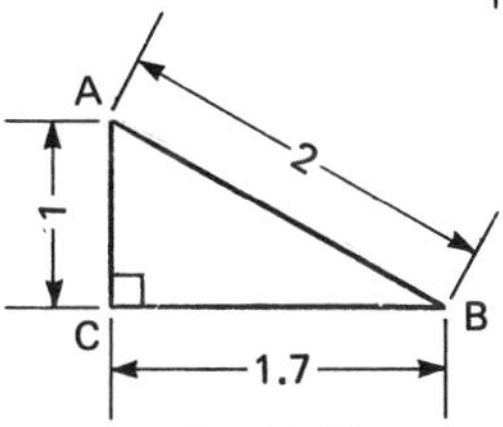

Fig. 11-18.

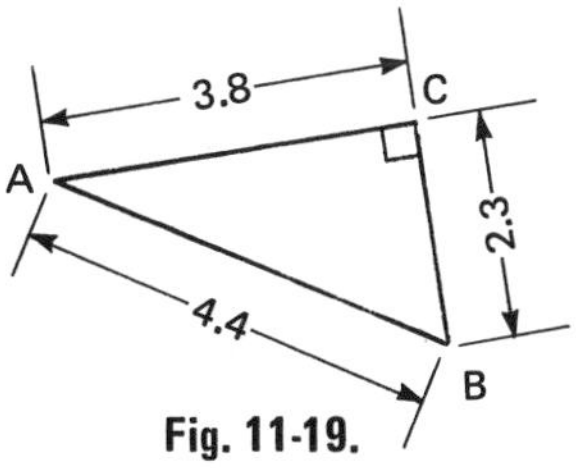

Fig. 11-19.

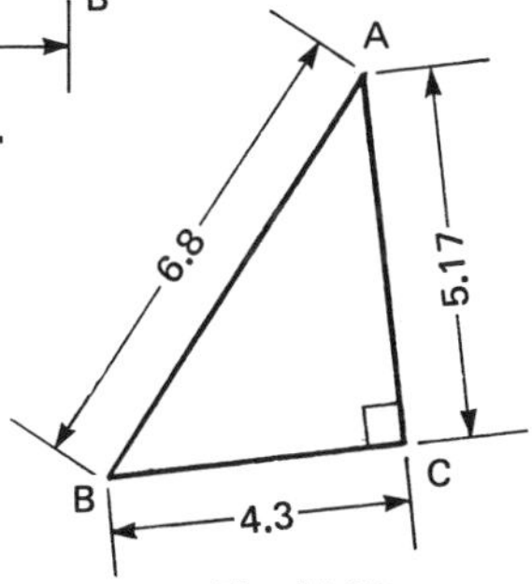

Fig. 11-20.

6-10. Write also the reciprocal of sin A and sin B (csc A and csc B) in the above exercises.

The cosine of an angle

In the right triangle (Fig. 11-21), the side adjacent to angle A is side b.

Fig. 11-21.

ADJACENT SIDE

The ratio, b/c, formed by dividing the measure of the side adjacent to A, side b, by the measure of the hypotenuse, c, is called the *cosine* of angle A. This is written briefly cos A.

We can write

$$\cos A = \frac{\text{measure of side adjacent to } A}{\text{measure of hypotenuse}} \text{ or } \frac{b}{c}$$

The cos $B = \frac{a}{c}$. Can you state why this is so?

Example. From Fig. 11-22, find the numerical value of the cosine of B.

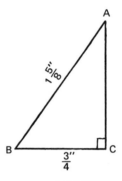

Fig. 11-22.

THINK	DO THIS
1. Since cos B is the ratio of the measure of the side adjacent to B and the measure of the hypotenuse, we (a) Measure BC. (b) Measure hypotenuse AB.	1. (a) $BC = 3/4''$ (b) $AB = 1\ 5/8''$
2. Write the formula for cos B.	2. cos $B = \dfrac{BC}{AB}$
3. Replace the variables BC and AB by their values.	3. cos $B = \dfrac{3/4}{1\ 5/8} = \dfrac{6}{13}$
4. Express the result as a decimal.	4. cos $B = 0.4615$ *Ans.* cos B = 6/13 or 0.4615
Summary	EXAMPLE
	Find cos P in Fig. 11-23.
1. The cosine of an acute angle in a right triangle is the ratio obtained by dividing the measure of the side adjacent to the angle by the measure of the hypotenuse.	1. Side adjacent to angle $P = 1\ 3/16''$ Hypotenuse = $3''$

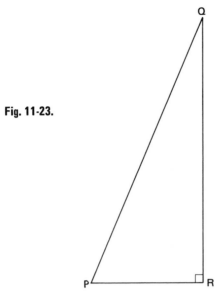

Fig. 11-23.

2. To find the cosine of an angle in a right triangle divide the measure of the side adjacent to the angle by the measure of the hypotenuse.

2. $\cos P = \dfrac{1\ 3/16}{3} = \dfrac{19}{48}$

Ans. $\cos P = 0.3958$

EXERCISES

1-5. Find $\cos A$ and $\cos B$ for each of the triangles in Exercises 1-5 on page 409.

The reciprocal of $\cos B$ is equal to $\dfrac{1}{\cos B}$. This is the ratio obtained by dividing the measure of the hypotenuse by the measure of the side adjacent to the angle. In Fig. 11-23,

$$\cos P = \frac{PR}{PQ}; \quad \frac{1}{\cos P} = \frac{1}{\dfrac{PR}{PQ}} = \frac{PQ}{PR}.$$

This ratio is called the *secant* of the angle P, written sec P. Hence,

$$\sec P = \frac{1}{\cos P} = \frac{\text{measure of hypotenuse}}{\text{measure of adjacent side}}$$

6-10. Write sec A and sec B for each of the Exercises 1-5 on page 409.

The tangent of an angle

In the right triangle shown in Fig. 11-24, page 412, the side BC opposite angle A is 2 inches long; the side adjacent to angle A is the side AC, also 2 inches long.

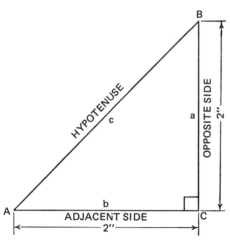

Fig. 11-24.

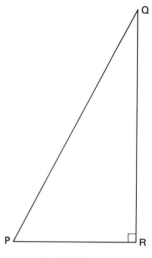

Fig. 11-25.

The ratio obtained by dividing the measure of the side opposite the angle A by the measure of the side adjacent to angle A is called the *tangent* of angle A. Written briefly, the tangent of angle A is tan A.

In Fig. 11-24,

$$\tan A = \frac{\text{measure of side opposite } A}{\text{measure of side adjacent to } A}$$

$$\tan A = \frac{BC}{AC}$$

$$\tan A = 2/2 \text{ or } 1$$
$$\tan A = 1$$

Example 1. Find the numerical value of the tangent of each of the acute angles in Fig. 11-25.

THINK	DO THIS
1. Since tan P is the length of the side opposite P divided by the length of the side adjacent to P, first measure QR and PR.	1. $QR = 2\,7/16''; PR = 1\,5/16''$
2. Write the formula for tan P.	2. $\tan P = \dfrac{QR}{PR}$
3. Replace the variables by the values obtained for them.	3. $\tan P = \dfrac{2\,7/16}{1\,5/16}$
4. Make the computations.	4. $\tan P = \dfrac{39}{21} = 1.8571$

5. Proceed similarly for tan Q.

5. $\tan Q = \dfrac{PR}{QR} = \dfrac{21}{39}$

$\tan Q = 0.5385$

Ans. $\tan P = 1.8571$;
$\tan Q = 0.5385$

The reciprocal of the tan P is $\dfrac{1}{\tan P}$. For Fig. 11-25 $\tan P = \dfrac{39}{21}$. The reciprocal of tan P is $\dfrac{1}{39/21}$ or $\dfrac{21}{39}$. The reciprocal of the tangent is called the *cotangent*. Then in this figure cotangent P, which is written cot P, equals 21/39 or 0.5385.

Notice that cot P is the same as tan Q in the right triangle PQR.

$\cot P = \dfrac{\text{measure of side adjacent to } P}{\text{measure of side opposite to } P} = \dfrac{PR}{QR}$. This equals tan Q.

Example 2. Find the numerical values of the tangent and the cotangent of the acute angles of the triangle in Fig. 11-26.

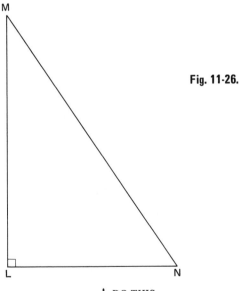

Fig. 11-26.

THINK	DO THIS
1. Since tan N is $\dfrac{\text{measure of opposite side}}{\text{measure of adjacent side}}$, or $\dfrac{ML}{LN}$, find the lengths of these sides.	1. $ML = 2\,5/8''$; $LN = 1\,13/16''$
2. Write the formulas for tan N and cot N.	2. $\tan N = \dfrac{ML}{LN}$; $\cot N = \dfrac{LN}{ML}$

3. Substitute these values in place of the variables in the formula.

3. $\tan N = \dfrac{2\ 5/8}{1\ 13/16}$; $\cot N = \dfrac{1\ 13/16}{2\ 5/8}$

4. Make the computations.

4. $\tan N = 1.4483$; $\cot N = 0.6905$

5. Proceed similarly for $\tan M$ and $\cot M$.

5. $\tan M = \dfrac{LN}{ML} = \dfrac{1\ 13/16}{2\ 5/8} = 0.6905$

$\cot M = \dfrac{ML}{LN} = \dfrac{2\ 5/8}{1\ 13/16} = 1.4483$

Ans. $\tan N = 1.4483$
$\cot N = 0.6905$
$\tan M = 0.6905$
$\cot M = 1.4483$

Summary

The tangent of an acute angle of a right triangle is the ratio obtained by dividing the length of the side opposite the angle by the length of the side adjacent to the angle.

Briefly, tangent of angle $A = \dfrac{\text{measure of side opposite to } A}{\text{measure of side adjacent to } A}$

Also, cotangent of angle $A = \dfrac{\text{measure of side adjacent to } A}{\text{measure of side opposite to } A}$

EXAMPLE

To find the numerical value of the tangent of an acute angle of a right triangle:

Find the numerical value of the $\tan A$ and $\cot A$ in the triangle of Fig. 11-27.

1. Write the formula for tangent of the angle. For Fig. 11-27,

$\tan A = \dfrac{CB}{AC}$.

1. $\tan A = \dfrac{CB}{AC}$

2. Find the lengths of the sides of the triangle.

2. $CB = 2\ 11/16''$
$AC = 13/16''$

3. Replace the variables of the formula by their values.

3, 4. $\tan A = \dfrac{2\ 11/16}{13/16} = 3.3077$

4. Make the computations.

5. Cotangent of an angle is the reciprocal of the tangent of the same angle. For Fig. 11-27,

$\cot A = \dfrac{1}{\tan A} = \dfrac{AC}{CB}$

5. $\cot A = \dfrac{AC}{CB}$

$\cot A = \dfrac{13/16}{2\ 11/16} = 0.3023$

Ans. $\tan A = 3.3077$
$\cot A = 0.3023$

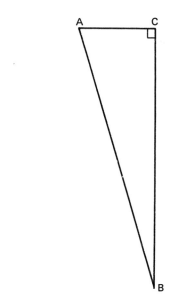

Fig. 11-27.

EXERCISES

Find the numerical value of the tangent of each of the acute angles in the following right triangles:

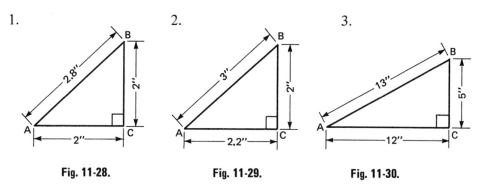

1.

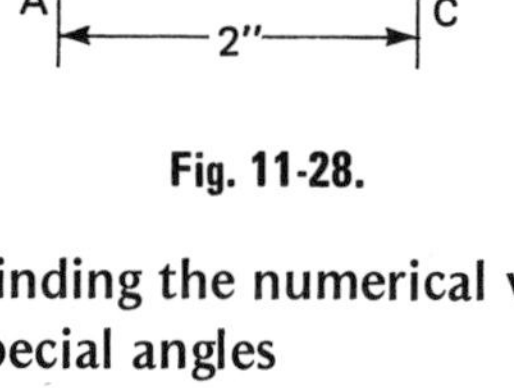

Fig. 11-28.

2.

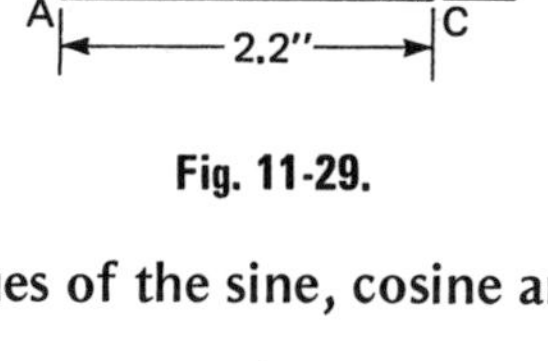

Fig. 11-29.

3.

Fig. 11-30.

Finding the numerical values of the sine, cosine and tangent of some special angles

Example. Find the numerical values of the sine, cosine and tangent of 45°.

THINK	DO THIS
1. Draw a right triangle of any convenient size that contains an angle of 45° (∠A). Make the sides 1″ each; then the hypotenuse must be 1.414″. Why?	1. See Fig. 11-31, page 416.

2. Write the formula for each function and substitute for each side the numerical value of that side.

2. $\sin 45° = \dfrac{BC}{AB} = \dfrac{1}{1.414} = 0.707$

$\cos 45° = \dfrac{AC}{AB} = \dfrac{1}{1.414} = 0.707$

$\tan 45° = \dfrac{1}{1} = 1.000$

Ans. $\sin 45° = 0.707, \cos 45° = 0.707,$
$\tan 45° = 1.000$

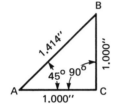

Fig. 11-31.

Summary

To find the sine, cosine, tangent of 45°, 30°, 60°:

1. Construct a right triangle that contains one of these angles as an acute angle. For the 45° angle use the fact that the sides of the right angle must be equal in length. For the 30°-60° triangle use the fact that the length of the side opposite the 30° angle is half the length of the hypotenuse.

2. Measure the lengths of the sides and of the hypotenuse.

3. Use these formulas to find the values required:

$\sin A =$

$\dfrac{\text{measure of side opposite } A}{\text{measure of hypotenuse}}$

$\cos A =$

$\dfrac{\text{measure of side adjacent to } A}{\text{measure of hypotenuse}}$

EXAMPLE

Find the values of sine, cosine, and tangent of 60°.

1. Construct a 30°-60° right triangle (Fig. 11-32).

First, construct a right angle *ABD*.

Second, construct a segment *AE* that is 2 x *AB*.

Third, with *AE* as a radius and *A* as center, make arc *RS* intersecting *BD* at *C*. In triangle *ABC* angle *A* = 60°, angle *C* = 30°.

2. If *AB* = 13/16, then *AC* = 1 5/8 or 13/8.

3. $\sin 60° = \dfrac{(13/16)\sqrt{3}}{13/8} = \dfrac{\sqrt{3}}{2}$

$= 0.8661$

$\cos 60° = \dfrac{13/16}{13/8} = \dfrac{1}{2} = 0.5$

$\tan 60° = \dfrac{(13/16)\sqrt{3}}{13/16} = \sqrt{3}$

$= 1.7321$

tan A

$$\frac{\text{measure of side opposite } A}{\text{measure of side adjacent to } A}$$

Ans. sin 60° = 0.8661,
cos 60° = 0.5,
tan 60° = 1.7321

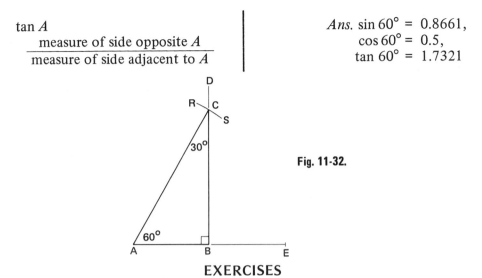

Fig. 11-32.

EXERCISES

1. Find the numerical values of the sine, cosine, and tangent of 30°. *Hint:* Construct an angle of 30°. Make the hypotenuse 2″ and the side opposite the 30° angle 1″ (p. 342). Measure length of the other side.
2. Using the same triangle as in Exercise 1, find the numerical values of the three ratios of 60°.

Finding the numerical values of the functions of any angle

Roughly speaking, a function is an expression by which one collection or set of numbers is matched, one for one, with another set of numbers. The sine, cosine, tangent, cotangent, secant and cosecant for angles between 0° and 180° each match this set of angles with the set of numbers. Take the case of sin A for each angle between 0° and 180° inclusive; 82° for example, if put in place of A, results in the sine of a special angle, sin 82°, which is a unique number, 0.9903 (see Appendix). So 82° is matched with the number 0.9903 by sin A. For this reason we call sin A a function and speak of it as a function of the angle A. We speak of 0.9903 as *the value of the function sin A for A = 82°*. Note especially that the A of sin A represents an angle, while sin A represents a number and sin A matches the angle with the number in a unique way.

The values of the functions of any angle may be determined from tables such as the one in the Appendix. This table gives values of sine, cosine, and tangent functions for every 1° angle from 0° to 90°. Other tables are available which include cotangent, secant and cosecant values and also smaller variations of angles, some in 10 minute intervals and some in one minute intervals. From this table, with a little practice, we can find the values of the functions of any angle.

Across the top are the names of the three functions. In the first column are the degrees from 0 to 45, and in the fifth column are degrees from 45 to 90. If you trace down the column headed "sine" you should notice that the angle starts at 0° and increases as you go down, and the numbers in the "sine" column also start at 0 and increase. When the values of the function increase as the values of the variable A, the number of degrees, increase, we say the function is an increasing function. Thus sin A is an increasing function. The cosine, however, starts at 1.0000 for cos 0° and as the angle A increases, cos A, the numbers in column headed "cosine," gets smaller and smaller so that when A is as large as 90°, cos A is zero. We call cos A a decreasing function.

In order to make these functions useful in solving problems, it is necessary to learn to use the tables to find the values of the functions of any angle between 0° and 90°.

Finding the numerical value of the function of an angle of a whole number of degrees

Example 1. What is the numerical value of sin 32°?

THINK

1. Since we are looking for the value of sin 32°, the value must be in the column headed "sine" and the angle must be in the line with 32° at the left.

DO THIS

1. Trace down degree column, the one headed "Degrees," until you meet 32°.

Degrees	sine	cosine	tangent
..	..	..	..
30	..	..	..
31	..	..	..
32	0.5299	..	..

2. Sin 32° must be in the column headed "sine."

2. Trace across this 32° line until you come to the sine column. Here we find the number 0.5299. Then sin 32° = 0.5299.

3. The number found in the 32° line and the sine column is sin 32°.

Ans. sin 32° = 0.5299

EXERCISES

Find the value of the sine of each of the following angles:

1. 42°
2. 87°
3. 2°
4. 51°
5. 76°
6. 32°

7. 75°
8. 61°

9. 32 1/2° (Can you do this?)
10. 57 1/2° (Can you do this?)

Example 2. Find the value of tan 58°.

THINK

1. Since 58° is greater than 45°, we must go to the fifth column of the table, the one headed "Degrees."

DO THIS

1. Start at top of the column headed "Degrees," trace down to the 58° line.

Degrees	sine	cosine	tangent
. .	. .	. .	. .
56	. .	. .	. .
57	. .	.	. .
58	. .	. .	1.6003

2. Since we seek the value of the tan 58°, we must find its value in the column headed "tangent" and the line of 58°.

2. Then trace to the right on the 58° line until you come to the column headed "tangent." The number in the tangent column and the 58° line is the value of tan 58°.

Ans. tan 58° = 1.6003

Summary

To find from the table the value of a function of an angle of a whole number of degrees:
1. Note whether the angle is greater or less than 45°.
2. If less than 45°, use column 1 of Degrees; if, greater than 45°, use column 5 of Degrees. Trace down the Degrees column until you find the given angle.
3. In the line of this angle trace across to the right to the column headed by the name of the function in question.
4. The number that stands in the column of the function and the line of the angle is the required value.

EXAMPLE A

Find the value of cos 84°.

1. 84° is greater than 45°.

2. Trace down column 5 to the line with 84°.

3, 4. Trace across the 84° line to the column headed "cosine." The number, 0.1045, which stands there is the value sought.

Ans. cos 84° = 0.1045

EXAMPLE B

Find the value of cot 63°.

5. Since the table does not contain the cotangents, secants, and cosecants, when the value of one of these is required first find the value of its reciprocal from the table and then compute the reciprocal of the number found. To find the reciprocal of the number found, divide 1.0000 by this number.

5. Cot 63° is not in the table. The cot 63° is the reciprocal of tan 63°.
Find the value of tan 63°.
tan 63° = 1.9626

$$\cot 63° = \frac{1}{\tan 63°} = \frac{1}{1.9626}$$

$$= 0.0510$$

Ans. cot 63° = 0.0510

EXERCISES

Find from the table of trigonometric functions the numerical value of each of the following:

1. tan 62°
2. sin 14°
3. cos 71°
4. cos 18°
5. tan 59°
6. cos 26°
7. sin 44°
8. cos 65°
9. sin 48°
10. tan 82°
11. cot 55°
12. sec 40°

Calculating the numerical values of the functions of angles of any number of degrees

We now learn how to find the ratios of angles that are not listed in the table. This is important; study carefully the examples and the summary that follow.

Example 1. Make use of the table of trigonometric functions to calculate the value of sin 56.8°.

THINK

Sin 56.8° is not given in this table. We know that sin 56.8° must lie between sin 56° and sin 57°. We make use of these to calculate sin 56.8° as follows:

1. Write the values of sin 56° and of sin 57°.

2. Write sin 56.8° between these two.

DO THIS

1, 2. sin 56° = 0.8290
sin 56.8° = ?
sin 57° = 0.8387

3. At this point we assume that since 56.8° is 0.8 of the difference between 56° and 57°, from 56°, the sin 56.8° will be 0.8 of the difference between 0.8290 and 0.8387, from 0.8290. This suggests:

(a) subtract 0.8290 from 0.8387.

(b) find 0.8 of this difference.

(c) since *sine is an increasing function*, add this result to sin 56°.

3. (a) 0.8387
0.8290
0.0097 difference between sin 56° sin 57°

(b) 0.8 x 0.0097 = 0.0078 difference between sin 56° and sin 56.8° Since the sine is an increasing function, 0.0078 is *added* to sin 56° to obtain sin 56.8°.

(c)   sin 56°   =   0.8290
      increase for 0.8°   =   0.0078
         sin 56.8°   =   0.8368

Ans. sin 56.8° = 0.8368

Example 2. Calculate the numerical value of cos 38.7°.

THINK

1. 38.7° is between 38° and 39°.

2. Write from the table the value of cos 38° and cos 39°.

Write cos 38.7° between these two.

3. (a) Find the difference from cos 38° to cos 39°.

(b) Since 38.7° is 0.7 of the difference between 38° and 39°, from 38°, we assume a proportional difference in the cosines.

(c) Since *cosine is a decreasing function*, subtract this difference from cosine of the smaller angle.

DO THIS

1, 2.   cos 38°   =   0.7880
      cos 38.7°   =   ?
      cos 39°   =   0.7771

3. (a) cos 38°   =   0.7880
      cos 39°   =   0.7771
         0.0109   *decrease*
      between cos 38° and cos 39°

(b) 0.7 x 0.0109 = 0.0076 decrease between cos 38° and cos 38.7°

(c)   cos 38°   =   0.7880
      decrease for 0.7°   =   0.0076
         cos 38.7°   =   0.7804

Ans. cos 38.7° = 0.7804

Summary	EXAMPLE
To find, by means of the table, the numerical value of the function of an angle, when that value is not written in the table:	Calculate the value of sec 69.4°.
1. Determine the two consecutive angles in the table between which the given angle lies.	1. 69.4° is between 69° and 70°.
2. Write down the values of the function in question that correspond to these two angles. Write the larger of the two values first.	2. Since secant is the reciprocal of the cosine, we write cos 69° = 0.3584 cos 70° = 0.3420
3. Find the difference between these two values. This will be the difference for one degree.	3. 0.3584 − 0.3420 = 0.0164 difference between cos 69° and cos 70°
4. Multiply this difference by the decimal or fraction in the given angle.	4. 0.4 x 0.0164 = 0.0066 decrease between cos 69° and cos 69.4°
5. If the function increases as the angle increases, add the result obtained in Step 4 to the smaller of the two values of the function; if the function decreases as the angle increases, subtract the result obtained in Step 4 from the larger of the two values of the function.	5. Since the cosine is a decreasing function, subtract cos 69° = 0.3584 decrease for 0.4° = 0.0066 cos 69.4° = 0.3518
6. If function desired is the reciprocal of the value obtained in Step 5, make the necessary calculations.	6. sec 69.4° = $\dfrac{1}{\cos 69.4°} = \dfrac{1}{0.3518}$ = 2.8425 *Ans.* sec 69.4° = 2.8425

EXERCISES

From table in the Appendix, calculate the numerical value of each of the following:

1. sin 47.2°
2. tan 26.6°
3. cos 41.5°

4. csc 70.4°
5. cot 65.9°
6. sec 20.1°

7. cos 46.9°
8. sec 78.2°
9. tan 17.4°

10. cot 36.7°   12. sin 82.1°   14. csc 14.5°
11. tan 53.6°   13. cos 16.9°   15. csc 30.7°

Degrees, minutes and seconds

Angles are sometimes expressed as 46° 18′ 15″. The single stroke, ′, is read *minutes*, the double stroke, ″, is read *seconds*. The given angle is read 46 *degrees*, 18 *minutes*, 15 *seconds*.

An angle of 1′ (one minute) is 1/60 of a degree, and one of 1″ (one second) is 1/60 of a minute. The table of angular measure is:

60′ = 1°; 60 minutes equals 1 degree
60″ = 1′; 60 seconds equals 1 minute

Example. Calculate the value of cos 46° 15′.

THINK	DO THIS
1. Express 15′ as a decimal part of a degree. Since 60 minutes equals 1°, 15′ would be 15/60 or 1/4 or 0.25 of a degree.	1. 15′ = 15/60 = 0.25° 46° 15′ = 46.25°
2. Calculate the value of cosine of this angle as summarized above.	2. cos 46° = 0.6947 cos 47° = 0.6820 difference for 1° = 0.0127 x decrease for 0.25° = 0.25 x 0.0127 = 0.0032 Since cosine decreases, subtract this result from cos 46° cos 46° = 0.6947 decrease for 0.25° = 0.0032 cos 46.25° = 0.6915 *Ans.* cos 46° 15′ = 0.6915

Summary

	EXAMPLE
To find the value of the function of an angle when the angle is expressed in degrees, minutes and seconds:	What is the numerical value of sin 18°13′30″?
1. Express the minutes and seconds as a fractional part of a degree.	1. 30″ = 30/60′ or 0.5′ 13.5′ = 13.5°/60 = 0.23° 18°13′30″ = 18.23°

2. Then find the value of the function of the angle by the procedure of p. 420.

2. sin 18° = 0.3090
sin 19° = 0.3256
difference for 1° = 0.0166
increase for 0.23° =
0.23 x 0.0166 = 0.0038
sin 18° 13′30″ = 0.3090 + 0.0038
= 0.3128

Ans. sin 18° 13′30″ = 0.3128

EXERCISES

Using the table of trigonometric functions, calculate the numerical value of each of the following:

1. sin 47° 45′
2. cos 20° 30′30″
3. tan 76° 24′
4. tan 16° 4′45″
5. cot 62° 16′15″

6. cos 8° 30′30″
7. sec 81° 24′45″
8. sin 31° 11′30″
9. csc 10° 52′15″
10. cot 45′45″

Finding the angle when a numerical value of a function of an angle is known

Example 1. Find angle A (briefly ∠A) when tan A = 0.8391.

THINK

1. Since 0.8391 is the tan A, this number or a number close to it must be found in the tangent column of the table.

2. If this number appears in this column, the angle A whose tangent equals 0.8391 is in the Degrees column in the line with 0.8391.

3. The angle in the same line with 0.8391 is the required angle.

DO THIS

1. Trace down the tangent column until a number near 0.8391 is found.

2. Trace the line in which 0.8391 is found across left to the degree column. The number here is 40.

3. 0.8391 = tan 40°.

Ans. ∠A = 40°

Example 2. Find $\angle B$ to the nearest tenth of a degree if sin $B = 0.8231$.

THINK	DO THIS

THINK

1. The value 0.8231 or a number close to it must be found in column headed sine. Since 0.8231 is not in the table, we must interpolate for the angle B. This means to proceed in a way similar to that of the summary on p. 422.

DO THIS

1. Trace down sine column to the numbers near 0.8231 (this value is between 0.8192 and 0.8290).

Degrees	sine	cosine	tangent
45	0.7071	..	..
..	..	..	..
54	0.8090	..	..
55	0.8192	..	..
56	0.8290	..	..

2 (a) Write down the values of the sines of the angles between which 0.8231 lies in the table.

2. (a) sin 55° = 0.8192
    sin B = 0.8231
    sin 56° = 0.8290

(b) Find the difference between the values for sin 55° and sin 56°. This is the difference in sine for 1° of angle.

(b) difference for 1° =
  0.8290 – 0.8192 = 0.0098

(c) Find the difference between sine of the smaller of the two angles (sin 55°) and sin B.

(c) sin B =   0.8231
    sin 55° =   0.8192
              0.0039 difference
    between sin 55° and the given sine value

(d) Divide last difference by the difference for 1°.

(d) $\frac{0.0039}{0.0098}$ = 0.39

The difference in Step c is approximately 0.4 of the difference in Step b. Thus angle B is 0.4 degree from 55°.

(e) Since sine is an increasing function, add this result to the smaller of the two angles.

(e) 55° + 0.4° = 55.4°
Ans. 0.8231   =   sin 55.4° (approx.)
             $\angle B$ = 55.4° (approx.)

Example 3. Find angle C to the nearest minute (') when $\cos C = 0.5270$.

THINK	DO THIS
1. The value 0.5270 must be in the column headed cosine.	1. Trace down column headed cosine until the number 0.5270 is found.

2. Since 0.5270 is not in the table, we find the two values in this column between which 0.5270 lies.

2.
Degrees	sine	cosine	tangent
45	..	0.7071	..
..	..	..	..
..	..	0.5446	..
58	..	0.5299	..
59	..	0.5150	..
..	..	0.5000	..

(a) Write down those two values.

(a) $\cos 58° = 0.5299$
$\cos C \ \ = 0.5270$
$\cos 59° = 0.5150$

(b) Find the addend to 0.5150 to equal 0.5299.

(b) difference for $1°$ =
$0.5299 - 0.5150 = 0.0149$

(c) Find the addent to 0.5270 to equal 0.5299.

(c) $0.5299 - 0.5270 = 0.0029$

(d) Divide this last result by the difference for $1°$.

(d) $\dfrac{0.0029}{0.0149} = 0.19$
$= 0.2°$ (approx.)

3. Since $1° = 60'$, we express this decimal part of a degree in minutes by multiplying it by $60'$.

3. $0.2 \times 60' = 12.0'$ (approx.)

4. Since the angle *increases* as the value of the cosine decreases, we *add* these minutes to the smaller of the two angles obtained from the table.

4. $58° + 12' = 58°12'$

Ans. $0.5270 = \cos 58°12'$ (approx.)
$\angle C = 58°12'$ (approx.)

Summary

To find, with the assistance of tables, an angle when the value of a function is known:

1. Determine from the tables the two consecutive values of the given function between which the given value lies.

EXAMPLE

Find D if $\cos D = 0.9000$. Express the result in degrees and minutes.

1. 0.9000 in the cosine column lies between 0.9063 and 0.8988.

2. Write down the angles that correspond to these two values.	2. cos 25° = 0.9063 cos D = 0.9000 cos 26° = 0.8988
3. Find the difference in values of the function for 1°.	3. 0.9063 – 0.8988 = 0.0075
4. Find the difference in values of the function of the smaller of the two angles and the given function.	4. 0.9063 – 0.9000 = 0.0063
5. Divide this last difference by the difference for 1°.	5. 0.0063 ÷ 0.0075 = 0.84 0.84 of 1° = 0.8° (approx.)
6. Add this to the smaller of the two angles in Step 2.	6. 25° + 0.8° = 25.8°
7. If the result is to be expressed in minutes, multiply the result of Step 5 by 60' and annex this result to the smaller of the two angles.	7. 0.8 x 60' = 48.0' 25° + 48' = 25°48' *Ans.* 0.9000 = cos 25°48' (approx.) $\angle D$ = 25°48' (approx.)

EXERCISES

Find the angle in each of the following. Express the angle in tenths of a degree and in degrees and minutes.

1. sin A = 0.2419	7. tan A = 0.6366	13. tan D = 4.6862
2. cos B = 0.3618	8. sin B = 0.5073	14. cos P = 0.7024
3. sin B = 0.1370	9. sin T = 0.4691	15. tan R = 1.2688
4. cos D = 0.8650	10. cos M = 0.3472	16. sin T = 0.3320
5. tan A = 3.5027	11. sin N = 0.7565	17. cos M = 0.0548
6. cos C = 0.3420	12. cos P = 0.6266	18. cos R = 0.2573

Using the trigonometric functions to solve problems that involve right triangles

The pitch of the roof of a house is by definition the rise divided by the span (see unit 12). One might be interested to know what angle a roof makes with the horizontal if the rise is 12' and the span 32'. This involves a right triangle (ABC) in which the angle A is wanted. It can be found by the methods learned in the last several pages. In fact, if we know the length of one side of a right triangle and one other part, such as another side or an acute angle, we can find any of the other parts.

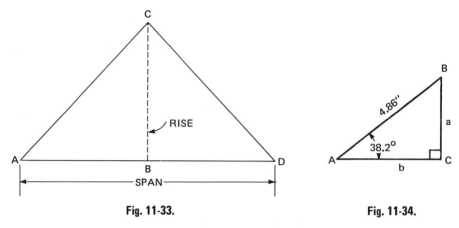

Fig. 11-33.   **Fig. 11-34.**

The following examples are illustrative of the thinking that must be done in such cases.

Example. Find all the parts of a right triangle (Fig. 11-34) in which the hypotenuse is 4.86″ (inches) and one of the acute angles is 38.2°.

THINK	DO THIS
Draw a carefully constructed figure showing the known parts and naming the other parts; indicate angles by capital letters and sides by small (lowercase) letters.	
1. To find side a:	1. Draw Fig. 11-34 on your paper.
(a) Ask what function is involved in the parts that are known and the part to be found.	(a) sin A involves a, AB and angle A.
(b) Write this function and make it equal to the proper ratio of sides. This results in an equation that involves the part to be found, here a.	(b) $\sin 38.2° = \dfrac{a}{4.86}$
(c) Solve this equation for the part to be found.	(c) $\sin 38.2° \times 4.86 = \dfrac{a \times 4.86}{4.86}$; $4.86 \times \sin 38.2° = a$

(d) Use the table of trigonometric functions to find
the value of the function
involved.

(d) sin 38° = 0.6157
sin 38.2° = ?
sin 39° = 0.6293
difference for 1° = 0.0136
0.2 x 0.0136 = 0.00272 or
0.0027
sin 38.2° = 0.6157 + 0.0027
= 0.6184

(e) Substitute the value of the
function thus found and
complete the computa-
tions for the part required.

(e) From Step c,
$a = 4.86 \times 0.6184 = 3.01$

Ans. Side $a = 3.01''$

2. To find side b:
(a) Function involved?
(b) Write the function equal
to the proper ratio.
(c) Solve this equation for
the part to be found.
(d) Use the table to find the
value of the function
involved.
(e) Substitute this value in the
equation and find the re-
quired part, b.

2. (a) $\cos A$ involves side b and AB.

(b) $\cos A = \dfrac{b}{4.86}$

(c) $b = 4.86 \times \cos A$
$b = 4.86 \times \cos 38.2°$
(d) $\cos 38.2° = 0.7880 - 0.0022$
$\cos 38.2° = 0.7858$

(e) $b = 4.86 \times 0.7858 = 3.82$

Ans. Side $b = 3.82''$

3. To find angle B:

Since the sum of the two
acute angles of a right triangle is 90° and we know one
of them, 38.2°, we ask what
angle added to 38.2° equals
90°.

3. $\angle A + \angle B = 90°; \angle A = 38.2°$
$90° - 38.2° = 51.8°$ or $51°48'$

Ans. $\angle B = 51.8°$ or $51°48'$

A second method for finding angle B may be used as a check on the
earlier work as well as the size of $\angle B$.

THINK

(a) To find $\angle B$, using functions,
we would need a ratio, either
$\dfrac{a}{AB}$ or $\dfrac{b}{AB}$.

DO THIS

(b) Write a function of B equal to one of these ratios.

(c) Substitute value of b found in Step 2.

(d) Find value of B using the table.

(b) $\sin B = \dfrac{b}{4.86}$

(c) $\sin B = \dfrac{3.82}{4.86} = 0.7860$

(d) $\sin 52° = 0.7880$
$\sin B = 0.7860$
$\sin 51° = 0.7772$
difference for $1° = 0.0108$
$\sin B = 0.7860$
$\sin 51° = 0.7772$
difference is 0.0088
$\dfrac{0.0088}{0.0108} = 0.807$ or $0.8°$

Ans. $\angle B = 51° + 0.8° = 51.8°$ or $51°48'$

Since this result agrees with the result in Step 3, we accept the work as correct.

Summary

To find a part of a right triangle of which one side and either a second side or one of the acute angles is known, first draw a reasonably accurate figure. Place the values on the known parts and variables (letters) on the ones not given.

1. When a side of the triangle is to be found:

   (a) Write a function of a known angle that involves the given side and the side to be found.

   (b) Solve the equation for the unknown side.

   (c) Substitute the numerical value for the function.

EXAMPLE

In a right triangle ABC with right angle at C, side $AC = 1.3125'$ (ft.) and angle $A = 68°34'$:

1. Find the length of side BC.

   (a) Known side $AC = 1.3125$

     tangent $68°34' = \dfrac{BC}{1.3125}$

   (b) $BC = 1.3125$ x tan $68°34'$

   (c) $\tan 68° = 2.4751$
     $\underline{\tan 69° = 2.6051}$
     diff. $1° = 0.1300$
     diff. $34' = 34/60$ x 0.1300
     $= 0.0737$
     $\tan 68°34'$
     $= 2.4751 + 0.0737 = 2.5488$

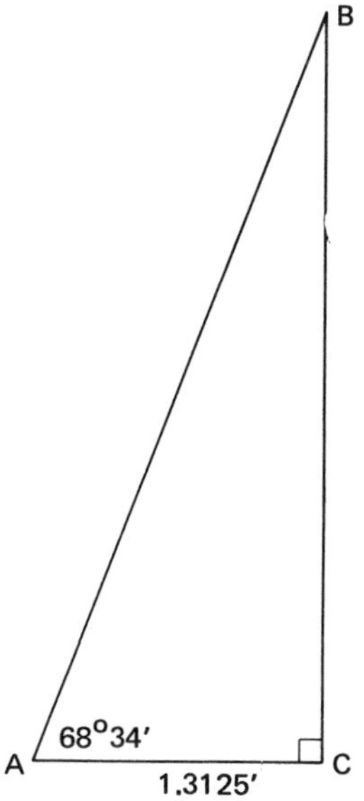

Fig. 11-35.

(d) Make the computations to find the length of the side.

(d) $BC = 1.3125 \times 2.5488$
$BC = 3.3453'$
Ans. $BC = 3.3453'$

2. If an angle is to be found:

2. Find angle B, if $BC = 3.3453'$ and $AC = 1.3125'$.

(a) Write a function of that angle which involves two known sides.

(a) $\tan B = \dfrac{1.3125}{3.3453}$

(b) Perform the division. Carry the result to as many decimal places as there are places in the table to be used, here 4 places.

(b) $\tan B = 0.3923$

(c) Find the angle by the method of p. 424.

(c) $\tan 21° = 0.3839$
$\tan B \quad = 0.3923$
$\tan 22° = 0.4040$
diff. for $1° = 0.0201$
diff. $21°$ to $B = 0.0084$
$\dfrac{0.0084}{0.0201} = 0.42$

0.42 of $1° = 0.4°$ (approx.)
$\angle B = 21.4°$ or $21°24'$

Ans. $\angle B = 21.4°$ or $21°24'$

Note: For convenience in computation, the figures in parts (b) and (c) may be arranged in a chart like this:

Tangent	Angle
0.3839	21°
0.3923	x
0.4040	22°

EXERCISES

Find all the parts of the following right triangles:

1.

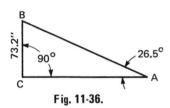

Fig. 11-36.

2.

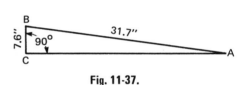

Fig. 11-37.

3.

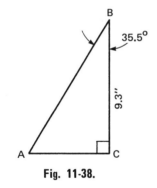

Fig. 11-38.

4.

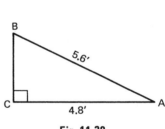

Fig. 11-39.

5. A right triangle in which the two sides (not the hypotenuse) are 16.2″ and 11.7″.
6. One in which the hypotenuse is 51.0′ and one angle is 74.3°.
7. One in which one side is 4.55″ and the angle opposite this side is 18°15′.
   Note: In 18°15′, 15′ is read "fifteen minutes." There are 60 minutes in one degree; hence, 15′ is 15/60 or 0.25 of a degree.
8. One in which the hypotenuse is 426 feet and one angle is 32°40′.
   Hint: Be sure to change 40′ to a decimal part of a degree.
9. One in which one side (not the hypotenuse) is 72.8 feet and the angle adjacent to this side is 54°48′.

PROBLEMS

1. Find the length of the rafter and the angle of elevation in Fig. 11-40 for a 12-ft. run and an 8-ft. rise.

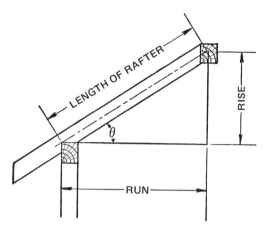

Fig. 11-40. Length of a rafter.

2. If the pitch of a gable roof is the total rise divided by the span, find θ, the angle of pitch, for each of the roofs in the following table:

Pitch	Rise	Span	Run*	Tan θ	θ
Full pitch	24	24	12		
3/4 pitch	18	24	12		
1/2 pitch	12	24	12		
1/3 pitch	8	24	12		
1/4 pitch	6	24	12		

*Run = 1/2 Span. See Fig. 11-3.

3. Figure 11-41 shows a sectional view of a dovetail. If $\angle\theta$ is 60°, h is 5/8″, and L is 3.5″, find w.

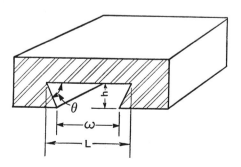

Fig. 11-41. Sectional view at dovetail.

4. The taper reamer in Fig. 11-42 has a flute length of 8″. The diameter D at the large end of the flute is 1.8″, while that of the small end is 1.4″. Find the included angle of taper.

Fig. 11-42. Taper hand reamer.

5. The carpenter's square (Fig. 11-43) is a convenient device for laying out angles. If the distance on the tongue is 9″ and that on the blade is 14″, what are the values of the angles θ and ϕ?

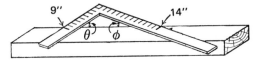

Fig. 11-43. Carpenter's square for laying out angles.

6. Nine holes are to be drilled in a plate of steel on the circumference of a circle having a diameter of 7.7 in. (Fig. 11-44). What will be the distance (c) between the centers of two adjacent holes?

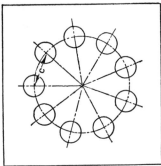

Fig. 11-44. Laying out nine holes to be drilled in steel plate.

Concerning oblique triangles

In this unit we have been working only with right triangles. There are, of course, many triangles that are not right triangles. These are called oblique triangles, but work with them is beyond the scope of this text.

Very often oblique triangles may be broken up into a number of right triangles and the parts found by the methods we have learned in this unit. If that cannot be done, the student is referred to a book such as *Algebra and Trigonometry, a Modern Approach*,* Book 2, by Peters and Schaaf, in which the solution of oblique triangles is discussed.

---

*Published by D. Van Nostrand Co., Inc.

REVIEW PROBLEMS

1. Find the length of the guy wire picture in Fig. 11-45.

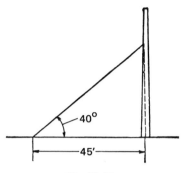

Fig. 11-45.

2. The distance the tailstock must be moved over to turn a taper is called the set-over. Whenever the taper is given in degrees, the set-over must be obtained from a table of set-overs or computed by trigonometry. Compute the set-over for the taper illustrated in Fig. 11-46.

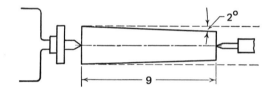

Fig. 11-46. Set-over of lathe tailstock for taper.

3. Find the taper angle θ for the taper in Fig. 11-47.

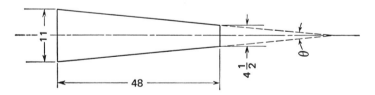

Fig. 11-47. Finding taper angle θ.

4. The sprocket wheel illustrated in Fig. 11-48 has ten teeth. Find the diameter of the pitch circle if the chordal distance C is 1.25".

5. The distance across the corners of the hexagonal nut illustrated in Fig. 11-49 is 1.75". Find the distance across the flats.

6. Find the distance (A) across the head of the flathead cap screw shown in Fig. 11-50 if $D = 7/16"$ and $H = 0.220"$.

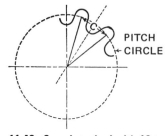

Fig. 11-48. Sprocket wheel with 10 teeth.

Fig. 11-49. Hexagonal nut.

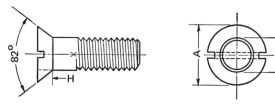

Fig. 11-50. Flathead cap screw.

7. The Acme thread in Fig. 11-51 has a pitch of 1/2″. If the depth of the thread (D) is 0.2600″, and the width of the top or flat (F) is 0.1853″, find R, the flat at the root of the thread.

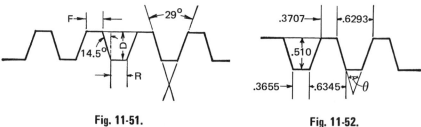

Fig. 11-51.   Fig. 11-52.

8. Find the angle of the thread in Fig. 11-52. What kind of thread is it?

9. The angle ϕ at which the 10″ sine bar in Fig. 11-53 is set measures 34°40′. If disk A just touches the surface plate, what should be the value of h when measured with a vernier height gage?

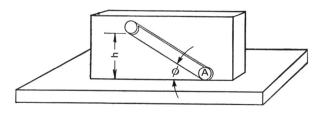

Fig. 11-53. Sine bar set for 34°40′.

10. What is the angle measured by a sine bar when the difference in the height between the disks of a 5″ bar is 3.4723″?

11. In Fig. 11-54, what is the center-to-center distance on a straight line between adjacent holes?

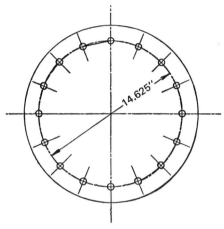

Fig. 11-54.

12. What is the diameter of the wire that will be flush with the top of a 7/8″-9 thread (Fig. 11-55). The wire will be tangent to the side of the thread at the pitch line.

13. Find the center-to-center distances missing in Fig. 11-56.

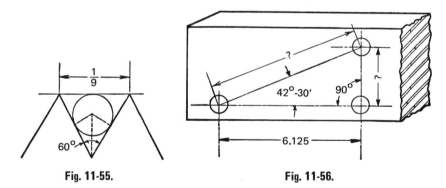

Fig. 11-55.   **Fig. 11-56.**

14. Find the helix angle of a 1 1/4″-7 N.C. thread. The pitch diameter D of the screw is 1.1572″.

    Hint: In a right triangle the side opposite the helix angle ϕ is equal to 1/pitch. The adjacent side is equal to the circumference at the pitch circle (Fig. 11-57, page 438).

15. In Fig. 11-58, page 438, find the $\angle\theta$ if the rise is 8′0″ for a span of 24′0″.

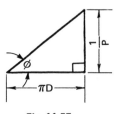

Fig. 11-57.

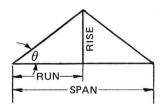

Fig. 11-58.

16. Find the length of each member of the roof truss (Fig. 11-59) if the span is 120 ft. and the rafters make an angle of 43° 15′ with the horizontal. The struts divide the span into three equal parts. The tie rods are perpendicular to the rafters.

17. Find the length of the rafter of the deck roof pictured in Fig. 11-60.

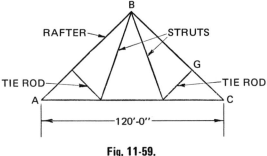

Fig. 11-59.

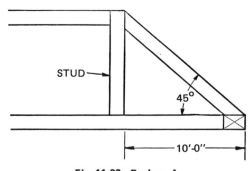

Fig. 11-60. Deck roof.

18. (a) Find the distance A across the base of the internal dovetail (Fig. 11-61) if the diameter of the plug P is 3/8″.

    (b) Find the distance B.

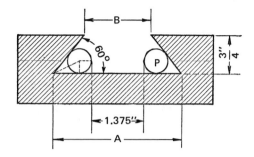

Fig. 11-61. Internal dovetail.

19. Find B for the external dovetail in Fig. 11-62.

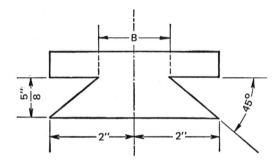

Fig. 11-62. External dovetail.

20. In Fig. 11-63, find the distances X and Y.

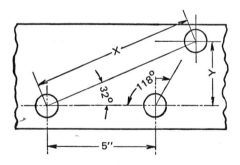

Fig. 11-63.

12 Mathematics of the Building Trades

The building trades include such fields as carpentry, plumbing, heating, electrical work, painting and decorating, masonry and concrete work. In this unit you will meet and learn to deal with some of the important problems that are met in some of these fields. Study the definitions until you know them thoroughly.

Board measure

The unit of measure that is used by the lumber trade is the board foot. A *board foot* (bd. ft.) is the amount of lumber contained in a board one inch thick, one foot wide, and one foot long, or its equivalent (see Fig. 12-1). A board two inches thick, one-half foot wide, and one foot long also contains exactly one board foot.

For example, a board 1 inch thick, 1 foot wide and 16 feet long would contain 16 board feet.

When a board is less than one inch thick, for purposes of computing board feet it is considered as though it were one inch thick. For example, a board 1/2 inch thick, 1 foot wide and 1 foot long contains 1 board foot. However, if a board is 1 1/2 inch thick, 1 foot wide and 1 foot long, it contains 1 1/2 board feet.

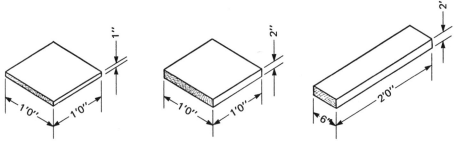

| Fig. 12-1. One board foot. | Fig. 12-2. Two board feet. | Fig. 12-3. Two board feet. |

Example 1. How many board feet are there in ten pieces of maple if each piece is 2" thick, 6" wide and 14' long?

THINK	DO THIS
1. Find the number of board feet in one piece. To do this we first express the thickness in inches and the width and length in feet. Then multiply to find board feet.	1. $2 \times \dfrac{6}{12} \times 14 = 14$
2. Multiply this result by the number of pieces.	2. $14 \times 10 = 140$ *Ans.* 140 board feet

Example 2. How many board feet are in 40 pieces of oak each 3/4″ x 10″ x 18′?

THINK	DO THIS
1. Find the number of board feet in one piece. Even though the thickness is 3/4″, we use 1″ for the thickness in the calculations.	1. $1 \times \dfrac{10}{12} \times 18 = 15$
2. Multiply this result by 40.	2. $15 \times 40 = 600$ *Ans.* 600 board feet

Summary	EXAMPLE Find the number of board feet in 38 pieces of hemlock, each 2 1/2″ x 4″ x 18′.
1. To find the number of board feet of lumber in one piece, multiply the number of inches in the thickness by the number of feet in the width by the number of feet in the length. (A thickness of less than one inch is taken as one inch.)	1. $2\frac{1}{2} \times \dfrac{4}{12} \times 18 = 15$
2. When several pieces are alike, multiply the number of board feet in one piece by the number of pieces.	2. $15 \times 38 = 570$ *Ans.* 570 board feet

EXERCISES

Determine the number of board feet in each of the following exercises:
1. A piece of oak 2″ x 10″ x 12′.*
2. A piece of poplar 1/2″ x 8″ x 16′.
3. In 12 pieces of yellow pine 8″ x 8″ x 16′.

*When dimensions are written as 2″ x 10″ x 12′, the first number is the thickness, the second the width and the third the length. In dimensions, the "x" is read "by."

4. In 30 pieces of hemlock 2″ x 6″ x 22′.
5. In 58 pieces of 7/8″ x 4″ x 16′ of clear white pine, D 4 S.*
6. In 10 pieces of clear white pine 7/8″ x 4″ x 12′.
   16 pieces of clear white pine 5/4″ x 3″ x 14′.
   10 pieces of clear white pine 5/4″ x 3″ x 18′.
7. Find the cost of the lumber in Example 6 when clear white pine is $438.00 per 1000 bd. ft.

PROBLEMS

1. How many board feet of lumber are there in a piece of white oak 1 3/4″ x 11″ x 18′?
2. A load of shelving consists of 45 pieces of 7/8″ x 10″ x 12′ yellow pine. How many board feet of lumber are there in this shipment? (First find the number of board feet in each piece.)
3. In estimating the amount of lumber in the rack of a certain lumber-yard, it was found that the lumber was stacked 5′ 2 1/2″ high and 6′4″ wide. Assuming each board to be 14′ long, find the number of board feet in the stack.
4. A cabinetmaker orders the following lumber in the rough. How many board feet of each size does he require? What is the total number of board feet in his order?
   12 pieces oak 3/4″ x 8″ x 8′
   2 pieces oak 2″ x 4″ x 12′
   9 pieces oak 1 1/2″ x 10″ x 14′
   1 piece oak 3″ x 3″ x 12′

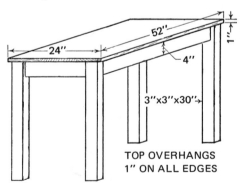

TOP OVERHANGS
1″ ON ALL EDGES

Fig. 12-4.

5. Find the number of board feet required to build the table shown in Fig. 12-4. Assume that dowel pins are used to fasten the apron to the legs. Top and apron boards are 1″ thick.

*The symbol "D 4 S" means "dressed on four sides."

Lumber terms everyone should know

Rough stock is lumber that is *not planed or dressed.* It is lumber in the rough as it leaves the saw. This type is not used very much.

Dressed stock is lumber that *has been dressed or surfaced* on one or more sides according to specifications. This type of lumber is used for most construction, cabinetwork, finished work or trim.

S 4 S means surfaced two sides and two edges.

D 2 S means dressed on two sides, D 4 S means dressed on 4 sides.

In dressing a board, the thickness is reduced from 1/4 to 1/8 inch depending upon the thickness of the board. It is necessary to make allowance for this reduction in thickness when dressed lumber is ordered. For example, a 2-inch board D 2 S will measure 1 3/4 inches when dressed. A "2 by 4" will measure 1 5/8 inches by 3 5/8 inches when D 4 S. You pay not for the lumber in the finished board but for what was in the rough stock from which it was made. When dressed lumber is figured, the full size of the rough material out of which it was made must be used. It is always better to specify the dimensions to which the finished lumber should be dressed if these dimensions are important; otherwise in dressing down lumber from a piece of larger dimensions the finished pieces may not be of the proper size.

Lumber is usually bought and sold by the *thousand board feet.* In specifying lumber quantities 1,000 board feet is indicated by the letter "M." The price quoted is usually the price per thousand board feet, as $375.00 per M. Certain stock sizes may be sold by the linear foot.

Lumber is supplied in even lengths unless otherwise specified, such as 10 feet, 12 feet, 14 feet, 16 feet, etc.

Example. Find the cost of the following order of lumber if redwood is selling at $400.00 per 1,000 bd. ft.:

| 100 pieces 1″ x 3″ x 16′ | 120 pieces 2″ x 4″ x 18′ |

THINK	DO THIS
For each kind of lumber in the order:	
1. Find the number of board feet in one piece.	1. $1 \times 3 \times 16$; $1 \times \frac{3}{12} \times 16 = 4$
	$2 \times 4 \times 18$; $2 \times \frac{4}{12} \times 18 = 12$
2. Multiply the number of pieces of each kind by the number of board feet in each piece of that kind.	2. $4 \times 100 = 400$ $12 \times 120 = 1,440$

3. Add the results to find the total number of board feet.

3. $400 + 1,440 = 1,840$

4. To obtain the number of 1,000 bd. ft., divide this sum by 1,000, and multiply by the cost of 1,000 bd. ft. (M) to find the cost of the entire lot.

4. $\dfrac{1840}{1000}$ x 400 = 736

Ans. $736

Summary

To find the cost (*C*) of a quantity of lumber:

EXAMPLE

Find the cost of 50 pieces 2″ x 4″ x 16′ white pine at $440 per M and 12 pieces 7/8″ x 2 3/4″ x 14′ oak at $760 per M.

1. Find the number of board feet (bd. ft.) required for each size.

2. Find the sum of the board feet in each of the required sizes.

1, 2. $2 \times \dfrac{4}{12} \times 16 \times 50 = 533.33$

bd. ft. of pine

$1 \times \dfrac{11}{48} \times 14 \times 12 = 38.5$

bd. ft. of oak

3. (a) If the cost (*C*) is quoted per board foot multiply the result of Step 2 by this price (*C*).
   (b) If the cost (*C*) is quoted per M bd. ft., divide the result of Step 2 by 1,000 and multiply by the cost (*C*) of M bd. ft.
   Use the formula

   $C = \dfrac{\text{bd. ft.}}{1,000} \times c,$

   in which *C* represents the total cost, bd. ft. represents the number of board feet, and *c* represents the cost of 1,000 bd. ft.

3. $c = \dfrac{533.33}{1,000} \times 440 = 234.67$

$c = \dfrac{38.5}{1,000} \times 760 = 29.26$

4. Add the costs of the individual sizes and kinds to obtain the total cost of the lumber in the order.

4. $234.67 + 29.26 = 263.93$

Ans. Total Cost = $263.93

EXERCISES

Find the cost of each of the following:

1. 30 pieces 1″ x 10″ x 12′ redwood at $400.00 per M.

2. 20 pieces 2″ x 8″ x 20′ white pine ⎫
   45 pieces 2″ x 4″ x 16′ white pine ⎬ at $438.00 per M
   6 pieces 6″ x 6″ x 18′ white pine ⎭

3. 24 pieces 7/8″ x 2 3/4″ x 18′ oak at $750.00 per M and 220 lin. ft. (linear feet) 3/4″ quarter round at $.04 per lin. ft.

PROBLEMS

6. At $750 per M bd. ft., what will be the cost of 4,800 bd. ft. of 1″ oak?

7. Estimate the cost of the following bill of material:

Quantity	Unit	Size	Cost per M Bd. Ft.	Total Cost
9	pieces oak	2″ x 8″ x 10′	$756.00	
8	pieces oak	1 1/2″ x 4″ x 12′	752.00	
22	pieces oak	3/4″ x 10″ x 14′	750.00	
7	pieces sugar pine	6″ x 8″ x 12′	477.50	

Total _____

8. How many linear feet of 4 x 6's are required to make 66 bd. ft.? At $.42 per linear foot, what will this lumber cost?

9. Find the cost of the material covered by the following specifications: California sugar pine "C" select 8″ wide and up in width, thoroughly kiln dried, D 2 S to sizes indicated:
   (a) 200 bd. ft., thickness 1″, at $438 per M bd. ft. $........
   (b) 300 bd. ft., thickness 1 1/2″, at $442 per M bd. ft.   ....
   (c) 200 bd. ft., thickness 2″, at $445 per M bd. ft.   ....
   Total $........

Floor tile

Floor tile is used extensively in many homes and buildings. Two standard sizes of floor tiles are the 12″ square and the 9″ square tiles. Material is asphalt, rubber, solid vinyl, vinyl-rubber, vinyl-asbestos, and linoleum.

Most 9″ tiles are packed in cartons of 80, for a coverage of 45 square feet. 12″ tiles are packed in cartons of 45, 50, 75, etc., depending on their weight. The total area to be covered is not a good guide for the number of tiles needed, as extra tiles must be purchased to allow for fitting and spoilage in application.

Tiles should be laid by starting in the center of the space to be covered. Centerlines should be chalked on the floor. If the distance from centerline to wall results in a space of less than one-half the tile width, the starting line should be shifted one-half tile width from the centerline. For example, if the room is 12'-8" x 13'-8", and the tiles to be used are 12" square, the short dimension would give a 4" space, so the starting line should be shifted 6" to make the last space 10" wide. Then it would take 13 tiles to cover the distance, rather than 12 tiles and 2 pieces of tile. If it is not required that the sides of the room have the same width rows (here, 10"), the starting line could be shifted only 4" off center, so that only one tile would need to be cut for each row. Or if 9" tile were to be used in this area, the 12'-8" distance would require 17 tiles, rather than 16 tiles and 2 pieces. Sometimes one full tile will make two part tiles.

Example. Determine the number of 12" square tiles needed for an area of 10'-9" x 12'-8".

THINK	DO THIS
1. For the 10'-9" side of the area, there would be required as many tiles in this direction as there are 12" sections in 10'-9". Hence change 10'-9" to inches and divide by 12.	1. 10'-9" = 129" $\dfrac{129}{12}$ = 10 3/4. Here 11 tiles are needed.
2. For the 12'-8" direction, do the same.	2. $\dfrac{12'\text{-}8''}{12} = \dfrac{152}{12}$ = 12 2/3 or 13 tiles
3. Multiply the whole number of tiles required in each direction.	3. 11 x 13 = 143 *Ans.* 143 tiles

Summary

	EXAMPLE
To find the number of whole tiles required to tile a rectangular area:	Determine the number of 9-inch square tiles needed to cover an area of 10'-9" x 12'-8".
1. Find the number of tiles required in each direction by dividing the dimension in inches by the width of the tile to be used.	1. $\dfrac{10'\text{-}9''}{9} = \dfrac{129}{9}$ = 14 1/3 tiles. This requires 14 1/2 tiles. $\dfrac{12'\text{-}8''}{9} = \dfrac{152}{9} = 16\dfrac{8}{9}$ or 17 tiles

2. Multiply the number of tiles in one direction by the number in the other direction. *Note:* When the division results in a fractional tile that is less than or equal to a half, we use 1/2. When the fraction is more than 1/2, a whole tile must be used.	2. 14 1/2 x 17 = 246 1/2 or 247 tiles *Ans.* 247 tiles

EXERCISES

Find the number of 12" square tiles needed to cover the following areas:

1. 21'-7" x 23'-2"   2. 16'-8" x 18'-9"

Find the number of 9" tiles needed to cover these areas:

3. 11'-7" x 13'-2"   4. 8'-4" x 7'-5"

What is the difference in cost of tiling the following rooms if 12" tiles are 38¢ each, and 9" tiles are 27¢ each?

5. 9'-8" x 14'-3"   6. 13'-4" x 13'-8"

Oak flooring

A simple way to find the amount of flooring to order for a room is to multiply the length by the width of the room. Since oak flooring is usually less than one inch thick, this result is the number of board feet of flooring that will actually be in the finished room. However, fitting, handling, cracking, and other causes make a certain amount of waste unavoidable, and this must be allowed for. This waste factor may be different for different sizes used.

The National Oak Flooring Manufacturers' Association recommends that the following percents of the floor area be added to cover waste:

Size	Add for Waste
25/32" x 1 1/2"	50%
25/32" x 2"	37 1/2%
25/32" x 2 1/4"	33 1/3%
3/8" x 1 1/2"	33 1/3%
3/8" x 2"	25%
1/2" x 1 1/2"	33 1/3%
1/2" x 2"	25%

An additional 5 percent should be added to these to cover damage in handling and cutting.

Example. Calculate the amount of oak flooring 25/32" x 2 1/4" to order for the floor of the bedroom shown in Fig. 12-5.

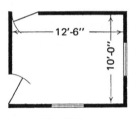

Fig. 12-5.

THINK	DO THIS
1. We must calculate the actual number of board feet (of 1" or less thick stock) required to cover the floor. Do this as on page 441.	1. 10 x 12 1/2 = 125
2. Use the table on page 447 to determine the percent, 33 1/3%, to be added for waste, and add 5% more for cutting and handling. This makes the total to be added 38 1/3%.	2. 33 1/3% + 5% = 38 1/3% to be added
3. Find this percent of the number of board feet needed to cover the floor.	3. 125 x 0.38 1/3 = 48
4. Add this result to the number of board feet needed to cover the floor.	4. 125 + 48 = 173 *Ans.* 173 bd. ft. of 25/32" x 2 1/4" oak needed for the floor shown

Summary	EXAMPLE
To find the number of board feet of oak flooring to order to cover a given floor:	Determine the amount of oak flooring 1/2" by 1 1/2" that must be ordered to cover a floor 13'-4" x 11'-3".
1. Find the area of the floor in square feet. This is the number of board feet of oak flooring needed to cover the entire floor.	1. Area of floor = 13 1/3 x 11 1/4 = 150 sq. ft. 150 bd. ft. are required to cover the floor.

2. Find from the table the percent of this number that must be added for waste and add to it the 5% needed for handling and cutting.

3. Find this last percent of the number of board feet needed to cover the floor.

4. Add this result to the number found in Step 1.

2. The table shows 33 1/3% to be added for waste and 5% for handling and cutting.
33 1/3% + 5% = 38 1/3% to be added.

3. 150 x 0.38 1/3 = 57.5 or 58 bd. ft.

4. 150 + 58 = 208 bd. ft.
Ans. Must order 208 bd. ft.

EXERCISES

Compute the amount of oak flooring to order for each of the following:

1. A living room 14'-6" x 16'-6", using 25/32" x 1 1/2" flooring.
2. Two bedrooms, one 10'-0" x 12'-6" and one 12'-0" x 14'-6", using 3/8" x 2" flooring.
3. A living room 16'-8" x 21'-6", from which a stairwell 3'-0" x 10'-4" is taken. The oak flooring is to be 1/2" x 1 1/2".

PROBLEMS

Unless otherwise specified, use the percentages for damage and waste as specified above.

10. It is proposed to cover the floor of a living room 12' x 15' with 25/32" x 2" matched oak flooring. If suitable allowance for waste and matching is made, how many board feet of lumber should be ordered?
11. If four rooms whose dimensions are 8'-6" x 8'-6", 5'-6" x 5'-8", 11'-0" x 12'-0", and 13'-6" x 28'-0", are to have oak floors laid with 25/32" x 2 1/4" flooring, how many board feet of flooring should be ordered?
12. If 50% is allowed for matching and 3% for waste, how many board feet of 7/8" x 3 1/4" flooring are required for the floor of a room 30 1/2' wide and 42' long?
13. The floor of a room 16'-3" x 22'-8" is to be laid with 25/32" x 1 1/2" flooring. At $285 per M, what will be the cost of the material in this floor?
14. How much will the material cost for the floor of a platform 50'-9" x 28'-4" if matched flooring 7/8" thick and 3" wide is to be used in its construction? Allow 25% for waste. The cost per M is $265.

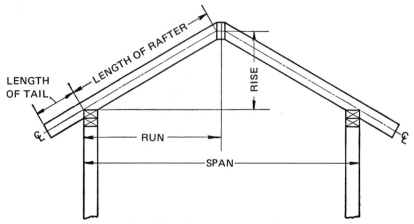

Fig. 12-6. Rise, run and span of a roof.

Roofs: rise, run and span

The *span* of a roof is the distance over the wall plates (see Fig. 12-6).

The *run* is the shortest horizontal distance measured from a plumb line through the center of the ridgeboard to the outer edge of the wall plate. For equally pitched roofs, the run is half the span.

While the rise of the roof is the distance from the top of the ridge to the level of the foot, in estimating rafters the *rise* is considered to be the *vertical distance from the top of the wall plate to the upper edge of the measuring line* (see Fig. 12-6).

The *pitch* of a roof is equal to the *rise divided by the span:*

$$\text{pitch} = \frac{\text{rise}}{\text{span}}$$

Also,

$$\text{rise} = \text{pitch x span}$$

The *cut* of a roof is the rise in inches *and* the unit of run (12″). A "cut" of 5/12 is very popular because the common rafter is 13 inches long for every foot of run.

PROBLEMS

15. A building is 28′ wide and has a roof pitch of 1/3. What is the rise?
16. A building is 22′ wide and has a roof with a 5 1/2′ rise. What is the pitch of the roof?
17. Find the pitch of the roof shown in Fig. 12-7.
18. How many square feet of plywood are needed to cover the five sections of the roof in Fig. 12-7? Allow 5% for waste. Assume common rafter to be 20 1/8′ long.

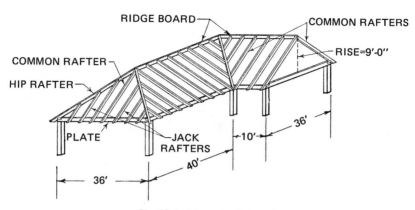

Fig. 12-7. Hip and valley roof.

19. A building 30' wide has a rise of 10'. What is the pitch of the roof?
20. A gable roof with 8'-6" rise, 18'-0" span, and 27'-6" length requires how many square feet of plywood sheathing?
21. What is the length of common rafter for a 5/12 cut roof with a 15' run?
22. Find the rise and the pitch of the roof shown in Fig. 12-8.

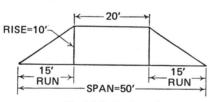

Fig. 12-8. Deck roof.

Fig. 12-9. Strip asphalt shingle exposed 4" to the weather.

Fig. 12-10. Plywood sheeting construction.

Shingling

Shingles of various materials such as wood, tile, asbestos, asphalt and other materials are used in the building trades.

Wood shingles come in bundles, Fig. 12-11, which contain approximately 250 shingles of an average width of 4″ and length of 16″. When the number of shingles necessary for a job is computed, the "square" is used as the unit of area. The *square* is an area equivalent to 100 square feet. The number of shingles needed to cover a square depends upon the length of the shingle "exposed to the weather." It takes about four bundles of wood shingles to cover a square. Wood shingles are used mostly on walls, for stained-shingle siding. Wood shingles are also used for shims and wedges in many other parts of construction.

Fig. 12-11.

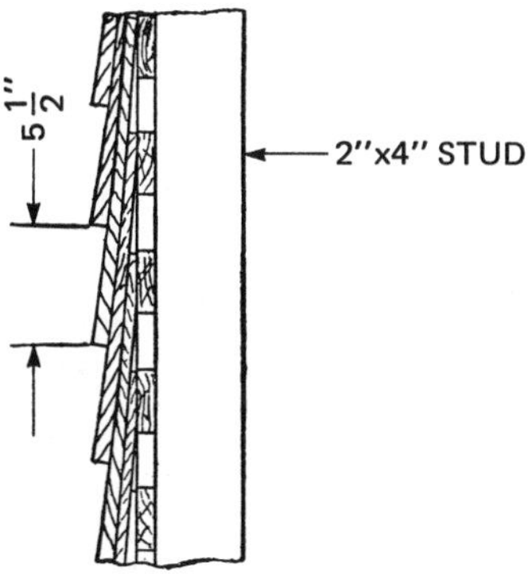

Fig. 12-12. Section through wall showing shingles exposed 5 1/2" to the weather.

2"x4" STUD

5 1/2"

Example. Estimate the number of bundles of wood shingles that will be required to cover the 8'-high walls of a 30' x 45' house. Make no allowance for openings.

THINK	DO THIS
1. To find the number of squares to be covered, we first find the area of the walls and divide the result by 100.	1. There are 2 walls 8 x 30 and 2 walls 8 x 45. $\dfrac{8 \times 30 \times 2}{100} = 4\,4/5$ $\dfrac{8 \times 45 \times 2}{100} = 7\,1/5$ $4\,4/5 + 7\,1/5 = 12$
2. Find the number of bundles at the rate of 4 bundles per square. Multiply the number of squares by 4.	2. $12 \times 4 = 48$ *Ans.* 48 bundles of shingles are needed.

The most popular asphalt shingle for roofs is the standard strip asphalt shingle, which is 12" wide and 36" long. One square requires 80 of these strip-shingles, when laid with a 5" exposure.

Summary	EXAMPLE
To find the number of shingles needed to complete a job:	Find the number of asphalt strip-shingles 12" x 36" needed to cover a roof 36' x 48'.
1. Find the number of square feet of area to be covered and divide by 100 to find the number of squares.	1. Number of sq. ft. to be covered: $36 \times 48 = 1{,}728$ $\dfrac{1728}{100} = 17.28$ or 17.3 squares

2. Multiply the number of squares by the number of shingles needed per square.

2. 17.3 x 80 = 1384

Ans. 1384 strip shingles required

PROBLEMS

23. How many asphalt strip-shingles laid 5″ to the weather are required to cover the upper roof illustrated in Fig. 12-13? Add 5% for waste.

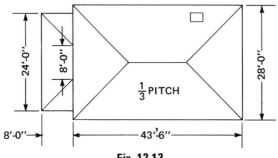

Fig. 12-13.

24. How many asphalt strip-shingles laid 5″ to the weather are required to cover the upper roof in Fig. 12-14? Add 5% for waste.

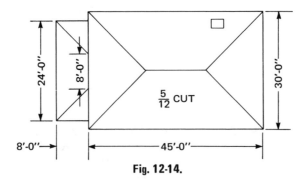

Fig. 12-14.

25. Estimate the number of shingles required to cover a gable roof having a ridge 55 ft. long and rafters 24 ft. on each side. Single-type 9″ x 12″ shingles which come in bundles of 95 each and require 380 shingles to the square when laid 4″ to the weather are to be used.

26. Estimate the number of single asbestos shingles 8″ x 16″ needed for the roof in Fig. 12-15. Assume that 260 such shingles will cover one square.

Fig. 12-15. Slant height of roof 18'-0", ridge length 41'-6".

Laths

When plaster is used as a wall surface, it is held in place by laths nailed to the studding. For many years, all laths were made of wood. They were 3/8" thick, 1 1/2" wide and 4' long. When installed with

Fig. 12-16.

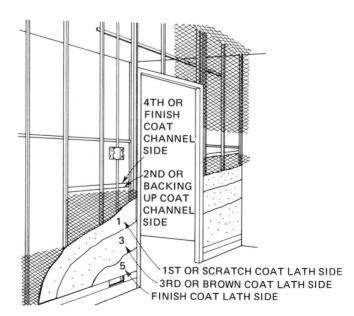

4TH OR FINISH COAT CHANNEL SIDE

2ND OR BACKING UP COAT CHANNEL SIDE

1ST OR SCRATCH COAT LATH SIDE
3RD OR BROWN COAT LATH SIDE
FINISH COAT LATH SIDE

spaces between, a bundle of 100 laths covered about seven square yards. Now metal lath is used. A common size is 2 feet by 8 feet.

Summary

To find the number of pieces of metal lath for a job:

1. Find the overall area of the surface to be lathed. First find the area of the two end walls, next the two sidewalls, then the ceiling. Add to get the overall area.

2. Find the sum of the areas of all the openings.

3. Subtract the result obtained in Step 2 from that obtained in Step 1.

4. Divide the result of Step 3 by 16, the area of each lath. Round off to the next whole number.

EXAMPLE

Estimate the number of pieces of 2' x 8' metal lath needed for the walls and ceiling of a room 12'-6" wide, 14'-6" long, 8'-4" high. There are 3 window openings — one is 5'-8" x 6'-3", and two are 3'-0" x 6'-3" each — and 2 door openings 3'-0" x 7'-2".

1. 12 1/2 x 8 1/3 x 2 = 208 1/3
   14 1/2 x 8 1/3 x 2 = 241 2/3
   12 1/2 x 14 1/2   = 181 1/4
                        631 1/4

2. Area of windows
   5 2/3 x 6 1/4 = 35 5/12
   3 x 6 1/4 x 2 = 37 1/2
   Area of doors
   3 x 7 1/6 x 2 = 43
   Total openings = 115 11/12

3. 631 1/4 – 115 11/12 = 515 4/12 or 516

4. $\frac{516}{16}$ = 32 1/4 or 33

Ans. 33 pieces of 2' x 8' metal lath

PROBLEMS

27. Find the number of 2' x 8' metal laths needed to cover the ceiling and sidewalls of a room 9 ft. wide, 12 ft. long and 8 ft. high. Deduct 50 sq. ft. for doors and windows.
28. A ceiling 25 ft. by 40 ft. has a rectangular opening for a skylight 6 ft. by 8 ft. How many metal laths will be required for the job?
29. How many metal laths are needed to cover the ceiling and walls of the study, living room, and dining room shown in Fig. 12-17? The ceiling height is 8 ft. Deduct 130 sq. ft. for openings.

Plastering

Plaster makes a very suitable interior wall finish. Plaster is estimated by the square yard. To determine the amount of plaster needed, it is customary to deduct one-half the area of the openings from the total surface to be plastered and express this result as the number of square yards of plaster needed.

Fig. 12-17.

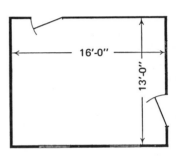

Fig. 12-18.

Summary

To determine the number of square yards to be plastered:

1. Obtain the total area to be plastered in the same way as for lathing.

2. Deduct 1/2 the area of the openings.

3. Express this result in square yards.

EXAMPLE

The room whose floor plan is given in Fig. 12-18 is 9 ft. high. It has 2 doors, each 3'-0" x 7'-0" and 3 windows, each 3'-0" x 6'-0". Estimate the square yards of plastering required for the 4 walls and ceiling of this room.

1. Area of ceiling 13 x 16 = 208
   Area of walls
   13 x 9 x 2 = 234
   16 x 9 x 2 = 288
   Area of walls and ceiling
   208 + 234 + 288 = 730

2. Area of openings
   Doors   2 x 3 x 7 = 42
   Windows   3 x 3 x 6 = 54
   Doors and windows   96
   1/2 x 96 = 48
   730 – 48 = 682 sq. ft.

3. $\frac{682}{9}$ = 75 7/9 or 76
   Ans. 76 sq. yds. of plastering

Dry wall

Plasterboard, rock lath, and other dry materials are often used in place of plaster. These are nailed to the joists and studs and the joints are then taped and filled to a smooth surface. Wood panels, applied vertically and horizontally, and some veneer sheets are also used for wall surfaces.

Ceilings are finished with acoustical tile. Sizes are 12" x 12", 12" x 24", 16" x 16" and 16" x 32", the most popular being the 12" x 12". These tiles are fastened to 1" x 3" nailing strips which are nailed to the ceiling joists. It takes one more nailing strip than there are rows of tile. That is, ten rows of tile require eleven nailing strips.

Suspended ceilings are also popular with the "do-it-yourself" home-owner. These consist of metal frames hung on wires, with loose panels dropped into place. The metal frame is made up of main tees, cross tees and wall angles. The main tees are 12' long (and can be joined together for longer lengths), the cross tees are 2' long and the wall angles are 10' long. The panels are 2 ft. x 2 ft. and 2 ft. x 4 ft. The cost of the material for the suspended ceiling and the permanently attached tile ceiling are practically the same.

PROBLEMS

30. In dividing a store to satisfy the requirements of a prospective tenant, it is necessary to build a partition wall 14 ft. high and 42 ft. long. How many square yards of plastering will be required to finish both sides of the partition?
31. The living room of a summer cottage is 11'-8" wide and 20'-0" long. The ceiling is 8' high. There are five openings for windows, each 2'-8 3/4" x 5'-6 1/2"; and two door openings, each 2'-8" x 6'-8". How many square yards of plaster are required for the walls and ceiling?
32. Find the number of square yards of plaster needed to cover all the walls and ceilings of the bedrooms in Fig. 12-19. The ceilings are 8'-6" high. Assume that each window is 2'-6" x 5'-6" and that each door opening is 2'-6" x 6'-8". Add 25% for hall and closets.
33. Estimate the number of square yards of plaster required for the walls of the bedrooms shown in Fig. 12-20. Make no allowances for openings. The ceilings are 8'-0" high.
34. At $2.30 per square yard, find the cost of plastering both sides of the wall shown in Fig. 12-21. Deduct for the openings.
35. Determine the number of 12 x 12 ceiling tiles needed for an area of 13'-8" x 16'-4". Find the number of lin. ft. of 1 x 3 nailing strips needed, if (a) the ceiling joists run the short dimension of the room and (b) the ceiling joists run the long way.

Fig. 12-19.

Fig. 12-20.

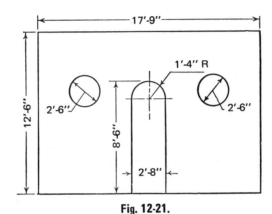

Fig. 12-21.

Painting

One of the most important problems for the painter is that of determining the amount of paint required for a given job.

The *covering power* (coverage) or *spreading rate* of a paint is the *number of square feet of surface one gallon of paint will cover with one coat.* This varies greatly according to the quality of the paint and the type of surface to be covered. Most paint manufacturers indicate an average coverage for their own products. In order to find the amount of paint required for a job, it is necessary therefore to know the area (in square feet) of the surface to be covered and the covering power of the paint to be used.

Interior work

To find the area of the walls of a room that must be painted:

1. Find the distance around the room (perimeter). This is the result of adding the lengths of the four walls.

2. Multiply this length by the distance from the baseboard to the picture molding, or to the ceiling.

3. Subtract from the product just found the total area of door and window openings.

The area of the ceiling is the product of the length and width of the room.

When the walls and ceiling are irregular in shape, it is well to divide the area into a number of convenient rectangles and triangles, find the area of each, and add the results.

For interior work one gallon of a latex flat paint will cover 450 square feet with one coat.

The number of gallons of paint required for a job may be found by the following formula:

$$\text{gallons} = \frac{\text{area to be painted}}{\text{coverage}} \times \text{no. of coats}$$

This means that the number of gallons required is the number of square feet to be painted divided by the number of square feet which one gallon will cover in one coat, and that this result is multiplied by the number of coats.

Exterior work

To find the area to be covered, measure the distance around the house and multiply this by the height to the eaves or cornice. If there are gables, multiply their widest part by one-half the height. Then add the results obtained.

For an exterior job, use the following formula for the number of gallons of paint required:

$$\text{gallons} = \frac{\text{area}}{700}$$

We divide by 700 because on the average a gallon of outside paint covers 700 square feet.

The following facts are sometimes useful for outside work:

1. For the priming coat a gallon of paint covers 500 to 700 square feet.

2. For a second or third coat one gallon covers 650 to 750 square feet.

3. For brickwork, stucco, concrete, stone, one gallon of priming coat covers 200 square feet; of second coat, 400 square feet; and of finishing coat, 600 square feet.

PROBLEMS

36. Estimate the number of gallons of paint required to paint a room 15 ft. long, 12 ft. wide, and 8 ft. high with two coats of paint having

a coverage of 600 sq. ft. per gallon. At $6.75 per gallon, what will be the cost of the paint? Estimate to the nearest quart, calculating the quart price as one quarter of the gallon price.

37. Estimate the quantity of paint required for a two-coat paint job on a building 18'-0" x 30'-0", which has two sides 18'-0" high and two gabled ends with an 8'-6" rise. Make no allowances for openings

38. The sidewalls and ceiling of a dining room 9' x 12' and a living room 11'-6" x 15'-8" are to be given two coats of latex paint. If the ceilings are 8' high, how many quarts of paint will be required?

39. How many quarts of spar varnish are needed to cover a hardwood hall floor with two coats if the floor measures 5' x 75'? The covering power of this varnish is to be assumed as 450 sq. ft. per gallon.

Concrete

This is a mixture of cement, sand and gravel. Different quantities of each ingredient are used for various purposes. When we speak of a 1:2:4 mixture of concrete, we mean that the concrete is composed of 1 part cement, 2 parts sand, and 4 parts stone or gravel. Most concrete is now ready-mixed at a mixing plant and delivered by truck to the building site.

The amount of concrete necessary for a job is usually expressed in cubic yards and costs are usually given as so much per cubic yard. The cost depends upon the amount purchased and the type of mixture required. For small jobs, one can purchase dry ready-mixed concrete in 90-lb. sacks. One sack will make about 2/3 cu. ft. of wet concrete.

The following table shows the quantities of cement, sand and stone or gravel that are needed to make one cubic yard of concrete for different mixtures.

Proportion by Parts			Proportion of Each Needed to Make 1 cu. yd. Concrete		
Cement	Sand	Gravel	Cement	Sand	Gravel
1	1	2	11.40 sacks	0.40 cu. yd.	0.80 cu. yd.
1	1 1/2	3	8.36 sacks	0.44 cu. yd.	0.88 cu. yd.
1	2	3	7.56 sacks	0.53 cu. yd.	0.80 cu. yd.
1	2	4	6.60 sacks	0.46 cu. yd.	0.92 cu. yd.
1	2 1/2	5	5.48 sacks	0.48 cu. yd.	0.96 cu. yd.
1	3	5	5.12 sacks	0.54 cu. yd.	0.90 cu. yd.
1	3	6	4.64 sacks	0.49 cu. yd.	0.98 cu. yd.

You read the table as follows: "A 1:1:2 mixture of concrete requires 11.40 sacks of cement, 0.40 cu. yd. of sand, and 0.80 cu. yd. of stone or gravel for each cubic yard of concrete."

Example 1. Estimate the quantities of cement, sand and gravel needed for a concrete wall 12" x 6' x 35', using a 1:2 1/2:5 mixture.

THINK	DO THIS
1. We must find the volume, in cu. ft., of concrete needed for the job ($V = l$ x w x t).	1. Volume = l x w x t $= 35 \times \dfrac{12}{12} \times 6 = 210$
2. Divide this result by 27 to find the volume in cu. yd.	2. $\dfrac{210}{27} = 7.78$
3. We must then determine the quantity of each ingredient needed for this number of cu. yd. of concrete.	3. 5.48 x 7.78 = 42.6 0.48 x 7.78 = 3.7 0.96 x 7.78 = 7.5 *Ans.* The amounts needed are 42.6 sacks of cement, 3.7 cu. yd. sand and 7.5 cu. yd. gravel. The amounts that would have to be purchased are 43 sacks of cement, 4 cu. yd. sand and 8 cu. yd. gravel.

Example 2. Estimate the quantity of concrete needed for a 4"-thick walk, 4 ft. wide and 150 ft. long.

THINK	DO THIS
1. We must first find the volume, in cu. ft., of the space to be filled ($V = l$ x w x t).	1. Volume $V = l$ x w x t $= 150 \times 4 \times \dfrac{4}{12} = 200$
2. Divide this result by 27 to find the volume in cu. yd.	2. $200 \div 27 = 7.40$ or 7 1/2 (approx.) *Ans.* 7 1/2 cu. yd.

Often, the object to be made of concrete is irregular in shape like that shown in Fig. 12-22. To find the volume of such a form, first find the cross-sectional area *ABCDEF* and multiply this area by the length or depth. Also in some cases the object to be built of concrete will have to be divided into smaller objects whose volumes must be found separately, then added.

The cross-sectional area can usually be broken up into squares, rectangles, triangles, trapezoids, and circles.

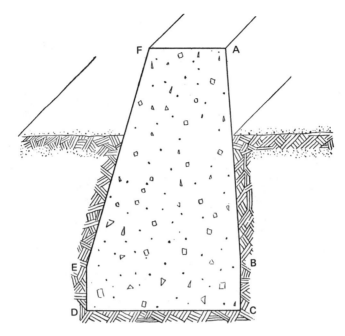

Fig. 12-22.

Summary

To estimate the amount of concrete, expressed in cu. yd., necessary for a particular job:

1. Find the volume of the space to be filled with concrete, expressed in cu. ft. ($V = l \times w \times t$).

2. Change this result to cubic yards (cu. yd.) by dividing by 27.

3. Round off this result to the next larger practical quantity.

EXAMPLE

How many cubic yards of concrete will be required for a rectangular column 24″ x 18″ and 16 ft. high?

1. Volume in cu. ft.
$$V = \frac{24}{12} \times \frac{18}{12} \times 16 = 48$$

2, 3. $\frac{48}{27}$ = 1 21/27 or 2

Ans.   2 cu. yd.

EXERCISES

Find the amounts of cement, sand and gravel to make each of the following:

1.   75 cu. yd. of 1:1 1/2:3 concrete.
2.   120 cu. yd. of 1:2:3 concrete.
3.   692 cu. ft. of 1:3:6 concrete
4.   584 cu. ft. of 1:3:5 concrete.
5.   475 cu. yd. of 1:2:4 concrete.

PROBLEMS

40. Figure 12-23 shows a section of curbing 250' long. How many cubic yards of concrete will be required? At a cost of $17.50 per cubic yard of ready-mixed concrete, what will be the cost? If this concrete is to be mixed on the job site, and is to be a 1:2 1/2:5 mixture, how many sacks of cement, how many cu. yd. of sand and how many cu. yd. of gravel will be required?

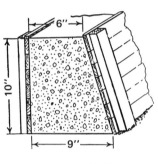

Fig. 12-23.

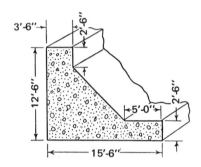

Fig. 12-24.

41. The concrete retaining wall in Fig. 12-24 is 72 ft. long. How many cubic yards of concrete will be required? At $16.80 per cu. yd., what will be the cost of the ready-mixed concrete needed to do this job?

42. The retaining wall in Fig. 12-25 is 125' long. Estimate the number of square feet of form work required for the outside finished face. What is the cross-sectional area? How many cubic yards of concrete are required?

43. The retaining wall in Fig. 12-26 is to be 75' long and made of 1:2:4 concrete. Estimate the amount of cement, sand and gravel needed to do this job. The cap is 2"-4" x 4" in cross section.

44. A driveway 12' wide and 56'-8" long is to be made of concrete 4" thick. How many cubic yards of 1:3:6 concrete will be required for this job? Find the quantity of each material in the concrete.

45. Estimate the quantity of sand, cement and gravel needed for the job illustrated in Fig. 12-27. The concrete mixture is to be 1:3:5.

46. Find the number of cubic yards of concrete needed to construct the foundation wall in Fig. 12-28. A section of this wall is shown in Fig. 12-29.

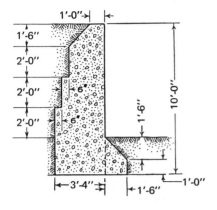

Fig. 12-25.

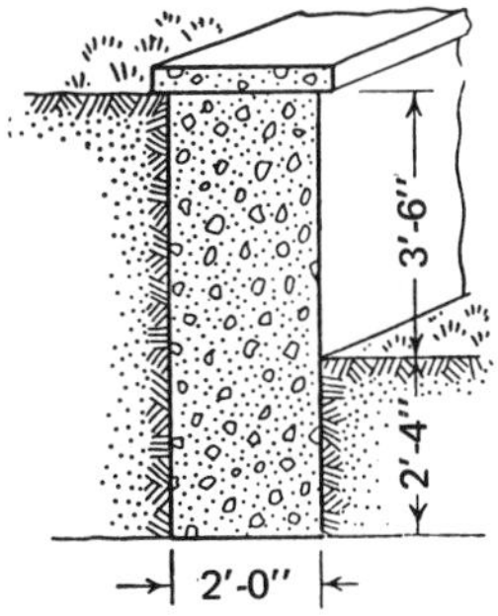

Fig. 12-26.

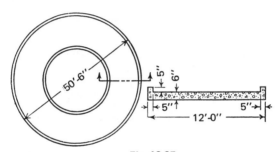

Fig. 12-27.

REVIEW PROBLEMS

1. How many square yards of plaster are required for each side of a blank partition wall 42'-8" x 14'-6"?

2. The foundation walls of a frame house, Fig. 12-30, are 12" thick and are made of poured concrete. The basement floor is 4" thick. How many cubic yards of concrete were required? Make no deduction for openings.

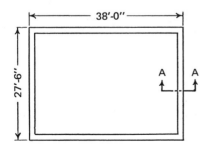

Fig. 12-28.

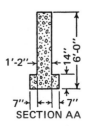

Fig. 12-29.

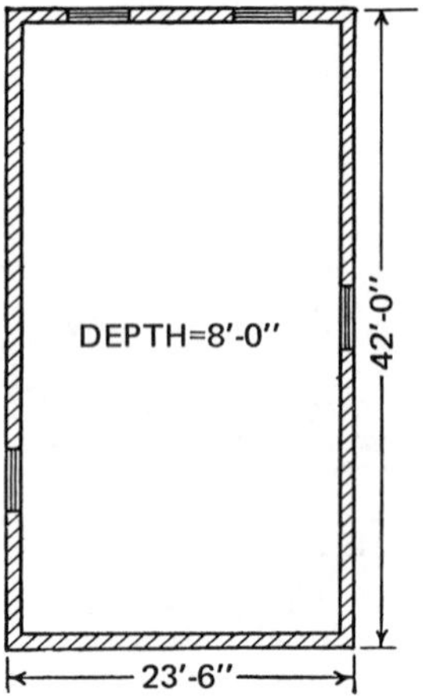

Fig. 12-30.

DEPTH=8'-0"

42'-0"

23'-6"

3. How many cubic yards of concrete were used to build the foundation in Fig. 12-31? Make no deductions for openings.

4. A stairway from the basement to the first floor of a one-family house has a rise of 8'-9". The riser is 7". How many treads are there in the stairway?

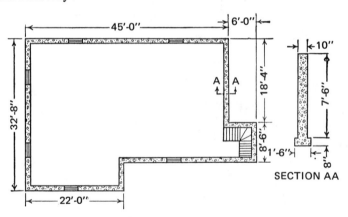

Fig. 12-31.

5. How many common bricks are needed to build the wall, which is 8'-6" deep and 16" thick, in Fig. 12-32? A 16" thick wall requires 30 bricks for each square foot of surface. Make no allowance for openings in the wall.

6. In excavating for the foundations shown in Figs. 12-33 and 12-34, it was agreed that the earth was to be removed for a distance of 1'-6" beyond the foundation walls in order that they might be waterproofed and inspected. If the depth of each excavation is to be 5'-6", estimate the number of cubic yards of earth to be removed. If a load is 4 cu. yd., how many truck loads were removed?

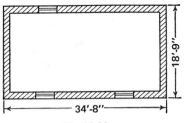

Fig. 12-32.

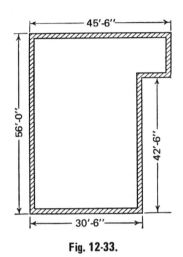

Fig. 12-33.   Fig. 12-34.

7. How many 12" x 36" asphalt shingles laid 5" to the weather are needed for the roof in Fig. 12-35? Assume 80 shingles per square. (Remember that a square is 100 sq. ft.)

8. Find the number of squares to be shingled on the 1/3 pitch hip roof in Fig. 12-36.

9. The span of a roof is 48'-6". What is the rise if the pitch is 1/4?

10. Find the length of the rafter in a 1/6 pitch roof if the span is 32'-0".

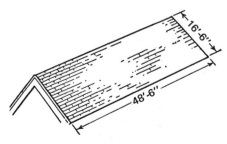

Fig. 12-35.

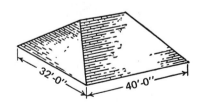

Fig. 12-36. Hip Roof.

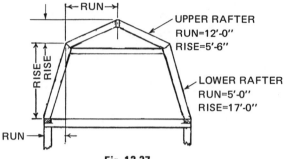

Fig. 12-37.

11. Find the length of the upper and the lower rafters in Fig. 12-37.
12. The outside face of the sill of a frame house measuring 36'-0" x 24'-0" at the foundation is set in 3/4" from the outside face of the foundation wall (Fig. 12-38). How many pieces of 2" x 6" x 16' lumber are required? At $380 per M bd. ft., what will this lumber cost?

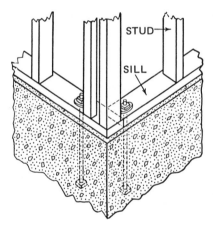

Fig. 12-38. Butting sills at corner.

13. Prepare a stock list and estimate the cost of the material required to make 55 open-shelf bookcases (Fig. 12-39) if the lumber used sells for $438 per M bd. ft.
14. Figure 12-40 shows a section of a building measuring 22'-6" x 52'-9". The sheathing is 2' x 8'. The distance from foundation to plate is 24'-6".
    (a) How many pieces of sheathing are needed? No allowance is to be made for openings. Add 1/6 of the surface area for waste.
    (b) Find the number of board feet of 10" siding laid 7" to the weather. The siding is 1/2" thick and tapers to 1/4". Add 10% for waste.
    (c) What will the lumber for this job cost if siding sells for $180 per M bd. ft. and sheathing for $120 per M bd. ft.?

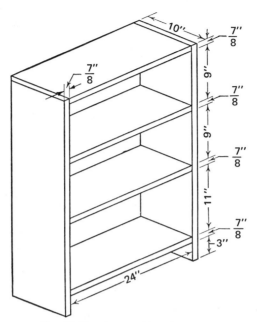

Fig. 12-39. Open-shelf bookcase.

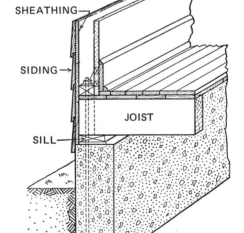

Fig. 12-40. Section of a wall.

15. How many square feet of plywood subflooring and how many board feet of finished flooring, respectively, are required for the rooms listed? The finished flooring is 3/8" x 2". Room size: (a) 13'-6" x 19'-4"; (b) 12'-6" x 14'-7"; (c) 16'-8" x 24'-6"; (d) 12'-0" x 14'-8"; (e) 12'-0" x 12'-6"; (f) 10'-8" x 15'-4".

16. Face bricks with 1/4″ joints are to be used in the construction of a one-family house 25′-8″ x 48′-6″ and 28′-9″ high (Fig. 12-41). If there are seven bricks per square foot, how many bricks are required for the job? Deduct 6% for openings.

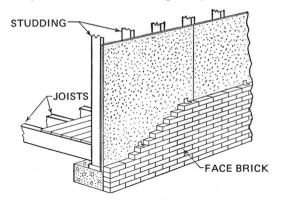

Fig. 12-41. Section of veneer brick wall.

17. How many gallons of paint are needed to paint the living room, alcove and bedroom in Fig. 12-42 if two coats of a latex flat paint are used? Walls are 8′ high. Make no allowance for openings.

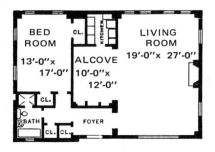

Fig. 12-42.

18. What is the cost of the material needed to tile a 12′ x 12′ ceiling if the nailing strips are 5¢ a linear ft., the molding 3¢ a linear ft., and the tiles 15¢ each? Use 12″ x 12″ tiles.

19. What is the cost of the material needed for a suspended 12′ x 12′ ceiling if the wall angles are 5¢ a foot, the 12 ft. main tees 75¢ each, the 2 ft. cross tees 18¢ each, and the 2′ x 4′ panels are $1.25 each?

13 Mathematics of the Electrical Shop

Some ideas about electricity

No one knows exactly what electricity is. However, a great deal is known about some of the things that it will do and the laws that govern its use.

The flow of electricity through a conductor (wire) is very similar to the flow of water through a pipe. The flow of water through a pipe is measured by the quantity of water that flows in a unit of time, as 1 gallon per minute, 5 gallons per minute, etc. (Fig. 13-1). Here the gallon of water is the unit of quantity and the minute the unit of time. The expression "gallons per minute" represents the rate at which water is flowing through the pipe. In a similar manner, the rate at which electricity flows through a circuit is measured by the quantity of electricity that flows in a unit of time. Here the unit of quantity is the *coulomb* and the unit of time the second. The rate of flow is expressed in terms of coulombs per second. The practical unit of electric current that is equivalent to the flow of one coulomb per second is called the *ampere* (amp.). If the flow is 15 coulombs per second, the current is 15 amperes.

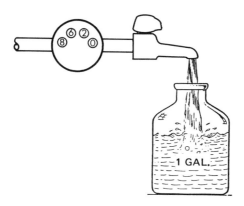

Fig. 13-1. Measuring the flow of water.

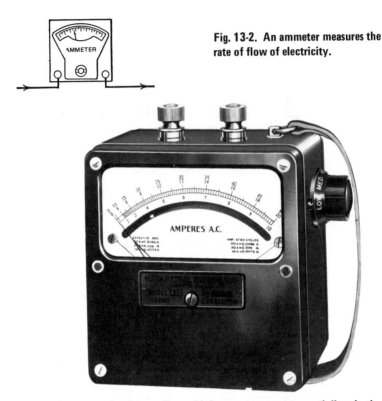

Fig. 13-2. An ammeter measures the rate of flow of electricity.

Fig. 13-3. Alternating-current ammeter. From this instrument you may read directly the number of amperes flowing through the circuit. *(Courtesy Weston Electrical Instrument Corporation)*

There are two kinds of electric current. *Direct current* is an electric current flowing in one direction only. *Alternating current* is an electric current which reverses its direction in a periodic manner.

The instrument used to measure the rate of flow of electricity is the ammeter (Figs. 13-2 and 13-3).

The rate of flow of water through a pipe, expressed in gallons per minute depends on the pressure on the pipe. The greater the pressure, the greater the flow of water. Similarly, an electric pressure known as an *electromotive force* (e.m.f.) causes a flow of electric current. The greater the electromotive force, the stronger the electric current. Electromotive force is measured in *volts*. The instrument used to measure the electromotive force is the voltmeter.

Electricity flows with less difficulty through some materials than others. A material that permits electricity to flow through it with ease is called a good conductor of electricity, whereas a material that hinders the flow of electricity to such an extent that it is almost impossible for a current to flow through it is called an insulator. That property of a material which hinders the flow of an electric current is called *resistance*.

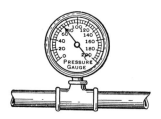
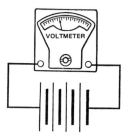

Fig. 13-4. A pressure gage measures water pressure. A voltmeter measures electric pressure.

The unit used to measure resistance is called the *ohm*. An ohm may be defined as the resistance offered to the flow of a constant current of electricity by a column of mercury weighing 14.4521 grams and having a length of 106.3 centimeters. Such a column of mercury would have a cross-sectional area of approximately one square millimeter.

Ohm's law

It has been determined that the rate of flow of current (amperes) passing through a circuit is equal to the pressure (volts) divided by the resistance (ohms). In electrical formulas the number of amperes is represented by I, the number of volts by E, and the number of ohms by R. The symbol "Ω" is sometimes used to represent the word "ohm." "Ω" is a Greek letter called "omega."

Fig. 13-5. Direct-current volt-ammeter. By using the upper terminals the instrument serves as an ammeter. By using the proper lower terminals the number of volts across the circuit may be read. *(Courtesy Weston Electrical Instrument Corporation)*

Using these letters, the statement, *Ohm's law*, is expressed in the formula

$$I = \frac{E}{R}$$

By solving this formula first for E and then for R, we obtain two other formulas

$$E = IR \text{ and } R = \frac{E}{I}$$

The student will find it helpful to repeat these formulas, using words for the letters.

Fig. 13-6. A simple circuit.

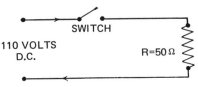

SWITCH

110 VOLTS
D.C.

R=50 Ω

Example. Find the current flowing in the circuit of Fig. 13-6 when the switch is closed.

THINK	DO THIS
1. Since we must find the amount of current, that is the amperes, I, we use the first formula.	1. Write the formula: $I = \dfrac{E}{R}$
2. From the statement of the problem, substitute the numbers given for their respective letters in this formula.	2. $E = 110$ volts, $R = 50$ ohms $I = \dfrac{110}{50}$
3. Make the computations.	3. $I = 2.2$ *Ans.* 2.2 amp.

Summary

To find the current (amperes) or the pressure (volts) or the resistance (ohms) in an electric current:

1. Use one of these three formulas:

Amperes	Volts	Ohms
$I = \dfrac{E}{R}$	$E = IR$	$R = \dfrac{E}{I}$

EXAMPLE

A current of 80 amp. is forced through a resistance of 4.4 ohms. What is the pressure across the circuit?

1. Formula to use: $E = IR$

2. Replace the letters by their known values.	2. I = 80, R = 4.4, E = 80 x 4.4
3. Make the computations to obtain the required units, in this case the pressure or voltage of the circuit.	3. E = 352

Ans. 352 volts

Fig. 13-7.

Fig. 13-8.

The chart shown in Fig. 13-7, if properly used, will help you to remember the various forms of Ohm's law. If you cover the I (Fig. 13-8), you get E over R. This shows that I (the one covered) = E/R.

Fig. 13-9.

If you cover the E (Fig. 13-9), the figure tells you that E (the one covered) = I x R.

The statement concerning R is left to the student.

EXERCISES

1. Find the missing value in each of the following:

	Amperes I	Volts E	Ohms R
(a)	?	220	4
(b)	?	110	5
(c)	25	220	?
(d)	45	220	?
(e)	10	?	4.5

2. How many amperes will a pressure of 110 volts send through a resistance of 3.5 ohms?

3. A current of 60 amp. is forced through a resistance of 3.8 ohms. What is the pressure across the circuit?

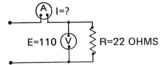

Fig. **13-10.** An ammeter and voltmeter connected into a circuit.

Fig. **13-11. Electrically heated glue pot.** *(Courtesy General Electric Company)*

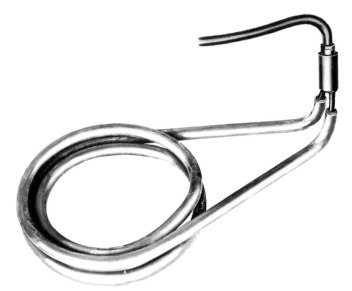

Fig. **13-12. Immersion heater.** *(Courtesy General Electric Company)*

PROBLEMS

1. An electric iron has a resistance of 22 ohms. Find the current drawn from the line when the applied voltage is 110 volts (see Fig. 13-10).

2. A jacketed electrically heated glue pot (Fig. 13-11) draws 6 amp. when connected to the 110-volt line. What is the resistance of the heater unit in this glue pot?

3. An immersion heater (Fig. 13-12) designed for use with an oil tempering bath has a resistance of 3.3 ohms. The normal operating current is 35 amp. What voltage is required to operate this unit?

4. How much current can pass through a field rheostat (Fig. 13-13) having a resistance of 600 ohms if the voltage impressed across it is 115 volts?

5. A 15,000-ohm voltmeter reads 125 volts. How much current passes through it at that voltage?

Fig. 13-13. Front and rear views of a field rheostat. *(Courtesy Ward Leonard Electric Company)*

The circular mil

The diameters of electric wires are usually small. It was found cumbersome to express these diameters in inches. Accordingly, a new unit was devised that is known as the *mil. The mil is a unit of length that is 1/1,000, or 0.001, of an inch.* It would, therefore, take 1,000 wires each one mil in diameter to equal one inch.

A unit of area was also devised that is helpful in measuring the cross-sectional areas of wires. This unit is the *circular mil* (C.M.). *A circular mil is the area enclosed by a circle that is one mil in diameter.*

The *area of a circle in circular mils* is obtained by *squaring the diameter (in mils) of the circle* (see Fig. 13-14).

$$\text{Formula: } A = D^2 \quad \text{or} \quad D = \sqrt{A}$$

Note that D is the number of *mils* in the diameter of the circle and A is the area expressed in *circular mils.*

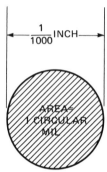

Fig. 13-14. Circular mil greatly enlarged.

Example. If a wire is 100 mils in diameter, find its area in circular mils.

THINK	DO THIS
1. Since we are to find the area in circular mils, we write the formula for A.	1. $A = D^2$
2. Substitute the value given in the problem for D.	2. $A = 100 \times 100$
3. Make the computations.	3. $A = 10,000$
	Ans. 10,000 C.M.

Summary	EXAMPLE
To find the cross-sectional area of a wire in circular mils:	Find the cross-sectional area of a 1/4″ wire in circular mils.
1. Express the diameter in mils.	1. There are 1,000 mils to an inch; then $1/4 \times 1,000 = 250 = D$.
2. Use the formula for A above.	2. $A = D^2$
3. Replace the letter by the value given.	3. $A = (250)^2 = 250 \times 250$
4. Make the computations.	4. $A = 62,500$
	Ans. 62,500 C.M.

EXERCISES

1. Express each of the following in mils: (a) 0.1 in., (b) 0.15 in., (c) 0.50 in., and (d) 0.625 in.
2. Express each of the following in inches: (a) 162 mils, (b) 460 mils, (c) 50.8 mils, and (d) 364.8 mils.
3. Find the area in circular mils of each of the following wires (the number given is the diameter of the wire): (a) 57 mils, (b) 41.9 mils, (c) 101 mils, (d) 64.08 mils, (e) 0.319 in., (f) 3/16 in., (g) 1/32 in., and (h) 0.364 in.

The square mil

The *square mil* is an area equivalent to that enclosed by a square that is one mil on a side (see Fig. 13-15). Figure 13-16 is an enlarged drawing that shows the relation between a square mil and a circular mil. The square mil is the area enclosed by the square, while the circular mil is the area enclosed by the circle. See definition of circular mil, p. 477.

It takes more circular mils than square mils to fill a given area.

The following formulas enable you to change numerical values from circular mils to square mils and from square mils to circular mils.

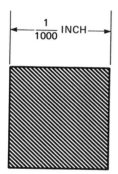

Fig. 13-15. Square mil greatly enlarged.

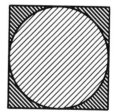

Fig. 13-16. Circular mil inscribed on a square mil greatly enlarged.

Summary

$$\text{circular mils} = \frac{\text{square mils}}{0.7854}$$

$$\text{square mils} = \text{circular mils} \times 0.7854$$

The first formula states: To find the number of circular mils in an area, divide the number of square mils by 0.7854.

The second formula states: To find the number of square mils in an area, multiply the number of circular mils by 0.7854.

EXERCISES

Use the following formulas to convert questions 1 through 6 to the unit indicated.

$$\text{circular mils} = \frac{\text{square mils}}{0.7854}$$

$$\text{square mils} = \text{circular mils} \times 0.7854$$

1. 7,500 circular mils to square mils.
2. 15,000 square mils to circular mils.
3. 62,500 circular mils to square mils.
4. 31,416 square mils to circular mils.
5. 45,000 square mils to circular mils.
6. 100,000 circular mils to square mils.

PROBLEMS

6. The diameter of No. 00 A.W.G.* wire is 0.3648". What is the area of this wire in circular mils?
7. The cross section of a solid cylindrical conductor is 66,560 C.M. What is its diameter in mils? In inches?
8. The diameter of No. 8 A.W.G. solid copper wire is 0.1285".
   (a) Determine the cross-sectional area in square inches.
   (b) Determine the cross-sectional area in circular mils.
9. The cross-sectional area of a square copper conductor is 0.60" on a side.
   (a) Determine the area in square mils.
   (b) Determine the area in circular mils.
10. The cross-sectional area of a 37-strand cable is 450,000 C.M. Determine the cross-sectional area of each strand and the diameter of the individual strands in mils and inches. The area of a stranded wire is the sum of the areas of the individual strands.

The circular mil-foot

A wire (electrical conductor) that is one foot long and one mil (or 0.001 inch) in diameter is a *circular mil-foot* of wire. The resistance that a circular mil-foot of wire offers is accepted as the unit of resistance of that wire. For example, a circular mil-foot of copper wire at 20 degrees Celsius has a resistance of 10.37 ohms. This is the unit of resistance for copper wire.

The resistance, R, of a conductor of any length, L, in feet and any diameter, D, in mils may be obtained by the formula

$$R = \frac{KL}{D^2}$$

*A.W.G. stands for American wire gage (see Fig. 13-17).

where K is the resistance in ohms per circular mil-foot for the type of wire under consideration. Note that D^2 is the cross-sectional area of the wire in circular mils. Can you express this formula in words? Start with, "The resistance of a conductor. . ."

Example. Find the resistance of a cable having a diameter of 0.229" and a length of 2,500'. Assume that K is 10.37.

THINK	DO THIS
1. Since the diameter is given in inches, we change the diameter to mils.	1. 0.229" = 0.229 x 1,000 = 229 mils
2. Write the formula for resistance.	2. $R = \dfrac{KL}{D^2}$
3. Substitute the numerical values given for K, L and D.	3. $R = \dfrac{10.37 \ \times \ 2,500}{229^2}$
4. Make the computations	4. $R = 0.494$ *Ans.* Resistance = 0.494 ohm

Summary	EXAMPLE
To find the resistance, R, in ohms in a conductor:	Find the resistance of a copper wire 0.204" in diameter and 1,500' long. Assume $K = 10.5$
1. Express the diameter in mils.	1. Diameter = 0.204 x 1,000 = 204 mils
2. Write the formula $R = \dfrac{KL}{D^2}$, in which L is the length of the conductor in feet, D is the diameter in mils, and K is the resistance of one circular mil-foot of the conductor expressed in ohms.	2. $R = \dfrac{KL}{D^2}$
3. Replace the letters by the values given for them.	3. $R = \dfrac{10.5 \ \times \ 1,500}{(204)^2}$
4. Make the computations.	4. $R = 0.378$ *Ans.* Resistance = 0.378 ohm

EXERCISES

Find R in each of the following exercises:

1. A conductor 1,875 ft. long, 0.125" in diameter. $K = 10.37$.

2. A conductor 2 mi. long, 0.1875″ in diameter. $K = 10.37$.
3. A conductor 1,000 yds. long, 0.4375″ in diameter. $K = 10.37$.

PROBLEMS

11. What is the resistance of a copper wire 0.204″ in diameter and 1,225′ long?
12. The diameter of copper annunciator wire in common use is 0.040 in. What is the resistance of a coil containing 100 ft.?
13. How many feet of copper wire are required to make a resistance of 18.2 ohms? The diameter of the wire is 15.9 mils.

Copper wire

A very interesting table that combines a number of the things which we have recently learned is the table of copper wire given in the Appendix. Part of a similar table is reproduced here. The table gives the gage numbers, the cross-sectional area in circular mils, the diameter in mils, and the approximate resistance in ohms per 1,000 feet at a certain temperature. Note that a copper wire gage No. 3 is 229.42 mils in diameter, 52,634 circular mils in area, and offers a resistance of 0.2 ohm per 1,000 feet. Notice also that as the gage number increases the size of the wire decreases. Can you suggest why the resistance increases as the gage number increases?

Gage No.	Area Circular Mils	Diameter Mils	Ohms per 1,000 Ft. (Approximate)
0	105,530	324.86	0.1
1	83,694	289.30	0.125
2	66,373	257.63	0.16
3	52,634	229.42	0.2
4	41,742	204.31	0.25
5	33,102	181.94	0.32
10	10,381	101.89	1.0
11	8,234	90.242	1.25
12	6,529	80.805	1.6

The gage most commonly used in the United States for determining the size of wires is the one shown in Fig. 13-17. It is the Brown and Sharpe (B & S) gage, also shown as the American Wire Gage (A.W.G.).

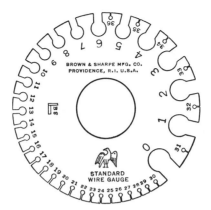

Fig. 13-17. B & S wire gage.

EXERCISES

Use the wire table in the Appendix for the information needed to solve the following exercises:

1. State the gage and the cross-sectional area of the wire that will have a resistance (a) of 1.25 ohms per 1,000 ft., (b) of approximately 0.85 ohm per mile.
2. What is the gage and the resistance per 1,000 ft. of copper wire that is approximately 1/10″ in diameter?
3. State the cross-sectional area and the approximate resistance per mile of each of the following gage wires: No. 4 and No. 11.

PROBLEMS

Use the wire table in the Appendix for these problems.

14. What size wire (A.W.G.) has a resistance of approximately 1 ohm per 1,000 ft.?
15. What size wire has a resistance of approximately 13.3 ohms per mile?
16. What size B & S copper wire should be used to construct a coil consisting of 500 turns of 16.5 in. average length, if the coil is to have a resistance of 114 ohms?
17. What is the resistance of a line consisting of two No. 6 conductors 275 ft. long?

Line drop

Whenever electricity flows through a wire, whatever the length or the diameter of the wire may be, that wire offers resistance to the flow.

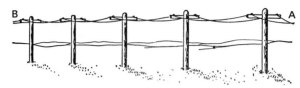

Fig. 13-18.

As the current flows from *A* to *B* in Fig. 13-18, some of the pressure (voltage) is lost. This loss in pressure due to the resistance which the conductor offers to the current is called *line drop*, or *volts lost in the line*.

The line drop is equal to the current (amperes) multiplied by the resistance (ohms) of the conductor.

Summary	EXAMPLE
To find (a) the line drop (loss of pressure) and (b) the terminal voltage of a generator:	Find, for the circuit of Fig. 13-19, (a) the voltage lost in the line and (b) the terminal voltage of the generator.
(a) 1. Multiply the current in amperes by the resistance for each line. Formula: *Ld* = *R* x *I* where *Ld* represents the line drop, *R* the resistance, and *I* the current.	(a) 1. Line drop in *AB* is 6 x 0.2 = 1.2 Line drop in *CD* is 6 x 0.2 = 1.2
2. Add these results to obtain the total line drop.	2. Total line drop is 1.2 + 1.2 = 2.4
(b) Find the terminal voltage of the generator by adding the line drop to the voltage across the circuit.	(b) Terminal voltage 112 + 2.4 = 114.4 *Ans.* Line drop = 2.4 volts. Terminal voltage = 114.4 volts

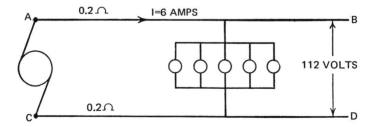

Fig. 13-19.

EXERCISES

1. Find the line drop in each of the following: (a) A to B; (b) D to E; (c) the entire circuit (see Fig. 13-20).

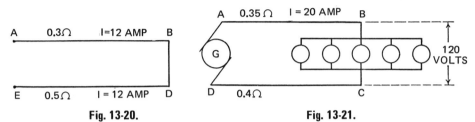

Fig. 13-20.   Fig. 13-21.

2. Find the line drop in a line that offers a resistance of 0.8 ohm to a current of 45 amp.
3. Find the line drop A to B, C to D, the total line drop and the terminal voltage of the generator in the circuit shown in Fig. 13-21.

PROBLEMS

18. Each of the line wires shown in Fig. 13-22 has a resistance of 0.5 ohm. If each of the bells requires 0.65 amp., how much voltage is lost in the line?

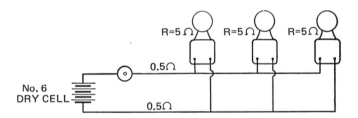

Fig. 13-22.

19. Figure 13-23 shows a method of wiring a circuit to make the voltage across several groups of lamps more nearly equal. If the generator voltage is 125 volts, and each lamp takes 1.25 amp., find (a) the current in sections AB, BC, DE, and EF, (b) the voltage drop in each section, and (c) the voltage across each group of lamps.

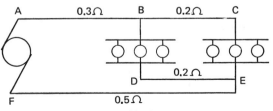

Fig. 13-23.

20. Find the line drop in a wire carrying 225 amp., 1 3/4 mi. long, if the wire is 0.460 in. in diameter.
21. A group of lamps draw 25 amp. from a line 4,000 feet from the generator. If the line drop is not to exceed 2.6 volts, determine the proper size of wire to be used.
22. How far can 22 amp. be transmitted through No. 8 A.W.G. copper wire with a line drop of 8 volts?

The series circuit

When an electric circuit contains several resistances that are connected end to end and make a single path through which the current must flow, the resistances are said to be connected *in series*.

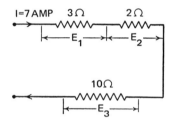

Fig. 13-24. A series circuit.

In a series circuit, remember these three rules:
1. The current (in amperes) is the same in all parts.
2. The total resistance is the sum of the separate resistances.
3. The total voltage is the sum of the voltages across the separate parts of the circuit.

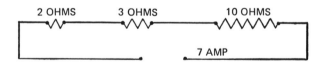

Fig. 13-25. Cast-iron grid resistor.

Example. Three cast-iron grid resistors, one of 2 ohms, one of 3 ohms, and one of 10 ohms, are connected in series and draw 7 amp. from the line. Find (a) the total resistance of the circuit, and (b) the voltage applied to the circuit (Fig. 13-25).

THINK	DO THIS
1. To find the total resistance in a series circuit, add the separate resistances.	1. $R = 2 + 3 + 10 = 15$

2. Apply Ohm's law to find E.	2. $E = IR$ $E = 7 \times 15 = 105$ *Ans.* Total resistance is 15 ohms. Voltage applied to the circuit is 105 volts.

Summary	EXAMPLE
To apply Ohm's law to a series circuit:	Four grid resistors of 2 ohms, 4 ohms, 5 ohms and 15 ohms are connected in series and draw 12 amp. from the line. Find the total resistance of the circuit and the voltage applied to the circuit.
1. Find the sum of the separate resistances.	1. Sum of resistances = $2 + 4 + 5 + 15 = 26$
2. Find the sum of the voltages if voltages across the separate parts of the circuit are given. Current is the same in all parts.	2. Circuit draws 12 amp.
3. To apply Ohm's law in the case of the series circuit, use the same current, I, throughout the circuit, use for R the sum of all the resistances, and use for E the sum of the voltages across the separate parts.	3. $E = IR$ $E = 12 \times 26 = 312$ volts *Ans.* Total resistance = 26 ohms Voltage applied = 312 volts

Ohm's law may be applied to a whole circuit or to a portion of a circuit. When Ohm's law is applied to an entire circuit, all data used must be for the entire circuit. When Ohm's law is applied to a part of a circuit, all data must be for that particular part of the circuit.

EXERCISES

Find the missing quantity in each of the following series circuits:

1. Resistances of 8 ohms, 20 ohms, 10 ohms; voltage, 55 volts. Find the current flowing in the circuit.
2. Resistances of 5 ohms and 2.6 ohms, current delivered 15 amp. Find the applied voltage.
3. Voltage, 37.5; resistances of 1.8 and 4.2 ohms. Find the current delivered to the circuit.
4. A pressure of 110 volts delivers 20 amp. through two resistances, one of which is 2.4 ohms. What is the other resistance?

PROBLEMS

23. The arc lamp shown in Fig. 13-26 requires 8 amp. The resistance of the regulating coils is 5 ohms, that of the solenoid 2 3/4 ohms, and that of the carbon arc 2 1/4 ohms. What is the line voltage?

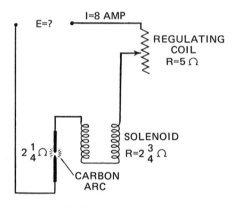

Fig. 13-26. Series arc lamp.

24. A 60-watt bulb is connected in series with a dimmer resistor across a 120-volt line. The resistance of the lamp is 240 ohms. The current drawn from the line is 0.25 amp. What is the value of the resistor in ohms? (See Fig. 13-27).

25. Three resistors are connected in series (Fig. 13-28). The voltages across them are 28, 42, and 56 volts, respectively. If the current through A is 0.25 amp., find (a) the line voltage, (b) the current through each part, and (c) the resistance of each part and the total resistance.

26. Three equal resistances connected in series have a combined resistance of 25 ohms. Find the current flowing through each and the voltage drop across each when connected to a 115-volt circuit.

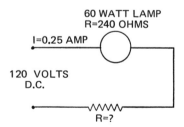

Fig. 13-27.

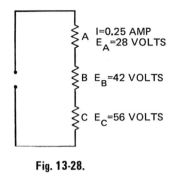

Fig. 13-28.

27. How much resistance would have to be connected in series with a 75-ohm circuit to limit the current to 0.4 amp. at 120 volts?

The parallel circuit

When a circuit is constructed in such a way that the current may flow through two or more paths, the circuit is called a *parallel circuit*.

The circuit shown in Fig. 13-29 is a parallel circuit. The current may pass from A to B to C to D to E to F to J, etc.; or it may flow from A to B to C to F to J, etc.

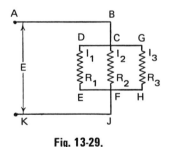

Fig. 13-29.

In parallel circuits, remember these three rules:

1. The voltage across each part of the circuit is the same as the voltage across the combination.

$$E = E_1 = E_2 = E_3$$

2. The current supplied to the circuit is equal to the sum of the currents flowing through all parts.

$$I = I_1 + I_2 + I_3$$

3. To obtain the total resistance of the combined circuits, find the reciprocal of each separate resistance, add these reciprocals, and finally obtain the reciprocal of this result.

$$R = \cfrac{1}{\cfrac{1}{R_1} + \cfrac{1}{R_2} + \cfrac{1}{R_3}}$$

Example 1. Find the line current and the current through each part of the circuit shown in Fig. 13-30. Assume the generator voltage to be 120 volts.

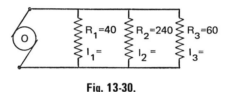

Fig. 13-30.

THINK	DO THIS
1. Since the generator voltage is 120, we apply Ohm's law to each of the resistances.	1. $I_1 = \dfrac{E_1}{R_1} = \dfrac{120}{40} = 3$ $I_2 = \dfrac{120}{240} = 0.5$ $I_3 = \dfrac{120}{60} = 2$
2. The line current then is the sum of the currents through each part.	2. $I = 3 + 0.5 + 2 = 5.5$ *Ans.* 5.5 amp.

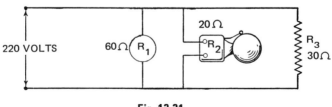

Fig. 13-31.

Example 2. Three resistances of 30 ohms, 20 ohms and 60 ohms are connected in parallel (Fig. 13-31). (a) What is the combined resistance of the circuit? (b) What is the line current?

THINK	DO THIS

(a) 1. To find the total resistance of the combined circuits in parallel, we employ the formula

$$R = \cfrac{1}{\cfrac{1}{R_1} + \cfrac{1}{R_2} + \cfrac{1}{R_3}}$$

2. Replace each letter by its known value.

3. Perform the computations.

(a) 1. Write the formula.

$$R = \cfrac{1}{\cfrac{1}{R_1} + \cfrac{1}{R_2} + \cfrac{1}{R_3}}$$

2. $R = \cfrac{1}{\cfrac{1}{30} + \cfrac{1}{20} + \cfrac{1}{60}}$

3. To find R here, first add the fractions in the denominator.

$$R = \cfrac{1}{\cfrac{1}{30} + \cfrac{1}{20} + \cfrac{1}{60}} = \cfrac{1}{\cfrac{6}{60}}$$

Use the principle of 1.

$$R = \cfrac{1}{\cfrac{6}{60}} \times \cfrac{\cfrac{60}{6}}{\cfrac{60}{6}} = \frac{60}{6} = 10$$

(b) Use Ohm's law to find the line current.

(b) $I = \dfrac{E}{R}$

$$I = \frac{220}{10} = 22$$

Ans. 10 ohms
22 amp.

Summary

In a parallel circuit:

EXAMPLE

Find the combined resistance and the line current if the resistances in Fig. 13-31 are 80, 50 and 70 ohms, respectively.

1. The voltage across each of the circuits is the same as that across the combined circuits:

$$E = E_1 = E_2 = E_3$$

1. E is the same in the entire circuit, 220 volts.

2. To find the resistance of the entire circuit use the formula:

$$R = \cfrac{1}{\cfrac{1}{R_1} + \cfrac{1}{R_2} + \cfrac{1}{R_3}}$$

3. The total current flowing equals the sum of the currents flowing in each part:

$$I = I_1 + I_2 + I_3$$

2. $R = \cfrac{1}{\cfrac{1}{R_1} + \cfrac{1}{R_2} + \cfrac{1}{R_3}} =$

$$\cfrac{1}{\cfrac{1}{80} + \cfrac{1}{50} + \cfrac{1}{70}}$$

$$R = \frac{2800}{131} = 21.37 \text{ ohms}$$

$$I = \frac{E}{R} = \frac{220}{21.37} = 10.29 \text{ amp.}$$

or

3. $I = \dfrac{220}{80} + \dfrac{220}{50} + \dfrac{220}{70}$

$$= 2.75 + 4.4 + 3.14$$

$$= 10.29 \text{ amp.}$$

Ans. Resistance = 21.37 ohms
Current = 10.29 amp.

EXERCISES

In the table below, find the combined resistance of A, B and C (Fig. 13-32) in cases 1 to 5. Then find the other missing item.

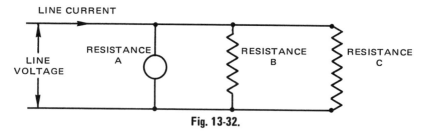

Fig. 13-32.

Case No.	Line Volts	Line Current	Resistance		
			A	B	C
1	110	?	35Ω	20Ω	25Ω
2	?	30	15Ω	25Ω	40Ω
3	220	?	40Ω	20Ω	55Ω
4	90	?	20Ω	20Ω	30Ω
5	?	22	15Ω	20Ω	25Ω

PROBLEMS

28. Six lamps are connected in parallel across a 115-volt line (Fig. 13-33). If the line current is 1.32 amp., what is the resistance of each lamp?

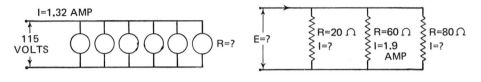

| Fig. 13-33. | Fig. 13-34. |

29. Three resistance units of 20 ohms, 60 ohms and 80 ohms are connected in parallel as shown in Fig. 13-34. If the current through the 60-ohm unit is 1.9 amp., (a) what is the current through each of the other two units, and (b) what is the line voltage?

30. Grid resistors of 4, 6 and 15 ohms are connected in parallel across a 120-volt line. Find the current through each resistor and the combined resistance of the circuit.

31. The current in the line feeding two coils connected in parallel is 10 amp. If the coils have resistances of 6 and 12 ohms, respectively, (a) what is the resistance of the circuit, (b) what is the voltage across each coil, and (c) what current flows through each coil?

32. Three resistances, A, B and C, are connected in parallel to a 120-volt line. The resistance of A is 2 ohms, and of B, 4 ohms. What must be the value of C if the current in the line is to be limited to 100 amp.?

The series-parallel circuit

Circuits, in practice, are usually combinations of the parallel and the series circuit. The circuit in Fig. 13-35 combines a parallel circuit *ABCEFD* with a series circuit. The resistance *HJ* is connected in series with the resistance *ABCEFD*. This type of circuit is called a *series-parallel* circuit.

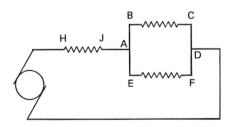

Fig. 13-35.

When Ohm's law is applied to series-parallel circuits, each part of the circuit must first be considered separately. Then the current, voltage, and resistance of the entire circuit may be determined.

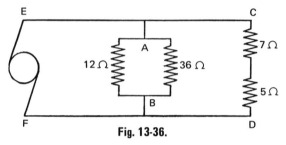

Fig. 13-36.

Example. What is the value of a single resistance that may be used in place of the combination shown in the circuit in Fig. 13-36?

THINK	DO THIS
1. Note whether the resistances are in parallel or in series.	1. In CD the resistances are in series, in AB they are in parallel.
2. Find the sum of the resistances in CD and use the formula $$R = \frac{1}{\frac{1}{R_1} + \frac{1}{R_2}} \text{ in } AB.$$	2. $R_{CD} = 7 + 5 = 12$ $$R_{AB} = \frac{1}{\frac{1}{R_1} + \frac{1}{R_2}}$$ $$= \frac{1}{\frac{1}{12} + \frac{1}{36}} = 9$$
3. Find the resistances in E to F.	3. We consider the parallel circuits CD and AB. $$R_{EF} = \frac{1}{\frac{1}{12} + \frac{1}{9}} = 5.14$$ *Ans.* 5.14 ohms

PROBLEMS

33. Find the resistance of the circuit in Fig. 13-37.

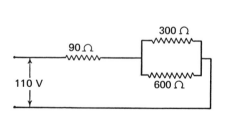

Fig. 13-37. Diagram of a series-parallel circuit.

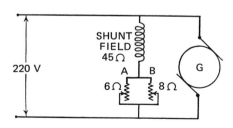

Fig. 13-38.

34. Two field rheostats, A and B, are connected in parallel. This combination is connected in series with a shunt-field winding of a generator as shown in Fig. 13-38. Find (a) the combined resistance of the two rheostats in parallel, (b) the resistance of the entire field circuit, (c) the current in each rheostat, (d) the voltage drop across each rheostat, and (e) the voltage across the field winding.

35. An electric furnace has a resistance of 6.5 ohms. A second furnace connected in parallel has a resistance of 8.5 ohms. If the line resistance is 0.05 ohm per wire, determine the current flowing through each furnace and the voltage at the furnace (see Fig. 13-39).

36. A series-parallel circuit is made up of two parallel groups connected in series. One parallel group consists of two resistances of 6 and 18 ohms each, whereas the other parallel group is made up of a 20- and a 30-ohm unit. If the line current is 5 amp., what is the voltage across the 6-ohm and the 30-ohm resistances?

37. In Fig. 13-40, find (a) the voltage across F, (b) the voltage across the line AB, and (c) the current through each unit.

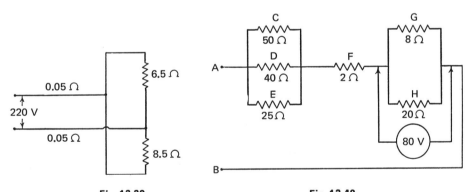

Fig. 13-39.   Fig. 13-40.

38. In the circuit shown in Fig. 13-41, resistance A is in parallel with resistance B. Resistance C is in series with this combination. If the voltage across coil A is 50 volts when 3.2 amp. flow through it, find (a) the voltage across the parallel combination, (b) the current through coil B which has a resistance of 5 ohms, (c) the current through the line, (d) the voltage across coil C which has a resistance of 4 ohms, and (e) the voltage across the line.

39. Find the combined resistance of the circuit shown in Fig. 13-42.

40. Each lamp in the circuit shown in Fig. 13-43 takes 0.85 amp. The motor draws 18 amp. If the generator voltage is 125 volts, find (a) the line drop of each wire, (b) the voltage across the lamps, and (c) the voltage across the motor.

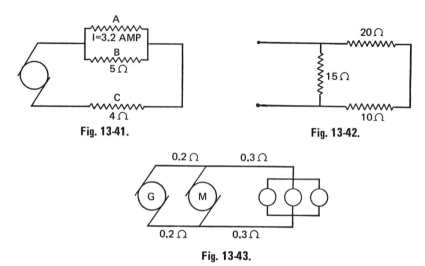

Fig. 13-41.

Fig. 13-42.

Fig. 13-43.

Power in a direct-current circuit

The power in watts or the rate at which electric energy is used in a circuit in operating a motor or a lamp may be found in two ways. First, if an ammeter (measures amperes) and a voltmeter (measures volts) are correctly connected in the circuit, the rate of power consumption in watts is the product of the readings on these two meters; or second, the number of watts may be read directly from a wattmeter (Fig. 13-44).

Fig. 13-44. Wattmeter. *(Courtesy Weston Electrical Instrument Corporation)*

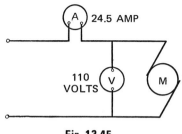

Fig. 13-45.

Example. The ammeter reading in a direct-current circuit (Fig. 13-45) in which a motor is wired is 24.5 amp., and the voltmeter reading is 110 volts. What power, in watts, is being consumed?

THINK	DO THIS
The number of watts equals the product of the number of amperes and the number of volts.	24.5 x 110 = 2,695
	Ans. 2,695 watts

Summary	EXAMPLE
In a direct-current circuit, to find the number of watts consumed:	If the ampere reading in a direct-current circuit (Fig. 13-44) in which a motor is wired is 41.8 amp., and the voltmeter reading is 220 volts, what power, in watts, is being consumed?
Multiply the number of volts by the number of amperes. With I the current in amperes, E the electromotive force or pressure in volts, and P the number of watts, we use the formula	Use the formula $P = I$ x E.
$$P = I \text{ x } E \text{ or } P = I^2 P.$$	P = 41.8 x 220 = 9,196 watts or 9.196 kilowatts
	Ans. 9,196 watts

Note: 1 kilowatt (k.w.) equals 1,000 watts.

EXERCISES

Find the power required in each of the following circuits:

1. The ammeter reading is 12.8, the voltmeter 45.
2. The ammeter reading is 5.9, the voltmeter 36.
3. A circuit delivers 40 amp. when the resistance is 8.2 ohms. (First use $E = IR$ to find E.)

4. A circuit delivers 32 amp. through a resistance of 14.5 ohms.
5. A circuit is under a pressure of 110 volts and a resistance of 6.5 ohms. (First find I, using $I = E/R$. Then find P, using $P = I \times E$. Note that this is equivalent to finding P, using $P = E^2/R$.)
6. The pressure is 220 volts and the resistance 5.2 ohms.
7. A current of 22.5 amp. under a resistance of 20.4 ohms.
8. The voltage is 55 and the resistance 12.6 ohms.

PROBLEMS

41. The shunt field of a direct-current motor (Fig. 13-46) has a resistance of 52 ohms and draws a current of 2.5 amp. from the line. How much power does the shunt field require?

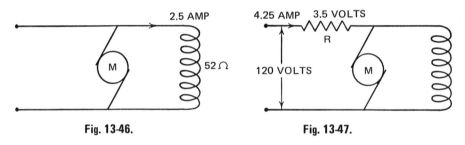

Fig. 13-46.   Fig. 13-47.

42. The motor shown in Fig. 13-47 draws 4.25 amp. at 120 volts. The resistance R requires 3.5 volts to force the current through it. Calculate the power required by (a) the resistor and (b) the motor.
43. (a) Calculate the power required by each of the resistors shown in Fig. 13-48. (b) What is the total power delivered to the circuit? (c) What is the voltage across the circuit?
44. Find the power required by each of the resistors in Fig. 13-49. Express your answer in watts and kilowatts.

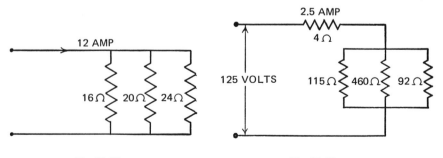

Fig. 13-48.   Fig. 13-49.

Alternating currents

When a single loop of wire is rotated in the magnetic field between two poles so as to cut lines of force at a changing rate, an electromotive force is induced in each conductor of the loop (see Fig. 13-50). The value of the induced electromotive force depends on the rate at which lines of force are cut. The direction of the induced electromotive force depends upon the direction of the lines of force and the direction of rotation of the conductors which make up the loop.

When conductor A is at position 0, it is moving parallel to the lines of force and no electromotive force is induced in the conductor. At position 1, some lines of force are cut and a small electromotive force is induced in A. As the conductor is moved from position 1 to position 3, 90 degrees away from the starting point, the rate at which lines of force are cut increases until position 3 is reached. At this position lines of force are cut at the greatest rate and the highest electromotive force is induced. As the conductor is moved from position 3 to position 6, the rate of cutting lines of force is decreased, resulting in a decreasing induced electromotive force until position 6 is reached, at which no lines of force are cut and no voltage is induced in the conductor. As the conductor is moved from this position to position 9, it comes under the influence of the opposite pole, and a small electromotive force is induced in the opposite

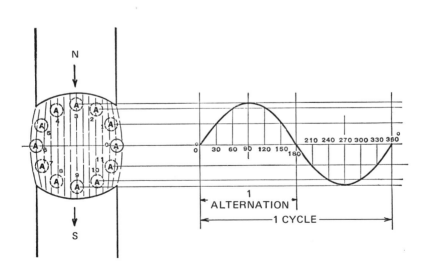

Fig. 13-50.

Fig. 13-51. A low-speed 20-pole A.C. generator or alternator. *(Courtesy General Electric Company)*

direction. This electromotive force increases until a maximum is reached when the conductor passes the center of the south pole. As the conductor is moved from the south pole to the original starting position, the electromotive force decreases until it reaches a value of 0.

The curve in Fig. 13-50 shows how the value of the induced electromotive force changes as the conductor makes one complete revolution. The complete set of values through which the induced electromotive force passes is called a *cycle*. If a current goes through these values 60 times in a second, it is said to have a *frequency of 60 cycles per second.* Electricity for light and power usually has a frequency of 60 cycles.

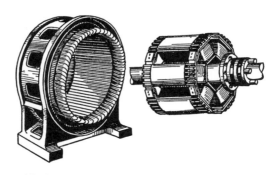

Fig. 13-52. Stator and rotor of a 6-pole synchronous motor.

The frequency of an alternating current may be calculated by the formula

$$F = \frac{P \times N}{120}$$

where F stands for the frequency, P for the number of poles the alternator has, and N for the number of revolutions per minute the alternator makes.

Summary

To find the frequency, F, of alternating current, use the following formula:

$$F = \frac{P \times N}{120}$$

in which F is the frequency, P the number of poles on the alternator, and N the speed or revolutions per minute of the alternator.

EXAMPLE

A 60-pole engine-driven alternator has a speed of 120 r.p.m. Find the frequency of the electricity generated.

Replace the letters in the formula by their values given in the problem.
$P = 60$, $N = 120$

$$F = \frac{P \times N}{120}$$

$$F = \frac{60 \times 120}{120}$$

$$F = 60$$

Ans. Frequency is 60 cycles

EXERCISES

Find the frequency in Exercises 1 through 4:

1. 40 poles at 1,200 r.p.m.
2. 60 poles at 1,500 r.p.m.
3. 50 poles at 1,800 r.p.m.
4. 20 poles at 1,000 r.p.m.

5. What are the revolutions per minute of a twenty-pole alternator that generates a 60-cycle current?

PROBLEMS

45. What is the frequency of a four-pole 1,800 r.p.m. alternator?
46. At what speed in revolutions per minute must a six-pole alternator rotate in order to develop 50 cycles?
47. A 25-cycle generator is driven at 75 r.p.m. How many poles does it have?
48. A twelve-pole turbo-alternator rotates at 600 r.p.m. Find the frequency in cycles per second.
49. A vertical turbo-alternator generating a 60-cycle electromotive force has ten poles. Find the speed of the alternator.

Power in a single-phase alternating circuit

The power (P) in a single-phase alternating current circuit is equal to the product of the number of amperes (I) times the number of volts (E) times the power factor of the circuit ($P.F.$). Expressed as a formula, this is

$$P = I \times E \times P.F.$$

This is also expressed as

$$P = I \times E \times \cos\theta$$

where the power factor becomes the cosine of the angle of lead or lag of the current.

The product $I \times E$ is the *apparent power*, while the *true power* is the reading obtained from the wattmeter in the circuit.

If you compare the apparent power $I \times E$ in the formula above, $P = I \times E \times P.F.$, with the true power P, you will see that

$$P.F. = \frac{P}{I \times E}$$

This shows that the power factor may easily be expressed as a percent. It may be expressed as the percent that the true power P is of the apparent power $I \times E$.

The power in an alternating circuit is also equal to $I^2 R$. Can you recall how this relation is derived? (Remember that $E = IR$.)

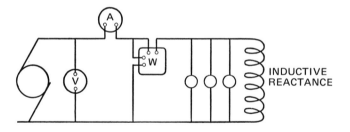

Fig. 13-53.

INDUCTIVE REACTANCE

Summary

The apparent power in an alternating current circuit is the product of E, the number of volts, multiplied by I, the number of amperes. Then the true power, P, is found by using the formula

$$P = I \times E \times P.F.$$

Here $P.F.$ is the power factor, sometimes given as percent, sometimes as the cosine of the angle of lead or lag (expressed as $\cos\theta$).

EXAMPLE

The ammeter in the single-phase circuit shown in Fig. 13-53 reads 25 amp. The voltmeter reads 220 volts and the wattmeter 4,675 watts. Find the true power, the apparent power, and the power factor.

1. Obtain the true power reading from the wattmeter.	1. True power = 4,675 watts or 4.675 kilowatts

1. Obtain the true power reading from the wattmeter.

2. To find the apparent power, use the formula

apparent power = $I \times E$

3. Power factor equals true power divided by apparent power:

$$P.F. = \frac{P}{I \times E}$$

1. True power = 4,675 watts or 4.675 kilowatts

2. $I \times E = 25 \times 220$
= 5,500 watts
or 5.5 kilowatts

3. $P.F. = \frac{P}{I \times E}$

$P = 4,675$
$I \times E = 5,500$

$$P.F. = \frac{4,675}{5,500} = 0.85$$

Expressed as a percent, 0.85 = 85%

Ans. True power = 4.675 k.w.
Apparent power = 5.5 k.w.
Power factor = 85%

PROBLEMS

50. A single-phase induction motor driving a small drill press draws 2.3 amp. at 110 volts. If the power factor is 0.65, what is the true power consumed?
51. At what power factor is a 60-cycle single-phase repulsion induction motor operating if it takes 375 watts and 5.2 amp. at 110 volts?
52. What power is lost in a 45-ohm series resistance connected to an alternating-current circuit through which 5 amp. are flowing?
53. An electrical refrigerator driven by an alternating-current motor draws 3.25 amp. from a 150-volt line. If the power factor is 65 percent, how much power is required to operate the refrigerator?
54. The power factor of a radio receiver is 91 percent, the line voltage 115 volts. If the receiver draws 120 watts, what current is supplied by the line?
55. If the apparent power is equal to 22 kva (kilovolt-amperes), and the power factor is 0.8, what is the true power in this circuit?
56. In a certain alternating-current circuit, the power factor is 95 percent. If the true power is 4,180 watts, what is the apparent power in kilovolt-amperes?

Finding the cost of electical energy

When you use electrical energy, you must pay for it the same as you do for any other commodity. If a store or a factory uses 10 lamps of 100 watts each, when all lamps are operating, there are 1,000 watts or a kilowatt of electricity being used. If these 10 lamps burn for one hour, we say that one *kilowatt-hour* (kw.-hr.) of electrical energy is consumed. The amount of electrical energy bought or sold is measured in kilowatt-hours. It is equal to the number of kilowatts being used multiplied by the number of hours the use continues.

Summary	EXAMPLE
To find the cost of electrical energy:	A motor when operating at full load requires 3,730 watts (or 5 horsepower). What does it cost to run this motor 8 hours if electricity costs 3.8¢ per kilowatt-hour.
1. Find the number of kilowatt-hours consumed.	1. Number of kilowatt-hours consumed: $\dfrac{3,730}{1,000}$ = 3.73 kw.-hr. per hr. For 8 hours: 3.73 x 8 = 29.84 kw.-hr. consumed
2. Multiply this result by the cost per kilowatt-hour.	2. 29.84 x 0.038 = 1.13392 = 1.14 *Ans.* $1.14

Note: The number of kilowatt-hours consumed was multiplied by 0.038, which is the cost per kilowatt-hour, 3.8 cents, expressed as dollars. Hence the product 1.14 is dollars.

EXERCISES

Find the cost or cost per month of each of the following:

1. 75 kw.-hr. of electricity at 4.5¢ per kw.-hr.
2. 4,675 kw.-hr. at 2.8¢ per kw.-hr.
3. The current consumed in one month by twelve lamps of 50 watts each if each lamp averages 90 hr. per month. The electricity costs 3.25¢ per kw.-hr.
4. A 3.5 h.p. motor (one h.p. = 746 watts) that averages 6 hr. per day, 22 days a month. Current costs 3.4¢ per kw.-hr.

The ordinary electric meter which you see in houses, stores, and factories is a kilowatt-hour meter (Fig. 13-54). Notice that such a meter has four dials, each with an indicator. The indicator on the right turns in the clockwise direction; the next one turns in the opposite direction. counterclockwise; the third clockwise; and the one on the left counterclockwise.

Fig. 13-54. Kilowatt-hour meter.

To read such a meter, write, beginning at the left, the number last passed by the indicator. The reading on Fig. 13-55 is 5 1 1 3 kilowatt-hours, or 5,113 kilowatt-hours. The reading from Fig. 13-56 is 5,069 kilowatt-hours. In this case, it is hard to tell whether the second indicator from the right has or has not passed the 7. To determine such a case correctly, look at the next dial to the right. If the indicator has passed 0, you would use the number 7; if the indicator on this first dial has not passed 0, use the number before the 7, or 6. Here the indicator has not yet passed the 0 mark. Therefore we take the reading on the second dial to be 6.

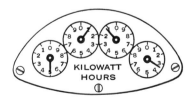

Fig. 13-55.

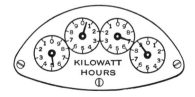

Fig. 13-56.

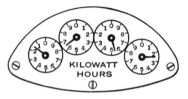

Fig. 13-57. June 1.

Fig. 13-58. July 1.

Summary

To read kilowatt-hour meters:

1. Start at the left and write the number on each dial that was last passed by the indicator of that dial. This number is the reading at the end of the period. To do this you must watch how the numbers increase.

2. To find the number of kilowatt-hours consumed during the period, subtract the reading recorded at the beginning of the period from the reading at the end of the period.

3. To find the cost of this energy, multiply the amount consumed in kilowatt-hours by the cost per kw.-hr.

EXAMPLE

Figures 13-57 and 13-58 show the readings given by an electric meter on June 1 and July 1, respectively. How much energy was consumed and what was its cost at 3.5¢ per kw.-hr. in the period June 1 to July 1?

1. Read the meters.
   Reading July 1 = 1,706
   Reading June 1 = 1,652

2. Subtract the earlier reading from the last reading.

   $1,706 - 1,652 = 54$ kw.-hr.

3. $54 \times 0.035 = 1.89$

   Ans. Cost is $1.89

EXERCISES

Make each of the readings indicated below:

1.
2.

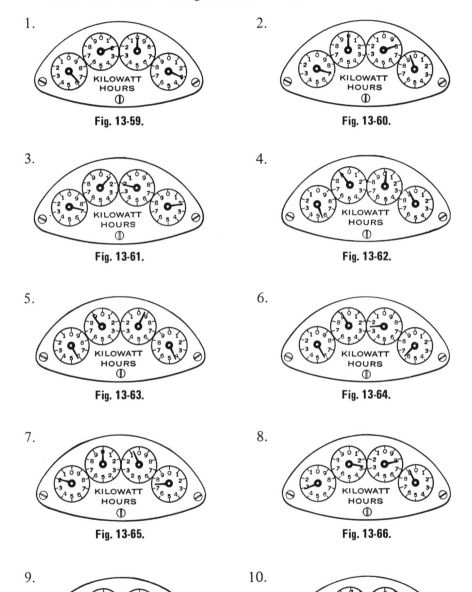

Fig. 13-59.   Fig. 13-60.

3.
4.

Fig. 13-61.   Fig. 13-62.

5.
6.

Fig. 13-63.   Fig. 13-64.

7.
8.

Fig. 13-65.   Fig. 13-66.

9.
10.

Fig. 13-67.   Fig. 13-68.

Find the cost of the electricity consumed in the period indicated in each exercise below. The cost of electricity was 2.6¢ per kilowatt-hour.

11.(a)   (b)

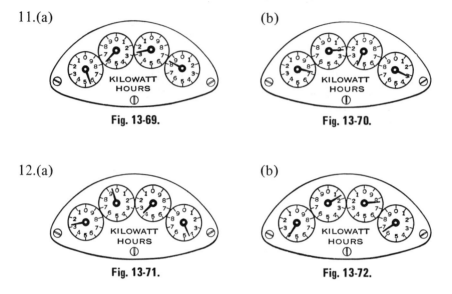

Fig. 13-69.   Fig. 13-70.

12.(a)   (b)

Fig. 13-71.   Fig. 13-72.

PROBLEMS

57. How much does it cost to operate an engine lathe per 8-hr. day if the lathe is driven by a 3-h.p. motor which operates at full load? Assume that the cost of electrical energy is 3¢ per kilowatt-hour.

58. A drill press driven by a 1 1/2 h.p. motor is operated 5 hr. per day. At $0.024 per kilowatt-hour, what is the cost of operation of this machine?

59. At 3.5¢ per kilowatt-hour, what is the cost of operation per month for using an electric iron which takes 5.6 amp. at 112 volts? The iron is used an average of 18 hr. per month.

60. A 7 1/2 h.p. general purpose integral-horsepower gear motor operating at full load is in use 40 hr. per week, 4 weeks per month. At 2 1/2¢ per kilowatt-hour, what is the amount of the monthly bill?

61. A bill for electric energy is $22.50 for 150 hr. If the cost of the electrical energy is $0.028 per kilowatt-hour, what was the average power used?

REVIEW PROBLEMS

1. How much current will flow through the windings of an electro-magnet when it is placed across a 115-volt direct-current circuit if the resistance of the wire on winding is 135 ohms?

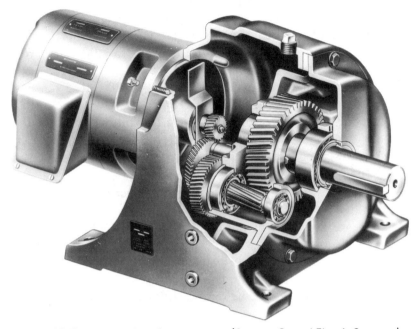

Fig. 13-73. Cutaway view of a gear motor. *(Courtesy General Electric Company)*

2. At 2 1/2¢ per kilowatt-hour, what will it cost to operate a 4,500 watt heater for 8 hr.?

3. A compound wound D.C. motor takes 48 amp. at 220 volts. The total losses at this load are 1,800 watts. Find the output and the efficiency of the motor.

4. An incandescent lighting installation consists of 46 lamps, each of which draws 0.85 amp. from the line at 112 volts, and 22 25-watt lamps. These lamps are used 4 1/2 hr. per day. At $0.026 per kw.-hr. and on a 30-day basis, what is the monthly bill for light?

5. The diameter of No. 6 B & S copper wire is 0.162". Without consulting the wire tables, find the resistance of a piece of this wire a quarter of a mile long.

6. Five resistances of 25, 42, 50, 35 and 64 ohms are connected in parallel (Fig. 13-74). Determine the resistance of the circuit.

Fig. 13-74.

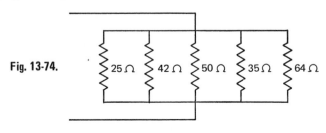

25 Ω   42 Ω   50 Ω   35 Ω   64 Ω

7. A heater resistance of 8.5 ohms dissipates 5,800 watts. What is the value of the voltage that must be supplied to the unit?
8. The electromotive force of a storage cell is 2.13 volts. When the cell delivers 18 amp., the terminal voltage drops to 2.00 volts. What is the resistance of the cell?
9. Find the current drawn from the line by a 220-volt, 5 horsepower motor if the efficiency at full load is 86%.
10. The electromotive force of a dry cell is 1.5 volts. The internal resistance is 0.12 ohm. How much current will flow through the circuit when a resistance of 2.25 ohms is connected across its terminals (Fig. 13-76)?
11. What is the resistance of 1 3/4 mi. of a copper cable having a cross section of 750,000 C.M.?
12. The average resistance of the human body is 10,000 ohms. What is the lowest voltage that will kill a person if a current of 0.1 amp. passing through the body is usually fatal?
13. The resistance of a telephone receiver is 75 ohms. What voltage must be impressed across the receiver if 0.006 amp. is required for its operation?
14. The incandescent lamp used in a certain slide projector is rated at 4.5 amp. at 32 volts. If it is desired to operate this projector from a 110-volt circuit, how much series resistance is necessary?

Fig. 13-75. *(Courtesy Atlas Supply Company)*

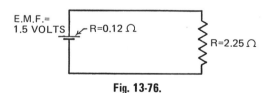

Fig. 13-76.

15. The output of an alternator is 325,000 watts at 10,000 volts. The corresponding input is 515 h.p. (a) Find the efficiency of the generator. (b) What is the value of the line current?
16. Find the cost of operating a 10-h.p. motor for 150 hr. if the efficiency of the motor is 85% and the energy is supplied at 3 cents per kw.-hr.
17. An electroplating bath takes 90 amp. at 25 volts for 2 3/4 hr. At 3 1/2¢ per kilowatt-hour, how much does it cost to operate this bath?
18. A 1/2" copper wire carries 110 amp. Find the voltage drop per mile.
19. What is the diameter in mils of a wire having a cross section of 250,000 C.M.? 750,000 C.M.? 62,500 C.M.?
20. A 1-h.p. single-phase induction motor draws 746 watts at full load. What is the power factor if the line current is 8.2 amp. at 110 volts?
21. What is the speed of a twelve-pole, 60-cycle synchronous motor?

14 Mathematics of the Machine Shop

In this unit we shall learn how to solve a number of the most common and yet most important problems of the machine shop. If a worker is to be more than a machine hand who repeats the same operation in the same way day after day, he must know how to solve certain machine problems. The three main subjects to be discussed are cutting speeds, tapers, and indexing.

Cutting speed

The length or distance in feet which the tool point of a machine cuts in one minute is the *cutting speed of the tool.* Since cutting tools are used in lathes, milling machines, vertical and horizontal boring mills, drill presses, and other shop machines, cutting speed applies to the tools of all these machines.

There are two problems connected with cutting speeds of the tool points of lathes, boring mills, drill presses, etc.: (1) that of finding cutting speed when revolutions per minute, r.p.m., and diameter of the work are known, and (2) that of finding revolutions per minute when cutting speed and diameter are known.

Finding cutting speed

Since the cutting speed is the distance in feet the tool cuts in one minute, it is found by multiplying the circumference of the work by the number of revolutions the work makes in one minute.

Example. Find the cutting speed in feet per minute of a tool that cuts 2 1/2-inch stock if the lathe makes 55 r.p.m.

THINK	DO THIS
1. The distance the tool cuts in one revolution is the circumference of the stock, so find the circumference.	1. $c = \pi D$ $c = \dfrac{3.14 \times 2\ 1/2}{12}$ feet

2. Since the lathe makes 55 r.p.m., the tool cuts 55 times the circumference in 1 minute.	2. $C = c$ x 55 $= \dfrac{3.14 \times 2\,1/2}{12}$ x 55
3. Use the slide rule to obtain the product.	$= 36$ *Ans.* Cutting speed = 36 ft. per min.

For the exercises and problems in this unit, use the slide rule for all calculations and, for the sake of uniformity in answers, use 3.14 as the value for π.

Summary	EXAMPLE
To find the cutting speed C in feet per minute:	Find the cutting speed in feet per minute of a tool that cuts 3 1/2-inch stock at 65 r.p.m.
1. Multiply the circumference, c, of the work in feet by the number of revolutions per minute, N. $C = \dfrac{c}{12}$ x N	1. $C = \dfrac{c}{12}$ x N
2. Substitute the values of the letters from the given problem.	2. $C = \dfrac{3\,1/2 \times 3.14 \times 65}{12} = 60$
3. Compute the result using the slide rule.	*Ans.* Cutting speed = 60 ft. per min.

EXERCISES

Find the cutting speed in feet per minute in each exercise.

1. 4" work, 36 r.p.m.
2. 3 1/2" work, 20 r.p.m.
3. 4.5" work, 48 r.p.m.
4. A piece of stock 2 1/2" in diameter and 16" long, 52 r.p.m.
5. A piece of stock 2 7/8" in diameter, 15" long, 45 r.p.m.

PROBLEMS

1. What would be the cutting speed of the cutting tool in a lathe if a steel rod 5 inches in diameter was revolving at the rate of 150 r.p.m.?
2. A 9-in. pulley on a countershaft revolving at the rate of 200 r.p.m. is belted to the 3-in. step of the spindle-cone pulley of a lathe. If the correct cutting speed for a cast-iron casting 4 1/2 inches in diameter is estimated at 35 ft. per min., is the casting turning too slowly or too fast, and how much?

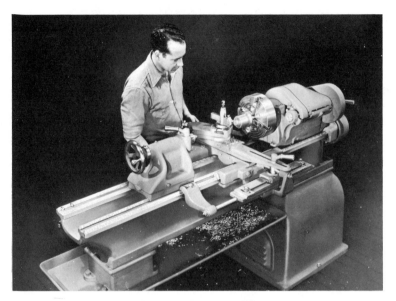

Fig. 14-1. Straight turning with a lathe using independent chuck. *(Courtesy South Bend Lathe)*

Fig. 14-2. Cutting off and turning a piece of stock in a turret lathe.

Finding the number of revolutions per minute to produce a given cutting speed

From the formula on p. 513 we note that

$$C = \frac{c}{12} \times N$$

If we solve this for N, we obtain

$$N = \frac{12C}{c}$$

That is, the number of revolutions per minute, N, is twelve times the cutting speed in feet, C, divided by the circumference in inches, c. Or

$$N = \frac{12C}{\pi \times D}$$

Here N is the number of revolutions per minute, C the cutting speed in feet per minute, and D the diameter in inches.

Summary

To find the number of revolutions per minute (r.p.m.), N:

1. Write the formula

$$N = \frac{12C}{\pi \times D}$$

in which C is the cutting speed, $\pi = 3.14$ and D is the diameter of the work.

2. Substitute in the formula the values given for the letters involved.

3. Find the value for N using the slide rule.

EXAMPLE

A piece of stock 5 1/4 in. in diameter is to be turned in a lathe at a cutting speed of 45 ft. per minute. At what speed, in revolutions per minute, should the work turn?

1. $N = \frac{12C}{\pi \times D}$

2. Substitue 45 for C, 3.14 for π and 5 1/4 for D.

$$N = \frac{12 \times 45}{3.14 \times 5.25} = 33$$

Ans. $N = 33$ r.p.m.

EXERCISES

Find the number of revolutions that the work must make per minute in each of the following:

1. To cut 3 in. stock, at 32 ft. per minute.
2. To cut 2 1/2 in. stock, at 48 ft. per minute.
3. To cut 4 1/2 in. stock, at 36 ft. per minute.
4. To cut a piece of stock that is 18″ long and 3 1/2 in. in diameter, at 40 ft. per minute.

PROBLEMS

3. If the cutting speed for bronze with a stellite cutting tool is 150 ft. per minute, what will be the correct speed, in revolutions per minute, for a bronze casting 1 7/8 in. in diameter?
4. How many revolutions per minute should a vertical boring mill make when boring a hole 6 1/2 in. in diameter in a cast-iron cylinder if the proper cutting speed is 35 ft. per minute?
5. If the cutting speed for a certain piece of soft steel 2 1/2 in. in diameter is 70 ft. per minute when using a high-speed steel tool, at what speed should the spindle of a lathe turn to produce this cutting speed?
6. Find the revolutions per minute for a 5/8 in. carbon-steel drill to cut mild steel containing 0.2-0.3 carbon. The cutting speed for this type of material is 40 ft. per minute.

Fig. 14-3. Drill press in operation. *(Courtesy South Bend Lathe)*

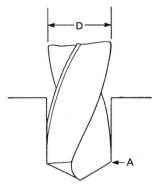

Fig. 14-4.

7. At how many revolutions per minute would the spindle of a 1 7/8″ drill have to revolve in order to drill a hole in a drop forging? Assume that the cutting speed is 60 ft. per minute.

Cutting speed and r.p.m. for drills

The *cutting speed of a drill* is the distance in feet traversed in one minute by the outer corners (*A*, Fig. 14-4) of the cutting edges of the drill. If *D* is the diameter of the drill, and *N* is the number of revolutions it makes per minute, then the cutting speed *C* is

$$C = \frac{\pi \times D \times N}{12}$$

Summary	EXAMPLE
To find the cutting speed, *C*, of a drill when the diameter, *D*, in inches, and the number of revolutions per minute, *N*, are given:	A 1 1/4″ high-speed drill has a speed of 225 r.p.m. when used to cut drill steel. What is the cutting speed of the drill in feet per minute?
1. Write the formula $C = \frac{\pi \times D \times N}{12}$	1. $C = \frac{\pi \times D \times N}{12}$
2. Substitute the values that are given for the letters. Remember to use 3.14 as the value for π.	2. Substitute 1 1/4 for *D*, and 225 for *N*.
3. Make the computations using the slide rule.	$C = \frac{3.14 \times 1.25 \times 225}{12} = 74$
	Ans. *C* = 74 ft. per min.

EXERCISES

Find the cutting speed of each of the following:

1. A 1 1/4″ drill making 350 r.p.m.
2. A 7/8″ drill making 250 r.p.m.

3. A 1 3/8" drill making 280 r.p.m.
4. Find the revolutions per minute which a 1 1/2-in. drill must make in order to have a cutting speed of 85 ft. per minute. (Round off the result to the nearest whole number.)

PROBLEMS

8. If a 7/8" carbon-steel drill is used to drill cast iron at a speed of 160 r.p.m., what is the cutting speed of the drill? What would be the r.p.m. if the cutting speed were 160 feet per minute?
9. A 1/8" high-speed drill used to drill brass revolves at a rate of 4,000 r.p.m. What is the cutting speed?
10. A 9/32" high-speed drill used to drill cold-rolled steel revolves at the rate of 550 r.p.m. What is the cutting speed?
11. A 1 3/8" high-speed steel drill in drilling malleable iron revolves at a rate of 236 r.p.m. What is the cutting speed?
12. A drill-press motor with a 5" pulley is running at 1,750 r.p.m. and is belted to a 2" pulley on the countershaft. What is the cutting speed of a 3/8" drill in this instance?
13. What is the cutting speed of a 5/16" drill when the 1 3/4" pulley of a countershaft is belted to a 4 1/2" pulley on a motor rotating at a speed of 1,725 r.p.m.?

Fig. 14-5. Milling with shell-end mill for face milling with arbor.

Cutting speed and r.p.m. of a milling machine cutter

The *cutting speed of a milling machine cutter* is the distance in feet traversed in one minute by a point on the outer edge of the cutter.

Fig. 14-6. High-speed plain milling cutter with straight teeth.
(Courtesy Brown & Sharpe Manufacturing Company)

To find the cutting speed, *C*, of a milling machine cutter, use the formula:

$$C = \frac{\pi \times D \times N}{12}$$

where *D* is the diameter of the cutter expressed in inches, and *N* the revolutions per minute of the cutter.

Summary

To find the cutting speed, *C*, of a milling machine cutter in feet per minute:

1. Write the formula

$$C = \frac{\pi \times D \times N}{12}$$

2. Substitute the given values for the letters.

3. Make the slide-rule computations. Estimate the result before placing the decimal point.

EXAMPLE

A 4 1/2″ milling machine cutter makes 88 r.p.m. What is the cutting speed?

1. $C = \frac{\pi \times D \times N}{12}$

2. Replace *D* by 4 1/2 and *N* by 88

$$C = \frac{3.14 \times 4.5 \times 88}{12} = 100$$

Ans. *C* = 100 ft. per min.

PROBLEMS

14. The 10" inserted tooth face milling cutter (Fig. 14-7) has a cutting speed of 75' per minute. At how many revolutions per minute should the cutter revolve?

Fig. 14-7. Carbide-tipped fine-pitch face milling cutter. *(Courtesy Brown & Sharpe Manufacturing Company)*

15. A high-speed helical plain milling cutter 2 1/2" in diameter (Fig. 14-8) has a cutting speed of 90' per minute when it machines cast iron. At how many revolutions per minute should the cutter revolve?

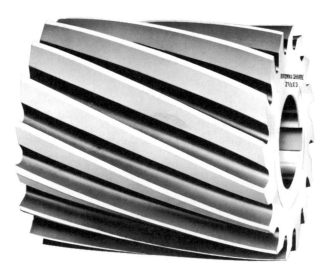

Fig. 14-8. High-speed plain milling cutter with 20° helix angle. *(Courtesy Brown & Sharpe Manufacturing Company)*

16. What will be the cutting speed of a 2 3/4" angular cutter (Fig. 14-9) used to cut cast iron revolving at 125 r.p.m.?

17. A 6 1/2" carbon-steel cutter used to cut brass revolves at 58.8 r.p.m. What is its cutting speed?

18. What diameter of milling cutter should yield a cutting speed of 45 ft. per minute at 91.7 r.p.m.?

Fig. 14-9. Left-hand angular cutter for milling angular grooves and dovetails.
(Courtesy Brown & Sharpe Manufacturing Company)

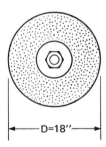

Fig. 14-10.

D=18″

Surface speed of a grinder

The *surface speed of a grinder* is the speed in feet per minute made by a point on the circumference or periphery of the wheel. To obtain the surface speed of a grinder, we multiply the circumference of the grinder by the number of revolutions per minute, N, it is making. Since the circumference of a circle is π times its diameter, D, the surface speed, S, equals $\dfrac{\pi \times D \times N}{12}$.

Summary	EXAMPLE
To find the surface speed, S, of a grinder in feet per minute:	An 18″ grinder is making 2,100 r.p.m. What is its surface speed?

1. Write the formula $S = \dfrac{\pi \times D \times N}{12}$

   where S is the surface speed, D the diameter in inches and N the number of revolutions per minute.

2. Substitute the given values for the letters.

3. Make the computations to find S.

---

1. $S = \dfrac{\pi \times D \times N}{12}$

2. Substitute 3.14 for π, 18 for D and 2,100 for N.

   $S = \dfrac{3.14 \times 18 \times 2,100}{12}$

   $= 9,900$

   Ans. $S = 9,900$ ft. per min.

PROBLEMS

19. What is the cutting speed of a grinding wheel that is 22″ in diameter and rotates at a speed of 3,800 r.p.m.?

20. A 7 1/2″ grinding wheel used for tool and cutter grinding has a cutting speed of 5,500 feet per minute. How many revolutions does it make a minute?

21. A grinding wheel 24″ in diameter used for wet tool grinding travels at 1,170 r.p.m. What is the cutting or peripheral speed in feet per minute?

22. A resinoid grinding wheel having a surface speed of 9,500 ft. per minute and revolving at 814 r.p.m. has a diameter of how many inches?

23. Find the peripheral or cutting speed of a 36″ grinding wheel revolving at 477 r.p.m.

Metal-working lathes

Turning is the process of revolving a piece of material and changing its shape to some form of a cylinder or "round" object. This operation is performed on a lathe.

Metal-working lathes are made for use on a bench, as in the toolroom lathe shown in Fig. 14-11, or for floor mounting, as in Fig. 14-11A. Some are of the standard-change gear type and some are of the quick-change gear type. Standard-change gear lathes require a set of independent change gears for obtaining various speeds and feeds. Figure 14-12 shows the gears at the head end of a bench model standard-change gear lathe. This type of lathe has the advantage of having a "setup" which is not too easily changed.

A quick-change gear lathe has the gearing between the "lead screw" and spindle so arranged that changes in speed and feed are made through a quick-change gear box. Figure 14-13 shows the inside of a quick-change gear box, looking from underneath.

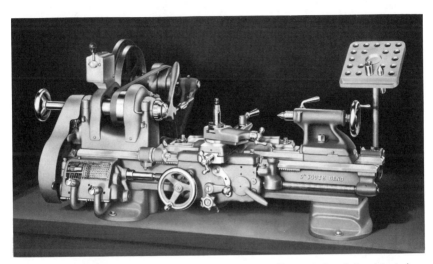

Fig. 14-11. Toolroom lathe with quick-change gears. *(Courtesy South Bend Lathe)*

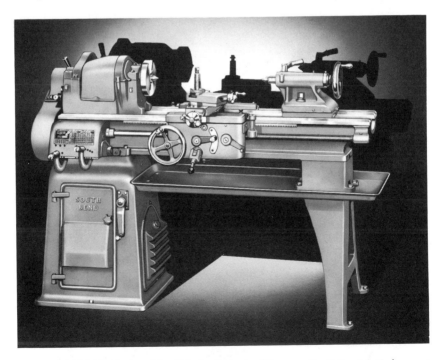

Fig. 14-11A. Lathe with quick-change gears. *(Courtesy South Bend Lathe)*

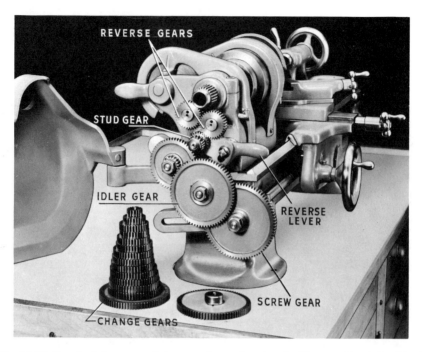

Fig. 14-12. Standard change-gear lathe setup for cutting threads. *(Courtesy South Bend Lathe)*

Fig. 14-13. Interior of quick-change box for lathe. *(Courtesy South Bend Lathe)*

The "size" of a lathe is determined by the double distance from the center of the headstock spindle to the lathe bed. For example, a 13″ lathe will accommodate a workpiece with a 6 1/2″ "swing."

Lathe change gears

In cutting threads with a lathe, the carriage, which carries the cutting tool, is moved along the bed of the lathe by means of a *lead screw*. As the work revolves in the lathe, the lead screw moves the carriage, and hence the cutting tool, along the work. If the work should revolve ten times while the lead screw moves the carriage one inch, the tool would cut ten threads per inch on the work.

Gears are used to give the proper ratio of speed of the carriage to the rate of revolution of the work in order that the proper number of threads per inch will be cut.

The gear on the spindle turns with the work. By properly selecting the gears, the lead screw can be made to move the tool so as to cut any desired number of threads per inch. We must determine how to select these *change gears*.

The change gearing may form either a simple or a compound train (see unit 7). Figure 14-14 shows a simple gear train connecting spindle-stud gear *A* with lead-screw gear *B*.

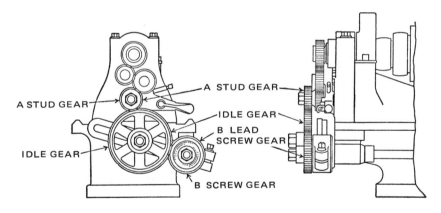

Fig. 14-14. Simple train of gears.

Lathe-screw constant

To determine the change gears required to cut a given thread, it is necessary to know the *lathe-screw constant* for the lathe to be used. This screw constant is the *number of threads per inch that will be cut by the given lathe when the gears (simply geared) on the spindle and lead screw have the same number of teeth.*

The constant is always the same for a given lathe, but may be different for different makes and sizes of lathes.

Change gears for simple gearing

In order to select the proper gears to cut a required number of threads per inch, we use the following formula:

$$\frac{\text{Lathe-screw constant}}{\text{Threads per inch}} = \frac{\text{No. of teeth in gear on spindle stud}}{\text{No. of teeth in gear on lead screw}}$$

Example. From the following set of gears select those suitable for cutting 16 threads per inch with a screw constant of 4.

Given Gears
24, 30, 36, 42
54, 60, 66, 72, 78
84, 90, 96, 102, 108

THINK	DO THIS
1. Write the formula.	1. $\dfrac{\text{Lathe-screw constant}}{\text{Threads per inch}} =$ $\dfrac{\text{No. of teeth on spindle-stud gear}}{\text{No. of teeth on lead-screw gear}}$
2. Substitute the numerical values in the proper places.	2. $\dfrac{4}{16} =$ $\dfrac{\text{No. of teeth on spindle-stud gear}}{\text{No. of teeth on lead-screw gear}}$
3. Select two gears from the set with teeth in the ratio 4/16, or 1/4.	3. Gear A, with 24 teeth for the spindle stud, and B, with 96 teeth for the lead screw, will do. Is there another pair that would work? *Ans.* Gears with 24 and 96 teeth.

Summary	EXAMPLE
To determine the gears to be used in a simple train of change gears:	From the set of gears above, select two that are suitable for cutting 18 threads per inch with a constant of 6.

1. Write the formula.

$$\frac{\text{Lathe-screw constant}}{\text{Threads per inch to be cut}} =$$

$$\frac{\text{No. of teeth on spindle-stud gear}}{\text{No. of teeth on lead-screw gear}}$$

2. Substitute the known values on the left. This will give a ratio.

3. Select two gears from the set so that the number of teeth on one divided by the number of teeth on the other will equal the same ratio.

1. Write the formula:

$$\frac{\text{Lathe-screw constant}}{\text{Threads per inch to be cut}} =$$

$$\frac{\text{No. of teeth on spindle-stud gear}}{\text{No. of teeth on lead-screw gear}}$$

2. Substitute 6 for the constant and 18 for threads per inch.

$$\frac{6}{18} =$$

$$\frac{\text{No. of teeth on spindle-stud gear}}{\text{No. of teeth on lead-screw gear}}$$

3. Select two gears which, when the number of teeth on one is divided by the number of teeth on the other, will give the ratio 6/18, or 1/3.

Ans. Gears with teeth numbering 24 and 72 or 30 and 90 will do.

EXERCISES

In solving these Exercises, select gears from the following set: 24, 30, 36, 42, 48, 54, 60, 66, 72, 78, 84, 90, 96, 102, 108. Find suitable change gears to cut the threads by means of simple gearing.

1. Cut 18 threads per inch with a lathe-screw constant of 5.
2. Cut 32 threads per inch with a lathe-screw constant of 8.

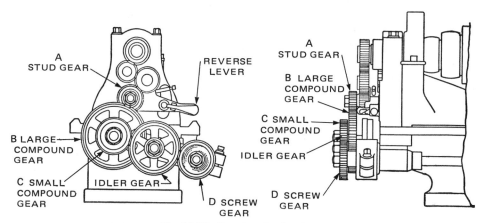

Fig. 14-15. Compound train of gears.

3. Cut 12 threads per inch with a lathe-screw constant of 6.

4. How many teeth should the gears for the stud and lead screw have in order to cut 24 threads per inch with a lathe screw having a constant of 6?

5. If the screw constant of a certain lathe is 6, find the number of threads per inch that will be cut with 35 teeth on the stud and 70 on the screw.

Change gears for compound gearing

It is not always possible to obtain gears that will make a simple train to cut desired threads. Then we must resort to compound trains.

In the train of Fig. 14-15, A and C are driving gears, B and D are driven gears. These four gears are selected as follows:

1. Write the formula

$$\frac{\text{Lathe-screw constant}}{\text{No. of threads per inch to be cut}} = \frac{\text{Product of teeth on driving gears}}{\text{Product of teeth on driven gears}}$$

2. Place the known numerical values on the left, which gives a ratio, such as 2/3 or 4/12 or 6/24.

3. Select the four gears from the set that is available in such a way that the product of the number of teeth on two, divided by the product of the number of teeth on the other two, will equal this ratio.

4. Use the first two as driving gears, the others as driven gears.

Example. Determine the change gears to be used to cut 24 threads per inch, when the lathe constant is 6. The gears available have from 30 teeth to 100 teeth in multiples of 5. That is, 30, 35, 40, 45, etc., up to 100.

THINK	DO THIS
1. Write the formula.	1. $\dfrac{\text{Lathe-screw constant}}{\text{No. of threads per inch to be cut}} =$ $\dfrac{\text{Product of no. of teeth on drivers}}{\text{Product of no. of teeth on driven}}$
2. Substitute the known values on the left. This will give a ratio.	2. $\dfrac{6}{24} =$ $\dfrac{\text{Product of no. of teeth on drivers}}{\text{Product of no. of teeth on driven}}$

3. We must now select from the set of available gears *four* gears: two for drivers, two for driven, so that the product of the number of teeth on the drivers divided by the product of the number of teeth on the driven equals 6/24.

To make this as easy as possible, we use the principle of 1 and first write $\dfrac{6}{24}$ as $\dfrac{2 \times 3}{4 \times 6}$ and then multiply by 1 as follows:

$$\frac{(2 \times 20) \times (3 \times 10)}{(4 \times 20) \times (6 \times 10)} =$$

$$\frac{40 \times 30}{80 \times 60}$$

4. Now choose 40- and 30-tooth gears as drivers, and 80- and 60-tooth gears as driven.

3. $\dfrac{6}{24}$ is the same ratio as $\dfrac{40 \times 30}{80 \times 60}$

4. 6/24 results if we use 40-tooth and 30-tooth drivers and 80-tooth and 60-tooth driven gears. *Ans.* If we use 40- and 30-tooth gears for the drivers, and 80- and 60-tooth gears for the driven, we have the proper compound train.

Summary

To select the proper gears for a compound train in order to cut a given number of threads:

1. Use the formula

$$\frac{\text{Lathe-screw constant}}{\text{No. of threads per inch}} =$$

$$\frac{\text{Product of no. of teeth on drivers}}{\text{Product of no. of teeth on driven}}$$

2. Substitute the known values on the left. This will give a ratio like 2/3, or 6/8, or 5/12, etc.

EXAMPLE

Determine the change gears to use to cut 18 threads per inch, when the lathe constant is 4.

1. Write the formula.

$$\frac{\text{Lathe-screw constant}}{\text{No. of threads per inch}} =$$

$$\frac{\text{Product of no. of teeth on drivers}}{\text{Product of no. of teeth on driven}}$$

2. $\dfrac{4}{18} =$

$$\frac{\text{Product of no. of teeth on drivers}}{\text{Product of no.of teeth on driven}}$$

3. Write this ratio as the product of two numbers divided by the product of two numbers.

$$\frac{6}{8} = \frac{3 \times 2}{4 \times 2}, \text{ etc.}$$

4. Use the principle of 1 to write this ratio in a form similar to

$$\frac{(3 \times 10) \times (2 \times 20)}{(4 \times 10) \times (2 \times 20)} =$$

$$\frac{30 \times 40}{30 \times 40}, \text{ where all are tooth}$$

numbers of available gears.

5. Select from the available gears two as driving and two as driven gears with the ratio of the product of the drivers to the product of the driven equal to the above ratio.

3, 4. $\dfrac{4}{18} = \dfrac{2 \times 2}{3 \times 6} =$

$$\frac{(2 \times 20) \times (2 \times 15)}{(3 \times 20) \times (6 \times 15)} =$$

$$\frac{40 \times 30}{60 \times 90}$$

5. The drivers have 40 teeth and 30 teeth, the driven 60 teeth and 90 teeth.

Ans. 40- and 30-tooth drivers, 60- and 90-tooth driven gears.

EXERCISES

Select four gears to form a compound train for each of the following exercises. Use the set of gears given in the example on p. 528.

1. With a lathe-screw constant of 6, cut 18 threads per inch.
2. With a lathe-screw constant of 5, cut 20 threads per inch.
3. With a lathe-screw constant of 6, cut 40 threads per inch.
4. With a lathe-screw constant of 4, cut 10 threads per inch.
5. With a lathe-screw constant of 8, cut 12 threads per inch.

PROBLEMS

24. The constant of a certain lathe screw is 6. The following gears are available: 30, 35, 40, etc., increasing by 5 up to 100. Find the gears for A, B, C and D in Fig. 14-16 to cut 24 threads per inch.

Fig. 14-16.

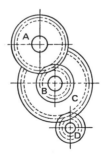

25. The following gears are furnished with a lathe: 25, 30, 35, 40, 45, 50, 55, 60, 65, and 70. Select the gears to be used to cut 20 threads per inch with a lathe-screw constant of 5.

26. The pitch of a thread is 3/8″. A lathe with a lathe-screw constant of 6 is to be used to cut the thread. If the compound gearing ratio is 2 to 1, select the compound gears to be used if the following gears are available: 20 to 90 teeth in steps of 5.

Tapers

Notice that the piece shown in Fig. 14-17 has a gradual and uniform change in size as it decreases from the larger (left) end toward the smaller (right) end.

When the dimensions of the cross section of a piece of work, or a part of the work, decrease gradually and uniformly from the large end to the smaller end, the work is said to be *tapered*.

The *taper* on a piece of round work is the difference between the diameter of the large end and that of the small end. For the piece of stock shown in Fig. 14-17, the large end is 1 3/4 inches in diameter, and the small end is 3/4 inch in diameter; then the taper is 1 3/4 inches − 3/4 inch, or 1 inch.

The piece of work in Fig. 14-17 is two inches long. The taper over the entire length of the piece is one inch. Then the *taper per inch* is the taper over the entire length divided by the length in inches, in this case the ratio 1 ÷ 2 = 1/2. The taper per inch of this piece of work is 1/2 inch per inch.

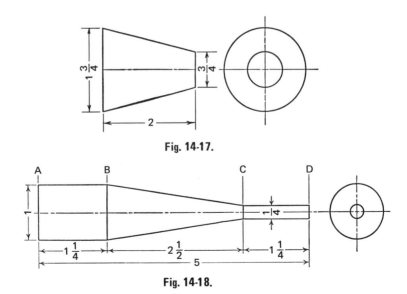

Fig. 14-17.

Fig. 14-18.

The piece shown in Fig. 14-18 has a taper only from B to C. The taper is 1 inch − 1/4 inch, or 3/4 inch. Notice that this taper is accomplished in the distance from B to C, which is 2 1/2 inches. The taper per inch of this piece is $3/4 \div 2\ 1/2$ or

$$\frac{3/4}{2\ 1/2} = \frac{0.75}{2.5} = 0.3, \text{ or } 0.3 \text{ inch per inch}$$

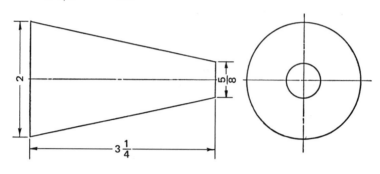

Fig. 14-19.

Example. Find the taper and the taper per inch of the piece of work in Fig. 14-19.

THINK	DO THIS
1. Subtract the diameter of the small end from that of the large end.	1. $2'' - 5/8'' = 1\ 3/8''$. This is the taper.
2. Divide the taper by the number of inches in the length of the piece that has the taper. The length is measured on a line parallel to the axis of the piece.	2. $\dfrac{1\frac{3}{8}}{3\frac{1}{4}} = \dfrac{\frac{11}{8}}{\frac{13}{4}} \times \dfrac{8}{8} = \dfrac{11''}{26}$
3. Express this result in decimal form.	3. $11/26 = 0.423$ *Ans.* The taper is 1 3/8''; the taper per inch is 0.423'' per inch.

The *taper per foot* is the taper per inch multiplied by 12.

Summary

To find the taper, t, of a piece of round work:

EXAMPLE

Find the taper, the taper per inch, and the taper per foot for a taper that is 8'' long, 2 1/2'' at the large end and 1 5/8'' at the small end.

1. Subtract the diameter, d, in inches, of the small end from the diameter, D, in inches, of the larger end. Formula: $t = D - d$.

2. To find the taper per inch, $t_{\text{per in.}}$, divide the taper by the number of inches, L, from the center of the large end to the center of the small end. This distance must be measured parallel to the axis of the work.

$$t_{\text{per in.}} = \frac{t}{L}$$

The operations performed in Steps 1 and 2 may be expressed as the formula:

$$t_{\text{per in.}} = \frac{D - d}{L}$$

3. To find the taper per foot, multiply the taper per inch by 12.

To find the taper per foot, $t_{\text{per ft.}}$, directly, use the formula:

$$t_{\text{per ft.}} = \left(\frac{D-d}{L}\right) \times 12$$

---

1. Find the taper. $t = D - d$

$$t = 2\ 1/2 - 1\ 5/8 = 7/8''$$

2. $t_{\text{per in.}} = \dfrac{t}{L}$

$$t_{\text{per in.}} = \frac{7/8}{8} = \frac{7/8}{8} \times \frac{8}{8} = \frac{7''}{64}$$

3. $t_{\text{per ft.}} = \dfrac{7}{64} \times 12 = \dfrac{84}{64}$

$$= 1\ 5/16''$$

Finding taper per foot directly:

$$t_{\text{per ft.}} = \left(\frac{D - d}{L}\right) \times 12$$

$$t_{\text{per ft.}} = \left(\frac{21/2 - 15/8}{8}\right) \times 12 = 1\frac{5}{16}''$$

Ans. Taper per inch $= 7/64''$
Taper per foot $= 1\ 5/16''$

EXERCISES

Find the taper, the taper per inch, and the taper per foot of each of the following pieces.

1.

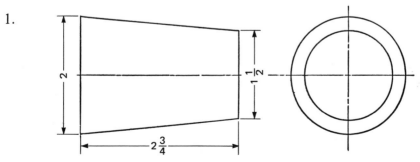

2 $2\frac{3}{4}$ $1\frac{1}{2}$

Fig. 14-20.

2.

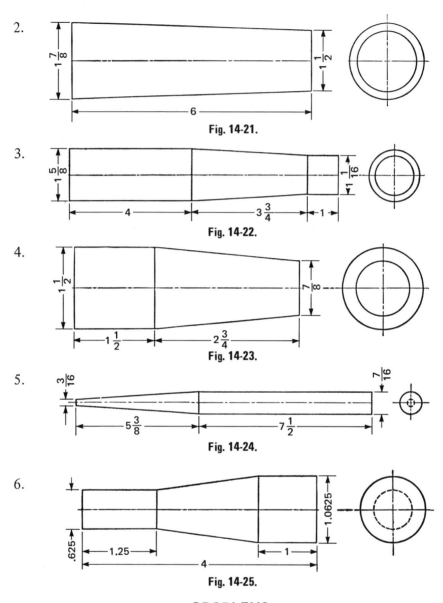

Fig. 14-21.

3.

Fig. 14-22.

4.

Fig. 14-23.

5.

Fig. 14-24.

6.

Fig. 14-25.

PROBLEMS

27. Figure 14-26 shows a taper pin. Find the taper per inch and the taper per foot.

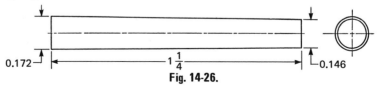

0.172

$1\frac{1}{4}$

0.146

Fig. 14-26.

28. The diameter of the large end of the No. 10 taper reamer shown in Fig. 14-27 is 0.7216"; that of the small end is 0.5799". If the flutes are 6 13/16" long, what is the taper per inch? What is the taper per foot?

Fig. 14-27. Taper reamer.

29. Find the taper per inch in the piece of work illustrated in Fig. 14-28. The diameter of the large end is 1 1/2".

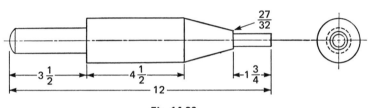

Fig. 14-28.

30. Find the taper, the taper per inch, and the taper per foot of the lathe center shown in Fig. 14-29.

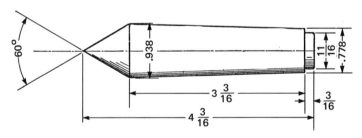

Fig. 14-29.

31. Figure 14-30 shows a journal that is to be turned to fit a bearing. What is the taper per foot? What is the taper per inch?

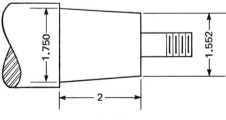

Fig. 14-30.

The diameter of one end of the taper

We often must find the diameter at the large end of a taper. We know the taper per foot, the diameter at the small end, and the length of the piece. To do this, we work out a new formula from the formula given on p. 533.

$$t_{\text{per ft.}} = \left(\frac{D - d}{L} \right) \times 12$$

When we solve this formula for the letter D, we obtain

$$D = \left(\frac{t_{\text{per ft.}} \times L}{12} \right) + d$$

This formula states that the large diameter equals the taper per foot divided by 12, then multiplied by the length of the work and added to the small diameter.

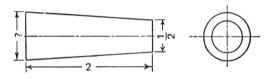

Fig. 14-31.

Example. Find the larger diameter of the piece shown in Fig. 14-31 if the taper per foot is 1 1/2".

THINK	DO THIS
1. To find the larger diameter, make use of the formula given above.	1. $D = \left(\dfrac{t_{\text{per ft.}} \times L}{12} \right) + d$
2. Replace the letters by their values.	2. $D = \left(\dfrac{1\ 1/2 \times 2}{12} \right) + 1/2$
3. Make the computations as shown in the formula.	3. $D = 1/4 + 1/2 = 3/4$ *Ans.* Larger diameter = 3/4"

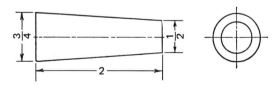

Fig. 14-32.

A nice check of this result is to use this answer and calculate the taper per foot of the piece.

The formula for the small diameter is: $d = D - \left(\dfrac{t_{\text{per ft.}} \times L}{12} \right)$

This formula is obtained from $D = \left(\dfrac{t_{\text{per ft.}} \times L}{12} \right) + d$ by subtracting $\dfrac{t_{\text{per ft.}} \times L}{12}$ from both sides of this formula.

Fig. 14-33.

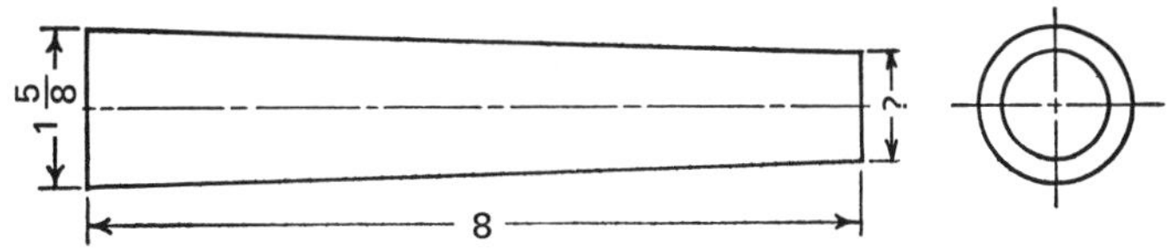

Summary

(a) To find the smaller diameter, use the formula

$$d = D - \left(\dfrac{t_{\text{per ft.}} \times L}{12} \right)$$

(b) To find the large diameter, D, of the work when we know the taper per foot, $t_{\text{per ft.}}$, the length, L, of the work, and the small diameter, d, use the formula

$$D = \left(\dfrac{t_{\text{per ft.}} \times L}{12} \right) + d$$

EXAMPLE

Find the small diameter of the piece shown in Fig. 14-33 if the taper per foot is 0.740''.

(a) 1. Write the formula.

$$d = D - \left(\dfrac{t_{\text{per ft.}} \times L}{12} \right)$$

2. Replace the letters by their values.

$$d = 1.625 - \left(\dfrac{0.740 \times 8}{12} \right)$$

3. Perform the calculations.
$$d = 1.625 - 0.493 = 1.132$$
$$\textit{Ans. } d = 1.132''$$

(b) Use 1.132'' for the smaller diameter and find the larger diameter as a check on the work in (a).

1. Write the formula.

$$D = \left(\dfrac{t_{\text{per ft.}} \times L}{12} \right) + d$$

2. Replace the letters by their values.

$$D = \left(\dfrac{0.740 \times 8}{12} \right) + 1.132''$$

3. Perform the calculations.
$$D = 0.493 + 1.132 = 1.625$$

$$\textit{Ans. } D = 1.625''$$

EXERCISES

Find the missing dimensions in each of the following drawings:

1. Taper per foot is 0.6" (Fig. 14-34).

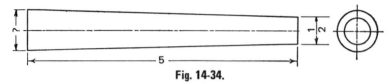

Fig. 14-34.

2. Taper per foot is 0.5" (Fig. 14-35).

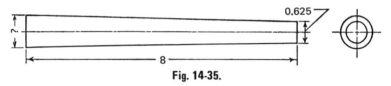

Fig. 14-35.

3. Taper per foot is 0.6" (Fig. 14-36).

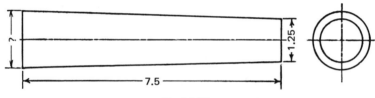

Fig. 14-36.

4. Taper per foot is 0.602" (Fig. 14-37).

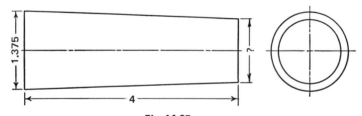

Fig. 14-37.

5. Taper per foot is 0.631" (Fig. 14-38).

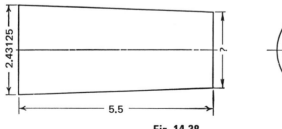

Fig. 14-38.

American standard machine tapers, self-holding (slow) series

There are several standard tapers in use in the United States, among which are the following:

1. Brown and Sharpe series, 0.500″ per ft., except No. 10, which is 0.5161″ per ft.

2. The Morse series

No. 1. . . . . 0.59858″ per ft.
No. 2. . . . . 0.59941″ per ft.
No. 3. . . . . 0.60235″ per ft.
No. 4. . . . . 0.62326″ per ft.
No. 5. . . . . 0.63151″ per ft.

3. The 3/4″ series, 0.740″ per ft.

4. The Jarno taper, 0.6″ per ft. See Appendix.

EXERCISES

Compute the missing dimension for each of the following drawings:

1. Brown and Sharpe series (Fig. 14-39).

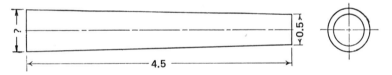

Fig. 14-39.

2. Morse series, taper No. 3 (Fig. 14-40).

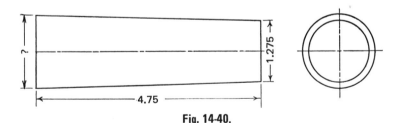

Fig. 14-40.

3. 3/4″ series taper (Fig. 14-41).

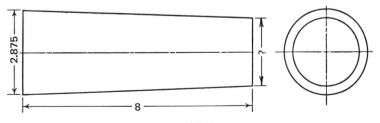

Fig. 14-41.

4. Jarno taper (Fig. 14-42).

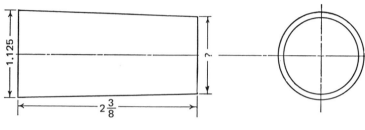

Fig. 14-42.

5. Morse series, taper No. 1 (Fig. 14-43).

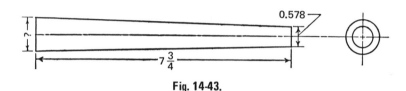

Fig. 14-43.

PROBLEMS

32. Compute the diameter at the large end of the taper of the 7/8″ Rose chucking reamer in Fig. 14-44. (This reamer has a No. 2 Morse taper.)

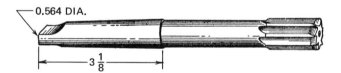

Fig. 14-44. Taper shank machine reamer.

33. The lathe mandrel in Fig. 14-45 is tapered 0.0005″ to one inch. What is the diameter of the small end?

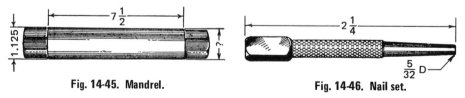

Fig. 14-45. Mandrel.   **Fig. 14-46. Nail set.**

34. The small diameter of the tapered section of the nail set in Fig. 14-46 is 5/32″. The length of the tapered section is 1 3/8″. Find the diameter of the large end. Taper per ft. = 1 3/8″.

35. Find the diameter of the small end of the special taper plug gage shown in Fig. 14-47.

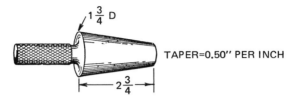

Fig. 14-47. Special taper plug gage.

36. Figure 14-48 shows an arbor for a face milling cutter having a No. 1 Morse taper. What is the diameter of the large end?

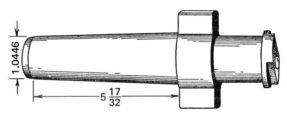

Fig. 14-48. Arbor.

The length of the tapered piece

In order to find the length, L, of a piece between two given diameters, D and d, that will have a given taper per foot ($t_{\text{per ft.}}$), we may use the following formula:

$$L = \left(\frac{D - d}{t_{\text{per ft.}}} \right) \times 12$$

This formula says that the length is equal to the difference of the diameters divided by the taper per foot, then multiplied by 12.

Example. Find the length of the following piece if the taper per foot is 0.740″.

Fig. 14-49.

542 • Unit 14

THINK	DO THIS

THINK

1. To find the required length of a piece when the taper per foot is known, use the formula
$$L = \left(\frac{D-d}{t_{\text{per ft.}}}\right) \times 12$$

2. Replace the letters by their values.

3. Make the calculations. First find the difference $(D - d)$, then divide by the taper per foot and multiply the result by 12.

DO THIS

1. Write the formula.
$$L = \left(\frac{D-d}{t_{\text{per ft.}}}\right) \times 12$$

2. $L = \left(\dfrac{2.875 - 1.5}{0.740}\right) \times 12$

3. $L = \dfrac{1.375}{0.740} \times 12 = 22.3$

Ans. Length of the piece is 22.3″.

Summary

To find the length, L, that a piece of work between two given diameters, D and d, must have in order to have a known taper per foot, $t_{\text{per ft.}}$:

1. Use the formula
$$L = \left(\frac{D-d}{t_{\text{per ft.}}}\right) \times 12$$

2. Replace the letters by their known values.

3. Make the calculations.

EXAMPLE

Find the length of a taper 2.5″ at one end and 1.25″ at the other when the taper per foot is 0.375″.

1. Write the formula.
$$L = \left(\frac{D-d}{t_{\text{per ft.}}}\right) \times 12$$

2. $L = \left(\dfrac{2.5 - 1.25}{0.375}\right) \times 12$

3. $L = 40$

Ans. Length = 40″

PROBLEMS

37. Find the length of the tapered part of the spindle in Fig. 14-50 if the taper per foot is 0.500″.

38. Find the length of the taper on the collet in Fig. 14-51 for a No. 5 Morse taper.

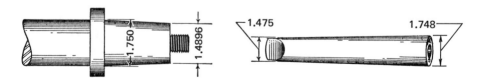

Fig. 14-50. Spindle.   **Fig. 14-51. Collet.**

39. The screw arbor shown in Fig. 14-52 has a No. 7 B & S taper. Find the dimension L.

Fig. 14-52. Screw arbor.

40. The chuck adapter in Fig. 14-53 has a diameter of 1 9/16″ at the small end of the taper. The diameter at the large end is 2 3/4″. What is the length of the tapered section if the taper per foot is 3 1/2″?

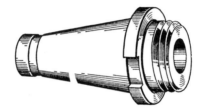

Fig. 14-53. Chuck adapter.

41. Compute the length of the bushing shown in Fig. 14-54. Taper per foot = 3/8″.

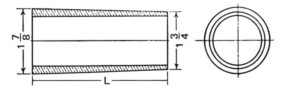

Fig. 14-54. Bushing.

Methods of turning tapers in a lathe

The three methods used in turning tapers are:
1. Offsetting the tailstock — only for outside tapers.
2. Using the taper attachment.
3. Using the compound rest — used for short tapers and tapers with a considerable angle.

Offsetting the tailstock

When the live center and the dead center of the tailstock are in alignment, then the cutting tool which is held in the tool post of the lathe will take a cut parallel to the line of lathe centers, and the piece of work will be turned "straight," or cylindrical (see Fig. 14-55).

Fig. 14-55.

Suppose the tailstock center is moved out of alignment with the live center by some definite distance which we shall call X. Then the center of the work at the dead-center end will be nearer the line of traverse of the tool than is the center of the work at the live-center end (see Fig. 14-56).

This will make the diameter of the work at the dead-center end smaller than that at the live-center end. Hence a taper will result.

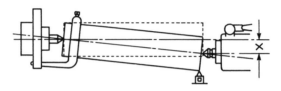

Fig. 14-56.

The farther off center the dead center is set, the smaller will be the small diameter. *The amount of taper* depends upon the *length of the work*, and the *"setover" of the tail center.*

To calculate the amount by which the tail center must be set over (S_o) in order to produce a required taper $(t_{per\ ft.})$ on a certain length piece (L), we use the following formula:

$$S_o = \frac{1}{2}\left(\frac{t_{per\ ft.} \times L^*}{12}\right)$$

Example. Find the setover of the tailstock required to put a taper of 3/4″ per foot in a piece 8″ long.

THINK	DO THIS
1. Write the formula for finding setover, S_o.	1. $S_o = \frac{1}{2}\left(\dfrac{t_{per\ ft.} \times L}{12}\right)$
2. Replace the letters by their values.	2. $S_o = \frac{1}{2}\left(\dfrac{0.75 \times 8}{12}\right)$
3. Make the calculations.	3. $S_o = 0.25$
	Ans. Tailstock must be set over 0.25″.

*Remember always that the length is measured parallel to the axis of the work, never along the tapered side.

Summary

To find the amount, S_o, by which the tailstock must be set over in order to cut a required taper ($t_{\text{per ft.}}$) on a piece of work of given length, use the formula

$$S_o = \frac{1}{2}\left(\frac{t_{\text{per ft.}} \times L}{12}\right)$$

EXAMPLE

Calculate the amount of setover for a 1/2″ per foot taper on a piece 8″ long.

1. Write the formula.

$$S_o = \frac{1}{2}\left(\frac{t_{\text{per ft.}} \times L}{12}\right)$$

2. Replace the letters by their values.

$$S_o = \frac{1}{2}\left(\frac{0.5 \times 8}{12}\right)$$

3. Complete the calculations.
$$S_o = 0.167''$$

Ans. Set over must be 0.167″.

EXERCISES

Calculate the amount of setover for each of the following:

1. A 3/4″ per foot series taper on a piece 10″ long.
2. A 3/4″ per foot series taper on a piece 18″ long.
3. A Brown and Sharpe series taper on a 7.5″ piece.
4. A Morse series No. 3 taper on a 12″ piece.
5. A Jarno system taper on a piece 4.0625″ long.

(a) *When the taper runs the full length of the stock and both diameters are known.* When the taper runs the full length of the stock and the diameters of both ends are known, a simpler formula is used. If you compare the formula of the last section with the formula on p. 536, you will notice that the quantity in parentheses,

$$\left(\frac{t_{\text{per ft.}} \times L}{12}\right)$$

is exactly $D - d$, where D and d are the diameters of the ends of the tapered piece.

Hence, when the taper runs the full length of the stock and these diameters are known, we use the following formula:

$$S_o = \frac{1}{2}(D - d)$$

This means that the setover in this case is one-half of the difference of the diameters.

(b) *When the taper runs only for a part of the length of the stock.*
Suppose you have to cut a taper for a distance of 2 1/2 inches on a piece
of stock 8 inches long (Fig. 14-57).

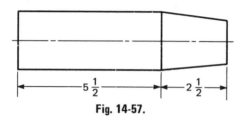

Fig. 14-57.

To set over the tailstock for such a job it is necessary to suppose that
this taper extends the entire length of the piece. That is, we must determine the taper in the whole length of the work and set the tailstock over
one-half of this amount.

TAPER PER FOOT =.600″

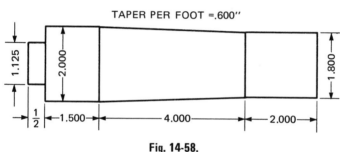

Fig. 14-58.

Example. Find the setover for the job shown in Fig. 14-58.

THINK	DO THIS
1. We must find the taper per inch.	1. Taper per inch = taper per foot divided by 12 0.600/12 = 0.050
2. To find the taper we multiply the taper per inch by the whole length of the work.	2. 0.050 x 8 = 0.400
3. Take 1/2 of this result.	3. 0.400/2 = 0.200 *Ans.* The tailstock must be set over 0.200″.

Summary	EXAMPLE
To find the setover of the tailstock:	Find the setover for a taper with large diameter 2.50″, small diameter 1 3/4″.

(a) When the taper runs the full length of the work, use the formula

$$S_o = \frac{1}{2}(D - d)$$

(a) When the taper runs the full length of the work:

1. Replace the letters in the formula by their values.

$$S_o = \frac{1}{2}(2.50 - 1.75)$$

2. Do the calculations.

$$S_o = 0.375$$

$$Ans. \; S_o = 0.375''$$

(b) When the taper runs for only part of the work, use the formula

$$S_o = \frac{1}{2}\left(\frac{t_{\text{per ft.}} \times L_t}{12}\right)$$

In this formula L_t is the total length of the work.

(b) When the taper per foot is 0.500'' and the total length of the work is 7.5'':

1. Replace the letters in the formula by their values.

$$S_o = \frac{1}{2}\left(\frac{0.500 \times 7.5}{12}\right)$$

2. Do the calculations.

$$S_o = 0.15625$$

$$Ans. \; S_o = 0.15625''$$

EXERCISES

Find the setover for each of the following jobs:

1.

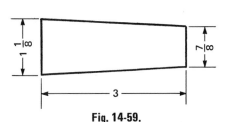

Fig. 14-59.

2.

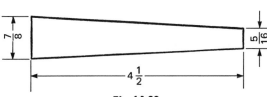

Fig. 14-60.

3.

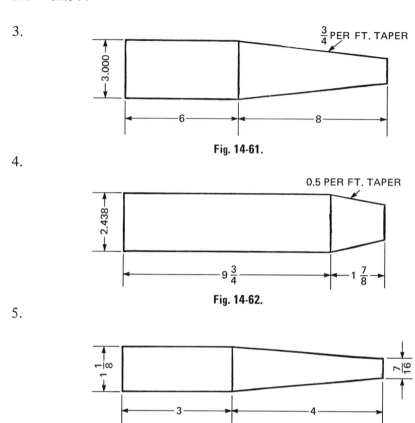

Fig. 14-61.

4.

Fig. 14-62.

5.

Fig. 14-63.

What can you say about the effect of the length of the work and the setover of the tail center on the amount of taper?

PROBLEMS

42. What should be the offset of the lathe tailstock to turn the brass bushing to the dimensions indicated in Fig. 14-64, if the length of the mandrel is 8 1/2"?

43. The taper plug gage shown in Fig. 14-65 has a Jarno taper. What would be the offset of a tailstock for a No. 16 Jarno taper?
   Note: The Jarno taper for all sizes is 0.6" per foot.

44. Find the tailstock offset to produce a No. 4 Morse taper plug having the dimensions shown in Fig. 14-65.

45. What should be the tailstock offset to turn the taper shown in Fig. 14-66? Assume total length of stock to be 7.844".

46. What is the amount of tailstock offset for cutting the taper shown in Fig. 14-67? Allow 3/4" for centering and rounding off the knurled end of the piece.

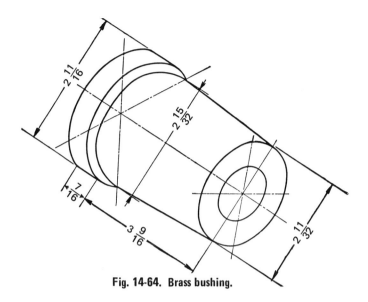

Fig. 14-64. Brass bushing.

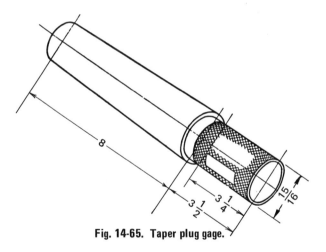

Fig. 14-65. Taper plug gage.

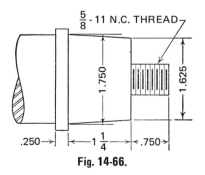

Fig. 14-66.

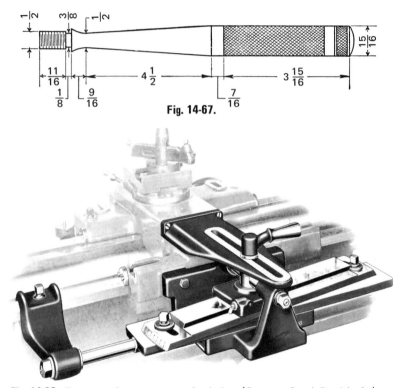

Fig. 14-67.

Fig. 14-68. Taper attachment on an engine lathe. *(Courtesy South Bend Lathe)*

Fig. 14-69. Taper attachment. *(Courtesy South Bend Lathe)*

Use of taper attachment in taper turning

When the taper attachment (Fig. 14-68) is used, it is necessary only to compute the taper per foot. When this value is known, the scale on the taper attachment (Fig. 14-69) is set for this amount.

The scales on the ends of the swivel bar are graduated in inches per foot of taper on one end and in degrees of the included angle of the taper on the other end.

The length of the stock and the length of the tapered section are required only when it is necessary to compute the taper per foot. When cutting a known taper, you need only set the scale on the taper attachment.

Summary	EXAMPLE
To determine the setting of the taper attachment:	Compute the setting for the taper attachment for the taper in Fig. 14-70.
1. Find the taper per foot, $t_{\text{per ft.}}$, of the tapered part of the work.	1. $t_{\text{per ft.}} = \left(\dfrac{D - d}{L}\right) \times 12$
2. Set the taper attachment at the reading just found. The taper per foot is the reading that is set on the taper attachment.	2. Replace the letters by their values. The length of the tapered part is 8″. $$t_{\text{per ft.}} = \left(\dfrac{2.500 - 2.0813}{8}\right) \times 12$$
	3. Make the computations. Work carefully and accurately. $t_{\text{per ft.}} = 0.628″$
	Ans. Taper attachment setting is 0.628″.

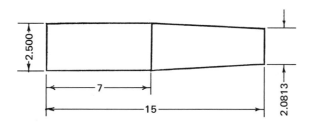

Fig. 14-70.

EXERCISES

Calculate the taper attachment setting for each of the following:

1.

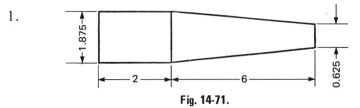

Fig. 14-71.

2.

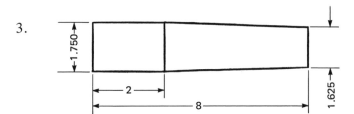

Fig. 14-72.

3.

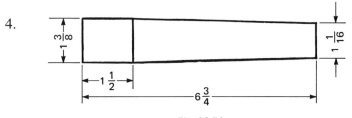

Fig. 14-73.

4.

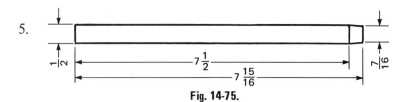

Fig. 14-74.

5.

Fig. 14-75.

PROBLEMS

47. Find the setting for the taper attachment used to cut a No. 3 Morse taper. *Note:* See Table of Tapers, p. 539.

48. What taper attachment setting is required to cut the taper in Fig. 14-76?

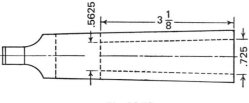

Fig. 14-76.

49. How should the taper attachment be set to cut the taper shown in Fig. 14-77? Find the total length of the piece of stock.

50. Compute the setting of the taper attachment to cut a Brown and Sharpe taper.

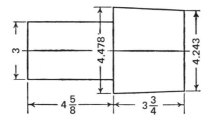

Fig. 14-77.

51. What taper attachment setting would be required to cut a Jarno taper?

Taper turning with compound rest

For cutting short tapers, the compound rest may be used. The compound rest swivel is adjusted to the proper angle and the compound rest feed screw is operated by hand to cut the taper (Fig. 14-78).

The carriage is locked in position on the "ways" of the lathe and the cross-feed is also locked in place for this operation.

Boring on the lathe

Boring is the operation of enlarging a hole with a single-edge cutting tool while the workpiece is being revolved in a lathe. Taper boring can be done by using the taper attachment or the compound rest, similar to taper turning, but since the work cannot be supported between centers, it is not possible to use the method of offsetting the tailstock.

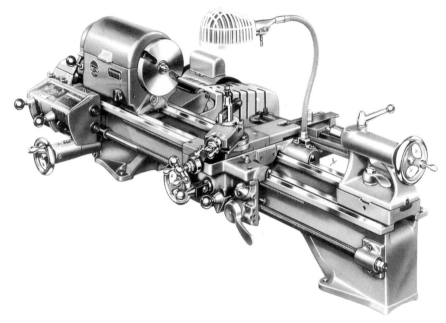

Fig. 14-78. Bench lathe with quick-change gears and compound rest.
(Courtesy Rockwell Manufacturing Company, Power Tool Division)

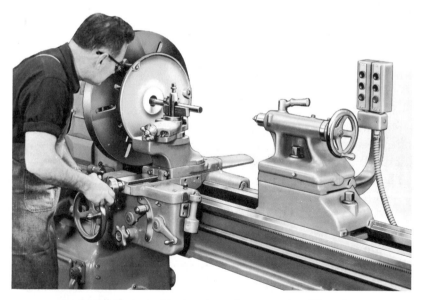

Fig. 14-79. Facing and boring a large-diameter pedestal floor flange.
(Courtesy South Bend Lathe)

Fig. 14-80. Universal spiral index center headstock.
(Courtesy Brown & Sharpe Manufacturing Company)

Fig. 14-81. Another view of a universal spiral index center headstock.
(Courtesy Brown & Sharpe Manufacturing Company)

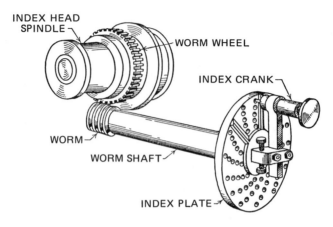

Fig. 14-82.

Indexing

In shop language, to *index* means to rotate the work on its axis to any desired position.

If a piece is to be machined on a milling machine and needs to be rotated to a special position for the operation, an attachment is needed for holding the work on the milling machine. This attachment consists of two parts: an index head, sometimes called a universal dividing head or spiral head (Figs. 14-80 and 14-81); and a footstock. These two together constitute the *index centers*. The index head is a complicated machine that contains the essential parts shown in Fig. 14-82. The work is attached to the index head by a "dog" securely fastened in a stationary position, so that the work will rotate as the index crank is turned. The footstock serves the purpose of supporting the workpiece, as does the tailstock of the lathe.

Methods of indexing

We will describe three methods of indexing: direct, angular, and simple.

(a) *Direct indexing.* Direct indexing is done by means of an index plate attached directly to the work spindle. An "index plate" is a circular plate containing a number of holes arranged in concentric circles, the number of holes being different in each circle — the inner, or smaller, circles have fewer holes; the outer, or larger, circles have the larger number of holes. The plate used for direct indexing usually has 24 equally spaced holes, but plates with 30 or 36 holes may also be used for direct indexing.

A plunger pin in the index head fits the holes in the plate and holds it in position. For indexing, the plate is turned by hand the required part of a revolution. Since the plate is attached to the work spindle the work also turns the required part of a revolution. This method is used to advantage when milling squares or hexagons, fluting taps, and in milling tools or cutters where the number of divisions of a circle is a factor of 24, 30, or 36.

PROBLEMS

52. What direct indexing is necessary to cut the four-fluted tap in Fig. 14-83?

Fig. 14-83. Four-fluted tap. *(Courtesy Morse Twist Drill & Machine Company)*

53. What direct indexing may be used to cut a reamer having eight flutes?
54. Give the direct indexing to be used to cut a pinion with 24 teeth.
55. What direct indexing would be used to cut a hexagon on the end of a round shaft?
56. What numbers can be indexed by direct indexing with a plate of 24 holes?

(b) *Angular indexing.* By using simple indexing it is possible to index to one-third of a degree on the Brown & Sharpe machine. With the Cincinnati millers, when certain of the extra plates are used, even smaller divisions than one-sixth of a degree may be indexed.

Since there are 360 degrees in a circle, one complete turn of the index crank will turn the work through 1/40 of a circle, or 1/40 of 360 degrees, or 9 degrees. If the eighteen-circle plate is used, two spaces will be equivalent to turning the work through one degree. If the index crank is turned through one space, one-half a degree will be indexed.

EXERCISES

Index for each of the following angles:

1. 30°	5. 0°30′	9. 8°
2. 25°	6. 270°	10. 0°20′
3. 56°30′	7. 65 1/2°	11. 16 1/2°
4. 2°30′	8. 63°20′	12. 5°40′

(c) *Simple indexing.* If the number of divisions of one revolution is not a factor of 24, 30, or 36, simple indexing is used.

The index plate is attached to a worm by a long worm shaft (Fig. 14-82). This worm turns the worm wheel which is attached to the spindle head.

When the index crank is turned, it turns the worm which in turn moves the worm wheel and the spindle head.

The index crank makes *40 complete* revolutions while the index-head spindle makes *one complete* revolution. Hence, when the index crank makes one complete revolution, the index-head spindle makes 1/40 of a revolution.

Thus, if it were required to cut a gear with 40 teeth, it would be necessary to turn the index crank one complete revolution for each cut that is made.

If it were required to cut a gear with 20 teeth, the index crank would need to be moved two complete revolutions for each cut.

The index crank is adjustable so that the pin will drop into any of the various hole circles. The sector arms are adjustable with reference to each other so that any number of spaces of any hole circle can be included between the sector arms. For instance, if there is a need to mark 12 equal divisions on the circumference of a cylinder, it would be necessary to rotate the index crank 40/12, or 3 1/3, turns. This can be accomplished by adjusting the index crank so that the pin will fit into some hole circle where the number of holes is a multiple of 3, such as 18, and by adjusting the sector arms so that there are 6 spaces on the 18-hole circle between them. The sector arms are locked together, and with the pin of the index crank in a hole, the sector arms are located as in Fig. 14-84. The pin is pulled out, the crank turned 3 complete turns and around to the other sector arm. The sector arms are then rotated until the follower arm reaches the pin, and the piece is then ready to be rotated to its next position (3 turns plus 6 holes more).

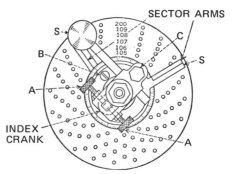

Fig. 14-84. Index plate.

Brown & Sharpe supplies a universal spiral index center-headstock, as well as index plates as follows:

Plate 1 – 15, 16, 17, 18, 19, 20 holes in each circle
Plate 2 – 21, 23, 27, 29, 31, 33 holes in each circle
Plate 3 – 37, 39, 41, 43, 47, 49 holes in each circle

Cincinnati Milling Machine Co. furnishes one plate drilled with holes arranged in circles as follows:

Side I – 24, 25, 28, 30, 34, 38, 39, 41, 42, 43
Side II – 46, 47, 49, 51, 53, 54, 58, 59, 62, 66

Other arrangements of holes are obtainable.

Example. It is required to cut a gear with 32 teeth. What simple indexing is required?

THINK	DO THIS
1. We know that 40 turns of the index crank produce one full turn of the work. So in order to cut 32 teeth, the cuts must be spaced 1/32 of a complete turn apart. Hence the index crank will be turned 1/32 of 40 turns, or 1 1/4 turns.	1. Find $40 \times \frac{1}{32}$, or $\frac{40}{32} = 1\ 1/4$ turns
2. From this, we know that the circle on the index plate must be chosen that has holes at each 1/4 turn.	2. The denominator here is 4. Hence the circle on the index plate must have holes 1/4 turn apart, or a multiple of 4 holes. The required index plate may have 16 holes, or 20 holes, or any other multiple of 4 holes.

Summary

To use simple indexing:

1. Divide 40 by the number of divisions required.

2. Express this result as a mixed number, which is the number of turns for the index crank.

EXAMPLE

It is required to cut a gear of 36 teeth on a Brown and Sharpe miller. What simple indexing is required?

1, 2. 40/36 = 1 1/9

3. Choose an index plate that contains a hole circle; the number of holes in the circle must be a multiple of the denominator of the fraction found in Step 2.

4. Adjust the sector arms to include the number of spaces as indicated by the numerator of the equivalent fraction.

5. Express the result as a number of full turns and a number of spaces on the particular hole circle being used.

3. Number of holes in the hole circle must be a multiple of 9, the denominator in 1 1/9. Choose Plate 1 with 18 holes, since 18 is 2 x 9.

4. *Note:* This operation is done on the index plate.

5. Result is one full turn and two spaces of the Plate 1, 18-hole index plate.

EXERCISES

Determine the number of turns and the number of spaces to be used for simple indexing for each of the following conditions:

1. 28 divisions − Brown & Sharpe plate 3.
2. 28 divisions − Cincinnati plate, side II.
3. 15 divisions − Brown & Sharpe plate 2.
4. 15 divisions − Cincinnati plate, side II.
5. 19 divisions − Brown & Sharpe plate.
6. 19 divisions − Cincinnati plate.
7. 22 divisions − Brown & Sharpe plate.
8. 22 divisions − Cincinnati plate.

Figure 14-85 shows a current model of the Brown & Sharpe Dynamaster Universal Milling Machine. This milling machine uses a universal spiral index center headstock (Fig. 14-80) which is supplied with one plate that provides for indexing all divisions up to 382 and many beyond this number.

PROBLEMS

57. Find the simple indexing to cut a 42-tooth gear on a Brown & Sharpe miller.
58. How would you index to cut the concave cutter shown in Fig. 14-86?
59. Find the simple indexing for milling the shell-end mill in Fig. 14-87.
60. How would you index to cut a gear with 48 teeth? 66 teeth? 84 teeth? 72 teeth? 30 teeth? 54 teeth?
61. The circumference of a piece of work is to be divided into 55 equal parts. Find the simple indexing to do this job.

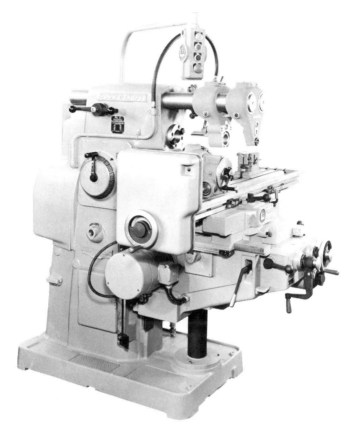

Fig. 14-85. Brown & Sharpe Dynamaster universal milling machine.
(Courtesy Brown & Sharpe Manufacturing Company)

REVIEW PROBLEMS

1. The diameter of the large end of a tapered cone is 9 3/4″. The diameter of the small end is 1 1/2″. If the work is 18″ long, find (a) the taper per foot and (b) the taper per inch.

2. A No. 3 Jarno taper is to be cut on a piece of stock 10″ long. What will be the offset of the tailstock?

3. How long will it take to turn a piece of steel 18″ long if the spindle revolves at the rate of 50 r.p.m. and the feed (the distance the tool moves per revolution) is 1/16″?

4. What will be the revolutions per minute of a 15/16″ high-speed drill having a peripheral speed of 58′ per minute?

5. The cutting speed for the face of a 6 1/2″ cast-iron pulley is 30 ft. per minute. What will be the revolutions per minute of the lathe spindle?

Fig. 14-86. Concave milling cutter. *(Courtesy Brown & Sharpe Manufacturing Company)*

Fig. 14-87. Shell mill for general-purpose face milling with arbor.
(Courtesy Brown & Sharpe Manufacturing Company)

6. How many revolutions per minute are necessary to drill a copper plate with a 1/4″ high-speed drill if the average cutting speed for this type of material is 200 ft. per minute?

7. Find the number of strokes per minute required to plane a cast-iron block in a shaper with a high-speed tool if the stroke is 5″, the cutting speed is 60′ per minute, and the number of strokes per minute is equal to seven times the cutting speed divided by the length of the stroke.

8. If brass were used in place of the cast iron in Problem 7, what would be the number of strokes if the cutting speed for brass is 160′ per minute when a high-speed cutting tool is used?

9. A 26″ grinding wheel revolves at 661 r.p.m. What is the peripheral speed?

10. (a) If 3/16″ is allowed for cutting off and facing, what is the total length of the stock required to make the jack shaft shown in Fig. 14-88?

    (b) Find the length of the tapered section if the taper per foot is 2 7/8″.

    (c) If the stock used to make the jack shaft is 3 3/4 in. in diameter, what are the revolutions per minute of the spindle if the cutting speed is 200 ft. per minute?

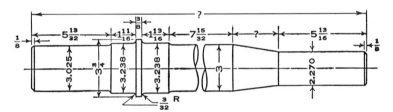

Fig. 14-88. Jack shaft.

11. In Fig. 14-89 find (a) the overall length and (b) the taper attachment setting for each of the tapered parts.

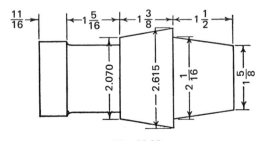

Fig. 14-89.

12. What would be the index setting for a 22-tooth spur gear?
13. What sector setting may be used to cut an 80-tooth gear? Can any other setting be used? If so, state the setting.
14. What indexing may be used to drill 15 holes equally spaced on a 10" bolt circle?
15. An engine lathe equipped with a set of gears having a progression of 6, starting with a 24-tooth gear, has a constant of 8. Find (a) the gears to be used on the stud and screw to cut a 1"-14 N.F. thread; (b) the pitch of the thread in thousandths of an inch; (c) the depth of the thread; and (d) the major and minor diameters of the thread.
16. A 1 1/8"-7 N.C. thread is to be cut on an engine lathe having a lead of 6 and equipped with a set of gears having a progression of 5 starting with a 25-tooth gear.
    (a) Find the gears to be used on the stud and screw to cut the thread.
    (b) Find the pitch of the thread in thousandths of an inch.
    (c) What is the depth of the thread?
    (d) What are the major and minor diameters of the thread?
17. How would you index to mill the ratchet in Fig. 14-90?

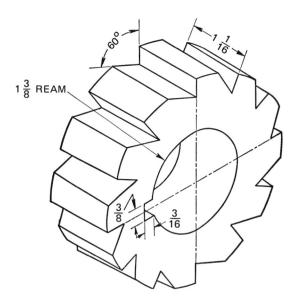

Fig. 14-90. Ratchet.

18. (a) Determine the setting of the plate and index crank to be used to index for the motor sprocket in Fig. 14-91.
    (b) Find the size of the drill needed for a 5/16"-18 N.C. thread.

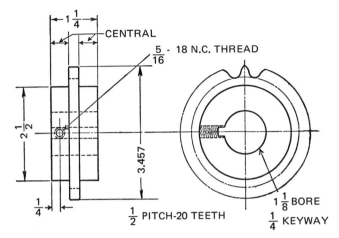

Fig. **14-91.** Blower drive motor sprocket.

15 The Steel Square

Purpose of the steel square

Among all the tools a carpenter uses, probably the most useful, the simplest and most indispensable is the *steel square* (Fig. 15-1). By means of this tool, the carpenter may quickly and easily solve problems in the laying out of his work which otherwise would be exceedingly difficult. The steel square is really a simple calculating machine.

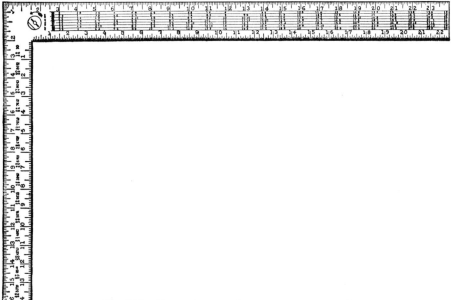

Fig. 15-1. The steel square. *(Courtesy Stanley Tools)*

When you look at the steel square with its many markings, figures and tables, you might get the impression that it is a complicated instrument that requires a knowledge of advanced mathematics to operate. It is in fact a comparatively simple tool to use, and the time saved by its use more than compensates for the little effort spent in learning the few

566

simple rules that are needed to use it effectively. These simple rules will enable the carpenter to use the steel square easily and quickly to determine the length of any common rafter (see p. 450) for any pitch of roof and to make the proper top and bottom cuts.

Information about the steel square

The steel square is often called a "framing square," and if it has rafter tables, it is also called a "rafter square." Some squares are made of aluminum to keep down the weight. Some are made of stainless steel, and others are polished, blued or have a royal copper finish. The longer arm, called the *body* or the *blade*, is usually 24 inches long and 2 inches wide. The shorter arm, the *tongue*, is usually 16 inches long and 1 1/2 inches wide. The *heel* of the square is the point where the outside edges of the square meet. The point where the inside edges of the square meet is also sometimes called the heel.

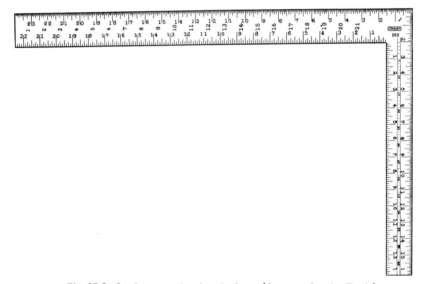

Fig. 15-2. Steel square, showing the face. *(Courtesy Stanley Tools)*

Compare Figs. 15-2 and 15-3 and you will observe that if any point on the tongue is connected to a point on the body, a right triangle is

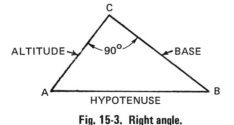

Fig. 15-3. Right angle.

formed (see page 336). The steel square as well as all roof framing is based on the principles of the right triangle.

The *face* of the steel square is the side upon which the manufacturer's name is usually stamped. It is the side that is visible when the body of the square is held in the left hand and the tongue in the right hand with the heel pointing away from the holder. The square shown in Fig. 15-2 has the face visible.

The back of the square is opposite to the face.

There are two kinds of markings on the steel square: scales and tables. To be able to use the square quickly and effectively, it is necessary for the carpenter to know these scales and tables thoroughly.

The scales on the inner and outer edges of both body and tongue (Fig. 15-2) are divided or graduated in inches and fractional parts of an inch. Note that all these scales begin at the heel of the square and extend outward. The graduations are usually in eighths, tenths, twelfths, sixteenths, and thirty-seconds of an inch. A scale in which the inch is divided into one hundred parts, each small division being one hundredth of an inch, is placed on the back of the square (see Fig. 15-6).

The scales and tables placed on the steel square vary according to the particular purpose for which the square is to be used. These are the ones usually available: inches, graduated in sixteenths, eighths, twelfths, and tenths; octagon scale; brace measure table; board measure table; hundredths scale; and rafter table.

Summary. One side of the steel square is the face, the other the back. On the face and back are scales and tables. The scales are graduated in inches and those fractional parts of an inch that are needed by and are helpful to the carpenter. The tables are useful in laying out cuts and making calculations.

The octagonal or eight-square scale

Octagonal means eight-sided. By means of the octagonal or eight-square scale, one is able to lay out accurately a regular octagon in a square.

The octagonal scale is found along the center of the face of the tongue (see Fig. 15-4).

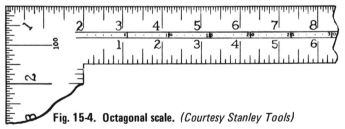

Fig. 15-4. Octagonal scale. *(Courtesy Stanley Tools)*

Example. Lay out the lines to make a regular octagon in a 12-inch square.

THINK	DO THIS
1. We must first draw lines through the center of the square that are parallel to the sides of the square.	1. In a 12-in.-square piece, mark points *A*, *B*, *C* and *D* each 6 in. from the vertices of the square. Connect *AB* and *CD*.
2. Using dividers, take as many spaces on the octagonal scale (Fig. 15-4) of the steel square as there are inches in the side of the square, here 12.	2. Open the dividers so that they will span just 12 spaces of the octagonal scale on the steel square.
3. Lay off this distance on both sides of each of the points *A*, *B*, *C* and *D*.	3. Keep the dividers with this opening and strike arcs at *E* and *M* from *A*, at *F* and *G* from *D*, *H* and *J* from *B*, and *L* and *K* from *C*.
4. Join these points: *EF*, *GH*, *JK* and *LM*.	4. Join *E* and *F*, *G* and *H*, *J* and *K*, and *L* and *M*. The figure *EFGHJKLM* is a regular octagon.
5. The figure *EFGHJKLM* is a regular octagon.	5. If this layout is on a square piece of timber and cuts are made along these lines *EF*, *GH*, *JK* and *LM*, the resulting piece will be in the shape of a regular octagon.

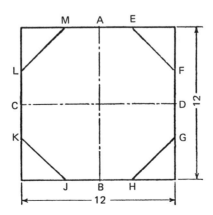

Fig. 15-5. Butt end of a 12-inch timber.

Summary

To lay out the lines to draw a regular octagon on a square:

1. Draw lines through the center of the square parallel to the sides of the square.
2. Take, on the dividers, as many spaces from the octagonal scale of the steel square as there are inches in the width of the given square.
3. Lay off this distance on both sides of the lines drawn on the square.
4. Draw lines joining these points so as to make a triangle at each corner of the square.
5. The figure thus formed within the square is a regular octagon.

EXAMPLE

Lay out a regular octagon on a 3″ square (*Note:* The student will first construct a three-inch square and perform this construction as he studies the example.)

1. Mark points *A, D, B* and *C*, each 1 1/2″ from the vertices, and draw *AB* and *CD*.
2. Open the dividers to 3 spaces of the octagonal scale.

3. Lay off points *E* and *M* from *A*, *F* and *G* from *D*, *H* and *J* from *B*, and *K* and *L* from *C*.
4. Draw *EF, GH, JK* and *LM*.

5. The figure *EFGHJKLM* is a regular octagon.

Porch columns, porch column bases, and newel posts are often made in an octagonal shape, and the octagonal scale is used in laying them out.

 Fig. 15-5A.

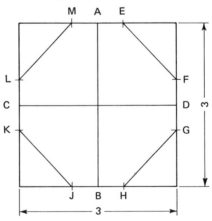

EXERCISES

Make five cardboard squares: (a) 6″ x 6″; (b) 8″ x 8″; (c) 10″ x 10″; (d) 16″ x 16″; (e) 18″ x 18″.

Using the steel square and dividers, make the layout for an octagon on each of the pieces above. Indicate the sections that are to be cut away.

Brace measure

The table by which a wooden brace may be laid out easily and accurately is located along the centerline of the back of the tongue (Fig. 15-6). The equal figures one above the other $\left(\dfrac{51}{51}\right)$ represent the run on the post and the run on the beam over which the brace must extend (Fig. 15-6).

Fig. 15-6. Brace table. *(Courtesy Stanley Tools)*

Notice that they are equal. The number to the right of this pair and between them, 72^{12}, represents the diagonal of a square of which 51 and 51 are the sides; it is the outside length of the brace. In this case, if the brace is to extend 51″ on the post and 51″ on the beam, the brace must be cut 72.12″ long. Note that the length of the brace is given in inches and hundredths of an inch.

Summary	EXAMPLE
To find the length of brace to extend equal distances, *AC* and *AB*, on the post and beam shown:	Find the length of brace to extend from *B* to *C* in Fig. 15-7.
1. Find the run on the post and beam.	1. Measure the distances *AB* and *AC*: 48″ each.
2. In the brace measure table on the back of the tongue, find the pair of numbers equal to the distances on post and beam.	2. On brace measure table find the pair of numbers $\left(\dfrac{48}{48}\right)$. Beside them is the number 67.88.
3. The number beside this pair is the length of the brace required. The length here is given in inches and hundredths of an inch.	3. The brace is 67.88 inches long.
	Ans. 67.88″

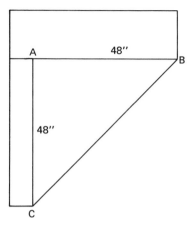

Fig. 15-7.

EXERCISES

Use the brace measure table to supply the length of brace for each of the following pairs of runs. *(Do not write in this book.)*

No.	Run on beam	Run on post	Length of brace
1.	24	24	
2.	27	27	
3.	30	30	
4.	33	33	
5.	36	36	
6.	39	39	
7.	42	42	
8.	45	45	
9.	48	48	
10.	51	51	
11.	54	54	
12.	57	57	
13.	60	60	

Board measure

On the back of the body, or blade, of the steel square may be found a table from which it is possible to read the number of board feet (see page 440) for many sizes of board or timber without having to make any calculations. This table consists of the columns of figures found directly under the whole inch divisions of the outer edge of the blade (see Fig. 15-8). The column under the 12″ division, *A* in Fig. 15-9, indicates the

Fig. 15-8. Essex board measure. *(Courtesy Stanley Tools)*

lengths of the boards to be measured. The other columns, such as *B*, give the number of square feet where the width of the board is the number at the inch graduation.

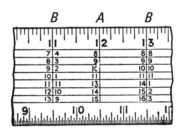

Fig. 15-9. Section of board measure table.

Example. Use the board measure table of the steel square to find the number of board feet in a piece of lumber 8 feet long, 7 inches wide, and 1 inch thick.

THINK	DO THIS
1. Find the 12-inch mark on the outer edge of the blade.	1. Find the 12-inch mark.
2. Trace down the scale under this mark to the line containing the length of the board to be measured.	2. Trace down the scale to "8," the first line, the length of the board we are measuring (Fig. 15-9).

3. Trace along the line to the inch graduation along the edge of the blade that is the same as the width of the board to be measured.

4. The number found here is the number of square feet in the board.

5. If the board being measured is one inch thick, the number of board feet is the same as the number of square feet.

3. Trace along this first line to the 7″ graduation on the edge of the blade (Fig. 15-8).

4. The number is 4|8, which means 4 8/12 square feet.

5. For a board one inch thick, the number of board feet is the same as the number of square feet.

Ans. A board 8 feet long, 7 inches wide, and 1 inch thick contains 4 8/12 board feet of lumber.

Note: The table on the blade of the steel square gives the number of board feet in a board that is 1 inch, or less than 1 inch, thick. If a board is more than 1 inch thick, the number found in the table must be multiplied by the thickness of the board being measured to obtain the number of board feet. If the board in the example above were 2 1/2″ thick, the number of board feet it would contain would be 4 8/12 x 2 1/2 board feet. If it were 1 3/4″ thick, it would contain 4 8/12 x 1 3/4 board feet.

Note that in the column of lengths, there is no 12-foot length given. This is omitted because the number of square feet in a 12-foot board is always the same as the number of inches in its width. A board 12 feet long and 10 inches wide contains 10 square feet; one 12 feet long and 8 inches wide contains 8 square feet, etc.

The board feet in a 6-foot piece equal 1/2 the board feet in a 12-foot piece; those in a 4-foot piece equal 1/2 the board feet in an 8-foot piece, etc.

Note also that this column of lengths extends only to 15 feet. For boards longer than 15 feet, divide the length into two smaller sizes, find the board feet in each, and add the results.

Summary

To find the number of board feet in a piece of lumber using the table of the steel square:

1. Find 12″ graduation on the blade of the steel square.

EXAMPLE

Find the number of board feet of lumber in a piece 2″ x 8″ x 14′.

1. 12″ graduation (Fig. 15-9).

2. Trace down the scale under 12″ to the number equal to the length of board being measured.

2. Under 12″ we find 14, the length of the board being measured.

3. Trace along this line to the graduation on the edge of the blade equal to the width of the board being measured. The number found here is the number of square feet in the board being measured.

3. Trace along the 14 line to the 8″ graduation. The number here is 9/4 or 9 4/12. The board being measured contains 9 4/12 square feet.

4. If the board being measured is 1″ thick or less, the number here is the number of board feet. If the board is more than 1″ thick, multiply the number found by the thickness of the board to find the number of board feet.

4. Since the board here is 2″ thick the number of board feet is 9 4/12 x 2 = 18 8/12.

Ans. 18 8/12 bd. ft.

PROBLEMS

1. Determine, using the board measure tables on the steel square, the number of board feet in each of the following:
   (a) a piece 1″ x 10″ x 10′
   (b) a piece 2″ x 10″ x 14′
   (c) a piece 1″ x 8″ x 7′
   (d) a piece 1″ x 7″ x 16′

2. How many board feet in each of the following lots?
   (a) 4 pieces 1″ x 8″ x 7′
   (b) 2 pieces 1″ x 7″ x 12′
   (c) 5 pieces 1″ x 9″ x 15′

3. Find the total number of board feet in the following shipment:
   (a) 40 pieces 1″ x 10″ x 15′
   (b) 25 pieces 1″ x 6″ x 6′
   (c) 10 pieces 1″ x 14″ x 13′

4. How many board feet of yellow pine are there in each of the following?
   (a) 80 pieces 4″ x 8″ x 11′
   (b) 65 pieces 1″ x 10″ x 20′
   (c) 15 pieces 2″ x 14″ x 12′

5. A contractor ordered the following materials to construct a cabinet. How many board feet are in the order?
   (a) 4 pieces 1" x 4" x 14'
   (b) 15 pieces 2" x 2" x 13'
   (c) 11 pieces 3" x 7" x 18'

To rip a board into a given number of strips of equal width

The layout for ripping a board into a given number of equal strips can be made easily with the steel square.

Example. Lay out the lines to rip a board into seven strips of equal width.

1. Lay the steel square diagonally across the board with the heel of the square on one edge of the board and the 7-inch graduation on the outside edge of the square on the opposite edge of the board (Fig. 15-10). (Pay no attention to the markings on the *other* arm of the square. They do not affect the layout.)
2. At each of the inch marks of the scale, used like a straightedge ruler, place a point on the board.
3. Through each of these points draw a line parallel to the edge of the board. If the board is ripped along these lines, the seven resulting strips will be of equal width.
   Compare this with the method of dividing a line into equal parts, p. 351.

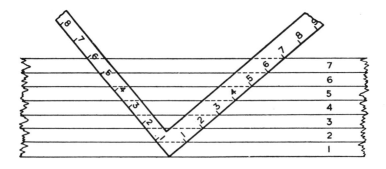

Fig. 15-10. Rip a board into seven equal strips.

Summary	EXAMPLE
To lay out lines for ripping a board into a number of strips of equal width:	Lay out the lines to rip a board into 5 strips, all of equal width.

1. Lay the steel square on the board so that its heel is on one edge of the board, and that number on the outside edge of the square that is equal to the number of strips desired is at the opposite edge of the board.

2. Place a point on the board at each whole inch division mark of the square.

3. Draw lines through these points parallel to the edge of the board. These lines are the ones along which the board is to be ripped.

1. Lay the steel square on the board with its heel on one edge and the number 5 on the other edge of the board (Fig. 15-11).

2. Place points on the board at the 1-, 2-, 3-, and 4-inch divisions of the scale of the square.

3. Draw lines through these dots, parallel to the edge of the board.

Ans. If ripped along these lines the board will be ripped into 5 strips, all of the same width.

Note: If the board is too wide to use inch marks, then use 2″ marks. That is, place the dots on the board at 2″, 4″, 6″, etc., of the steel square scale.

An easy way to draw the parallel lines required in Step 3 is to mark off a second set of points following the directions in Steps 1 and 2. Lines joining the first with the second set of points will be parallel.

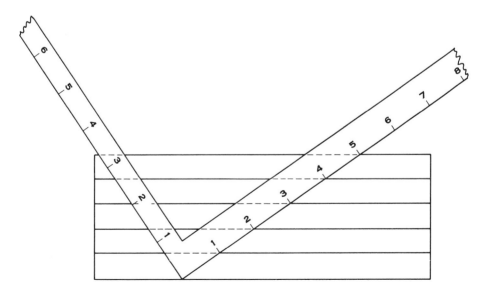

Fig. 15-11.

EXERCISES

Use several pieces of cardboard 6, 8, and 10 inches wide and lay out the lines for ripping these into:

1. 4 equal strips.
2. 5 equal strips.
3. 7 equal strips.
4. 9 equal strips.
5. 3 equal strips.
6. 11 equal strips.

Hundredth scale

Some squares contain a hundredth scale, Fig. 15-12, which is usually located on the back of the square, in the corner. The hundredth scale is one inch long and is divided into 100 equal parts. Also, on some squares there is another scale one inch long, graduated in 16ths, just below the hundredth scale. This scale makes the conversion from hundredths to sixteenths a very simple matter.

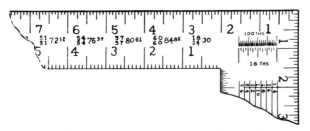

Fig. 15-12. Hundredth scale. *(Courtesy Stanley Tools)*

EXERCISES

Use the hundredth scale on the steel square and a pair of dividers to transfer the following lengths to a piece of cardboard.

1. 0.42″
2. 0.86″
3. 0.57″
4. 0.19″
5. 0.77″
6. 0.25″
7. 0.92″
8. 0.59″
9. 0.46″
10. 0.65″
11. 0.89″
12. 0.96″

To find the hypotenuse of a right triangle with the steel square

By laying a 6-foot rule across the steel square, the hypotenuse of a right triangle may easily be found.

Example. By means of the steel square, find the length of the hypotenuse of a right triangle in which one side is 13 feet and the other 19 feet long.

1. Place rule so that its zero point is at the 19-inch mark on the outer edge of the body of the square and the rule passes through the 13-inch division on the outer edge of the tongue (Fig. 15-13).

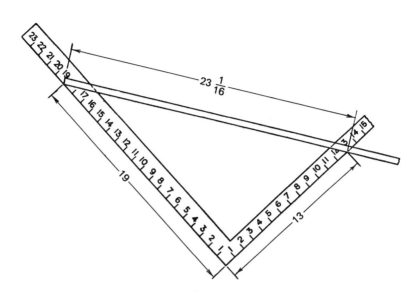

Fig. 15-13. Length of hypotenuse.

2. Read the measurement on the rule between these two points. This measurement, 23 1/16, is the length of the hypotenuse.
3. We may let the units on the square represent any units of measurement. Since the dimensions 19 and 13 in this example are in feet, then 23 1/16 is also in feet. This is approximately 23.1 feet.

Summary. To determine the hypotenuse of a right triangle by means of the steel square:

1. Lay a rule across the square so that its zero point is at the point that represents one of the dimensions on the body of the square and the rule passes through the graduation that represents the other dimension on the tongue.
2. Read the measurement on the rule between these two points.
3. Express this in the same unit in which the dimensions were given. This is, approximately, the length of the hypotenuse.

EXERCISES

Use the steel square and a rule to find the hypotenuse of the right triangles with dimensions as follows:

1. 3″, 4″	4. 8′, 15′	7. 21′, 7 1/2′
2. 5″, 12″	5. 12 1/2′, 16′	8. 19 1/2″, 8 1/2″
3. 10′, 20′	6. 4 1/2′, 15′	9. 10 1/2′, 14 1/2′

To bisect an angle with a steel square

An angle may easily be bisected using the steel square. Compare the method below with the method on page 341.

Summary	EXAMPLE
To bisect an angle using a steel square:	Bisect angle *MON* (Fig. 15-14) using the steel square.
1. Mark off, from the vertex of the angle, some convenient equal distance on each side of the angle.	1. Mark off from the vertex of the angle a length, say 4 inches, on each side of the angle.
2. Lay the square on the angle so that the heel of the square is away from the vertex of the angle and is the same distance from the two marks made on the sides of the angle. (If you use the outside graduations, use the outside heel of the square. If you use the inside graduations, use the inside heel of the square.)	2. Place the square as in Fig. 15-14 with the 3-inch points on the outside lines of the scale directly over the marks on the sides of the angle.
3. Place a point, *P*, on the work at the heel of the square.	3. Place dot *P* at the outside heel of the square.
4. Remove the square and draw a line through the vertex of the angle and *P*. This line bisects the angle.	4. Draw line *OP*. This lines bisects angle *MON*. *Ans.* Line *OP* bisects angle *MON*.

PROBLEMS

6. Draw five angles of different sizes and bisect each of them, using the steel square.

The cuts in floor bridging

The square may very conveniently be used in the processes of framing a building. It may be used to determine many different angles, including the right angle, along which a carpenter must make cuts when framing a building. These angles are called bevels or cuts.

The first of these cuts which we shall consider are those made in floor bridging.

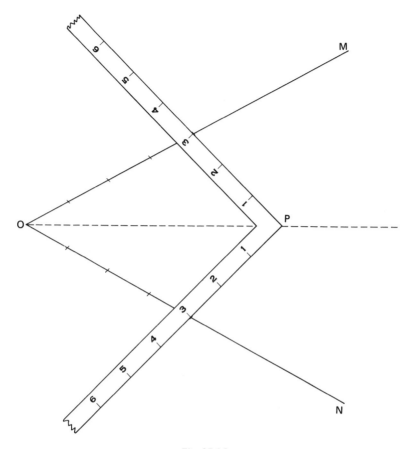

Fig. 15-14.

Example. Lay out the cut for bridging floor joists as indicated in Fig. 15-15.

1. Determine the width of the joists and the distance between joists. The joists in this floor are 11 5/8 inches wide, 1 5/8 inches thick, and are set 16 inches to centers. The distance between them is 14 3/8 inches.

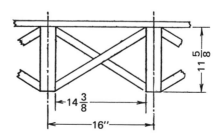

Fig. 15-15. Floor to be bridged.

2. Lay the square (Fig. 15-16) across the board with heel toward you, with the width of the joist extending from the heel to the far edge of the board, A, and the distance between joists extending from the heel to the near edge of the board, B.

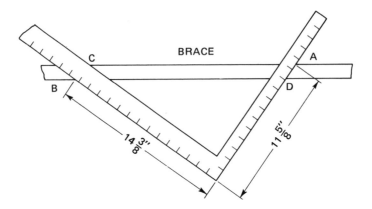

Fig. 15-16. Position of square to obtain bridging cuts.

3. Draw a line along the outer edge of the square, as AD, and place a point on the board at B. Remove the square and draw a line through this point parallel to the line AD.

    Ans. The cuts must be made along lines AD and BC.

Summary	EXAMPLE
To lay out bridge cuts:	Lay out the cuts for bridging floor joists 11 5/8″ wide, 1 5/8″ thick, and 24″ on centers.
1. Determine the width of the joists and the distance between joists.	1. Width of joists 11 5/8″, distance between joists 22 3/8″.
2. Place the square with heel toward the worker so that the width of the joists terminates at the outer edge of the board and the distance between joists terminates at the near edge (Fig. 15-17).	2. Place square on edge of the work, with heel toward worker so that 11 5/8″, the width of joists, just reaches the outer edge of the work at A, and the distance between joists, 22 3/8″, extends from the heel to the near edge of the board at B (Fig. 15-17).

3. Draw a line on the wood along the outer edge of the square at the first position, and place a dot on the board at the second one.

3. Draw line *AD* on the work and place a dot on the work at *B*.

4. Remove the steel square and draw a line through this dot parallel to the first line drawn. The cuts for the bridging will be made along these two lines. Use a T-bevel for this marking. See page 576 for a way to draw parallels.

4. Remove the steel square and draw *BC* parallel to *AD*. The cuts for the bridging are made along these lines.

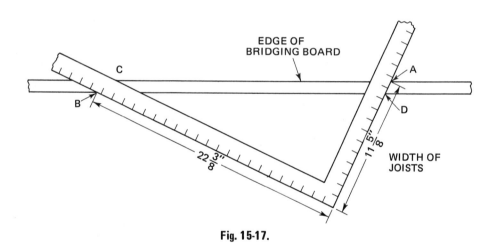

Fig. 15-17.

When one piece of bridging has been laid out, the others can be marked off by using a T-bevel. The T-bevel is set to the proper angle, and the lengths of the pieces are measured off between the cuts.

Fig. 15-18. T-bevel. *(Courtesy Stanley Tools)*

EXERCISES

Lay out the proper cuts for floor bridging under each of the following conditions:

1. Joists 9 5/8" wide, 1 5/8" thick, 14 3/8" apart.
2. Joists 7 5/8" wide, 1 5/8" thick, 16" on centers.
3. Joists 5 5/8" wide, 1 5/8" thick, 16" on centers.
4. Joists 7 5/8" wide, 1 5/8" thick, 24" on centers.
5. Joists 9 5/8" wide, 1 5/8" thick, 16" on centers.
6. Joists 11 5/8" wide, 1 5/8" thick, 16" on centers.

Layout of steps for a stairway

The rise of a stair step is the distance from the top of one tread to the top of the next tread (Fig. 15-19). The total rise of the stairs is the vertical distance from the floor line of the floor at which the stairs start to the floor line of the floor at which they end. Risers may vary from 5 to 9 1/2 inches, depending upon the angle of the stairs. For the same stairs, treads may vary from 12 1/2 down to 8 inches. The preferred dimensions are 6 1/2 to 7 1/4 inches for risers with the corresponding dimensions of 11 to 10 1/4 inches for treads.

It makes a "comfortable" set of stairs to have the sum of the dimensions of the tread and riser equal about 17 to 17 1/2 inches, exclusive of the nosing (Fig. 15-19), and to have the product of the dimensions equal from 70 to 75.

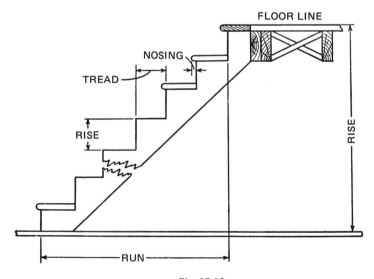

Fig. 15-19.

The following fundamental steps are suggested for laying out a stair stringer:

1. Determine the total rise of the stairs.
2. Determine the total run of the stairs.
3. Estimate the number of steps to be used in the stairs. Remember that there is always one less tread than there are risers, because the top tread is part of the upper floor.
4. Determine the exact dimension of each riser by dividing the total rise by the number of steps.
5. Determine the width of the tread, excluding the nosing. (The step boards are wider than the tread.)
6. Lay out each step.
7. The lowest riser is to be different from the others, according to the thickness of the tread board.
8. The end cut on the upper end of the stringer is parallel to the riser cuts and the length of the upper section is the amount necessary to reach the floor joists.
9. It is essential that the heights of *all* finished steps be the same, including the lowest step and the highest step.

Example. It is determined that each step of a set of stairs is to have a rise of 7 1/4 inches and a tread of 10 1/4 inches. What is the procedure for laying out such a step?

1. Lay the steel square on the stair stringer so that the reading on the tongue from the heel to the edge of the stair stringer is 7 1/4 inches (*A* in Fig. 15-20) and the reading on the body from the heel to the same edge of the floor stringer is 10 1/4 inches (*B*).

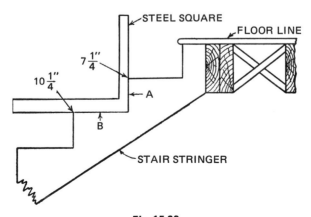

Fig. 15-20.

2. Lines drawn on the stringer along the outer edges of the tongue and body will be the layout for the step.

Summary	EXAMPLE
To lay out a step when the dimensions of the riser and the tread are known:	Lay out a step of a set of stairs that is to have a rise of 6 1/2″ and a tread of 11″.
1. Lay the steel square on the stair stringer so that the reading on the tongue from the heel to the edge of the stringer is the height of the riser, and the reading on the body of the square from the heel to the same edge of the stringer is the same as the dimension of the tread (Fig. 15-21).	1. Square Q is laid on the stringer with heel toward the worker, with (a) the 6 1/2″ point, A (Fig. 15-21), of the tongue at the further edge of the stringer, and (b) the 11″ point of the body of the square at the same edge of the stringer, B (Fig. 15-21)
2. Draw lines on the work along the outer edges of the tongue and body of the square.	2. Draw lines AD and BD on the work along the outer edges of the tongue and body.
3. The step is cut out along these lines. The bottom riser must be shorter than the others by an amount equal to the thickness of the stair tread board so all the finished steps have same height.	3. The step is cut out along these lines.

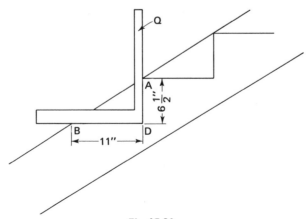

Fig. 15-21.

EXERCISES

On a 12-inch board or a 12-inch strip of cardboard lay out steps with the following dimensions:

	Tread	Riser		Tread	Riser		Tread	Riser
1.	10 1/2″	6 3/4″	4.	10 3/4″	6 7/8″	7.	10 7/8″	6 1/2″
2.	11″	7 1/2″	5.	11″	7 5/8″	8.	12 1/2″	5″
3.	10 1/4″	7 1/4″	6.	10 1/8″	8 1/8″	9.	9 1/2″	8″

Common terms in rafter framing

To lay out the various cuts of a rafter, it is necessary to have a knowledge of roof pitches and their relation to the figures on the steel square.

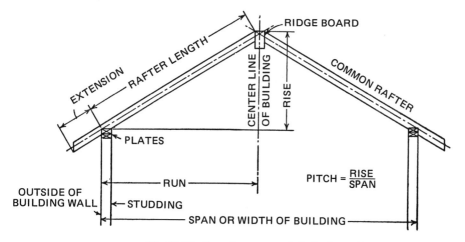

Fig. 15-22. Run, rise, span, and pitch of a roof.

The *plate* is the member of the wall to which the rafters are framed at the lower end.

The *span* of a roof is the distance between the outside of the wall plates. This is always the width of the building.

The *rise* of the rafter is the vertical distance from the top of the plate to the top of the ridge.

The *run* of the rafter is the horizontal distance from the outside of the plate to the center of the ridgeboard. On an equal-pitch gable roof, it is one-half the span.

The *pitch* is the ratio of the rise to the span.

Roof pitches are expressed as fractions, such as 1/6, 1/3, 5/8, etc. A 1/6 pitch roof may have a run of 12 feet and a rise of 4 feet. Since the run is 12 feet, the span will be 24 feet, and the pitch is 4 divided by 24, or 1/6.

Principal or common roof pitches

1/6 — in which the rise is 1/6 the span of the building.

1/4 — in which the rise is 1/4 the span of the building.

1/3 — in which the rise is 1/3 the span of the building.

5/12 — in which the rise is 5/12 the span of the building.

1/2 — in which the rise is 1/2 the span of the building.

5/8 — in which the rise is 5/8 the span of the building.

3/4 — in which the rise is 3/4 the span of the building.

Full pitch — in which the rise is equal to the span of the building.

The steel square is used by the carpenter to lay out the cuts of the various rafters used in the framing of a roof. Figure 15-23 shows the principal roof pitches laid on the steel square. The outer edges of the square are used for this purpose. Note that in each instance the run is assumed to be 12 inches. The reason for this will be discussed later.

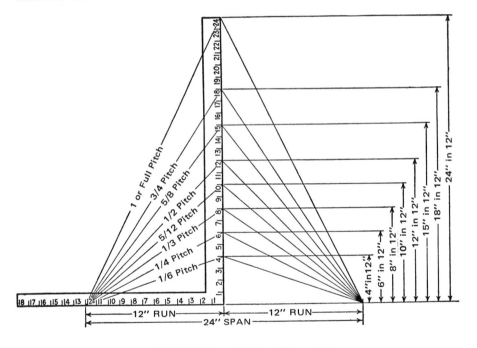

Fig. 15-23. Principal roof pitches laid on steel square.

Note: It is the accepted practice to have the full length of the body of the square, or 24 inches, represent the rise and 12 inches on the tongue represent the run of a full-pitch roof. All of the fractional pitches are determined on this basis. Roof pitches do not correspond to degree of an angle.

The pitch of a roof is also expressed in terms of inches of rise per foot of run. A roof that rises 6 feet, or 72 inches, over a span of 24 feet has a pitch of 6/24, or 1/4. The run of this building is 12 feet. Since the rise is 72 inches for a run of 12 feet, there are 72/12, or 6, inches of rise for each foot of run. This is expressed as a pitch of 6 inches per foot run (Fig. 15-24). For 12 feet of run and a rise of 60 inches there are 60/12, or 5, inches of rise for each foot of run. This is a pitch of 5 inches per foot of run.

 Fig. 15-24.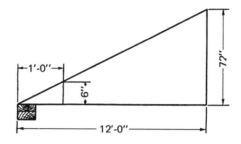

Cuts for a common rafter

The cut at the top end of a common rafter (Fig. 15-25) — the end which rests against the ridgeboard, or against the opposite rafter — is called the *top cut*, the *ridge cut*, or the *plumb cut* [see Fig. 15-26(*A*)]. The cut at or near the lower end of the rafter, the cut made where the rafter rests against the plate [Fig. 15-26(*B*)], is the *bottom* or *heel cut.* Both the ridge and the heel cuts may be laid out easily with the steel square.

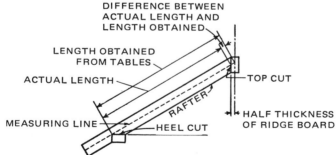

Fig. 15-25. Common rafter.

Example. Lay out the plumb or top cut and the bottom cut of a rafter for a roof with a pitch of 1/4, or 6 inches to the foot run.

1. Lay the square on the rafter with the heel toward the worker, the 12-inch mark of the body coinciding with the nearer edge of the rafter, and the 6-inch mark on the tongue coinciding with the same edge of the rafter.

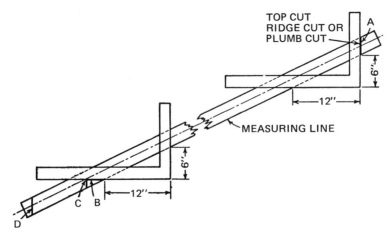

TOP CUT
RIDGE CUT OR
PLUMB CUT

A

6″

12″

MEASURING LINE

6″

12″

C B

D

Fig. 15-26. Applying the square for ridge and heel cuts.

2. Draw a line on the rafter along the outer edge of the tongue. The plumb or top cut is made along this line [Fig. 15-26(*A*)].

Note: The reading on the tongue must be the same as the number of inches per foot run. If the pitch is given only as a fraction, as 1/4 above, multiply this fraction by 24 to get the number of inches per foot run.

The heel cut is laid out in the same way. Lay the square so that the nearer edge of the body passes through the point which marks the end of the rafter (*C*), with the 12-inch division mark at the edge of the rafter and the 6-inch mark of the tongue coinciding with the same edge of the rafter (Fig. 15-26). Draw a line along the nearer edge of the body of the square (*B*). This will be the seat cut of the heel. At *C* make a right angle to complete the heel cut.

The rafter may be cut as at *D* by repeating the top cut here.

Summary	EXAMPLE
To lay out the top and heel (or seat) cuts of a common rafter:	Lay out the top and seat cuts of a rafter for a roof with a pitch of 1/3.
Top cut	
1. Lay the square with the heel toward the worker so that the 12″ mark at the outer edge of the body coincides with the nearer edge of the rafter and the mark of the tongue that is the number of inches per foot run coincides with the same edge of the work.	1. 1/3 pitch means 8″ of rise for each 12″ or 1 foot of run. Place the square with the 12″ mark of the outer edge of the body on the nearer edge of the work. Also place the 8″ mark of the outer edge of the tongue on this same edge of the work.

2. For the top cut draw a line on the rafter along the outer edge of the tongue. This is the line along which the top cut must be made.

Heel cut

1. Through the point on the work which marks the length of the rafter, lay the outer edge of the body of the square in the same way that it was laid for the top cut; that is, place the 12″ mark of the body on the nearer edge of the rafter. Also, place that number on the tongue that is the same as the number of inches per foot run on the same edge of the rafter.

2. Draw a line on the rafter along the outer edge of the body of the square to the measuring line which marks the end of the rafter. This is the line along which the heel cut is made.

3. At the point where this line meets the measuring line, another line at a right angle to the heel cut line is drawn. This completes the layout of the heel cut.

2. Hold the square firm and draw the line *AB* on the work along the outer edge of the tongue. This is the line of the top cut.

1. A 1/3 pitch means 8″ of rise for each 12″ of run.
Through the point *R* (Fig. 15-28) which marks the length of the rafter, lay the square in the same way as for the top cut. Place the 12″ mark of the outer edge of the body on the nearer edge of the work.

2. Draw line *RS* on the work along the outer edge of the body of the square.

3. A line through *P* at a right angle with *RS* completes the layout of the heel cut.

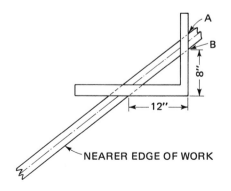

Fig. 15-27.

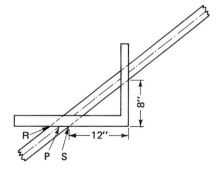

Fig. 15-28.

EXERCISES

For each of the following select two points, one near each end of a board or 4-inch strip of cardboard. Through one, lay out the top or ridge cut, and through the other, lay out the heel cut for common rafters to be used with the following roof pitches:

1. 8″ per foot run
2. 6″ per foot run
3. 15″ per foot run
4. 24″ per foot run

5. 1/6 pitch
6. 5/12 pitch
7. 1/2 pitch
8. 3/4 pitch

Length of rafters

There are several methods in use for obtaining the lengths of rafters. A practical and accurate method is the use of tables on the steel square. These tables enable the carpenter to determine the exact length of a rafter in a minimum amount of time and without errors.

There are two types of tables given on steel squares that are in common use. One type gives the following:

(a) Length of common rafters per foot run.
(b) Length of hip or valley rafters per foot run.
(c) Difference in length of jacks, 16-inch centers.
(d) Difference in length of jacks, 2-foot centers.
(e) Side cut of jack rafters.
(f) Side cut of hip or valley rafters.

The second type of table in common use gives the length of a rafter for a given span and pitch.

Length of rafter using table of length per foot run

Figure 15-29 illustrates the table of length per foot run as it appears on the steel square. Note that the first line of the table gives the length of main (or common) rafters, the second line the length of hip or valley rafters, etc.

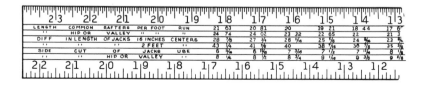

Fig. 15-29. Rafter table. *(Courtesy Stanley Tools)*

Example. Find the length of a common rafter for a building 18 feet wide where the rise of the roof is 6 inches per foot run.

1. Since the rise is given as 6 inches per foot run, find the 6-inch mark on the outer edge of the body of the square.
2. Under this number are given the lengths of rafters, as shown in Fig. 15-30. The first line gives the length of the common rafter. The first number under 6 is 13.42. This is the length in inches of a common rafter for one foot of run when the pitch is 6 inches to the foot run.

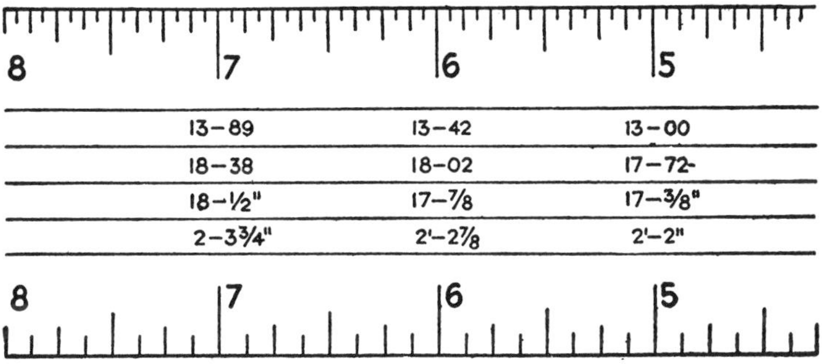

8	7	6	5
13–89	13–42	13–00
18–38	18–02	17–72
18–½"	17–⅞	17–⅜"
2–3¾"	2'–2⅞	2'–2"

Fig. 15-30.

3. Since the building is 18 feet wide, the run is 18/2, or 9 feet.
4. Since the table gives the length of 13.42 inches for one foot of run, multiply 13.42 x 9 = 120.78 inches.

    To express this in feet divide by 12: 120.78/12 = 10.065 feet.

Ans. A common rafter for a building 18 feet wide with a pitch of 6 inches per foot run would have to be 10.065 feet, or 10 feet 25/32 inches long.

Summary	EXAMPLE
To find the length of a common rafter using the table based on inches per foot run:	Find the length of a common rafter for a building 20 feet wide, where the rise of the roof is 6″ per foot run.
1. Find the inch mark that corresponds to the number of inches per foot of run on the body of the square that contains the table.	1. Mark on square = 6″

2. Trace down the table given under this mark to the line of the table that gives common rafter lengths. The number in this line is the number of inches in the length of the rafter for each foot of run.

2. Number under 6″ at common rafter line is 13.42 (Fig. 15-30).

3. Obtain the number of feet in the run.

3. Run = 1/2 the width
1/2 x 20 = 10 feet

4. Multiply the number from the table by the number of feet in the run. This is the length of the rafter in inches.

4. 13.42″ x 10 = 134.20″

5. Divide the result of Step 4 by 12 to express the length of the rafter in feet.

5. $\dfrac{134.20}{12}$ = 11.18′, or 11 2/12′

Ans. Rafter is 11 ft. 2 in.

Note: The length of rafter given here is measured along the measuring line from the plumb line (see Fig. 15-25).

Length of common rafter using table based on span and pitch

The second method used for obtaining the lengths of rafters is based on the table shown below (Fig. 15-31). Below the 3-inch and 4-inch marks appear three columns with numbers like 12 4 1/6, 12 6 1/4, etc. The first of these numbers represents the run; the second column gives the rise in inches per foot run; and the third column, the common pitches.

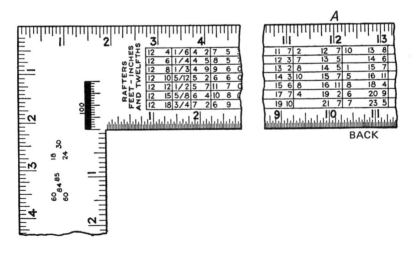

Fig. 15-31. Rafter table giving rise per foot of run.

Thus, in the first line, 12 4 1/6, the run is 12, which may be considered as either feet or inches, the second number means 4 inches per foot run, and the third means a 1/6 pitch.

The graduations along the upper edge of the square may be considered as either inches or feet, and since there are 12 smaller divisions in each large space, each small division may be considered as an inch or 1/12 of an inch. Under each whole division are found columns of numbers such as the ones under 4, which are 4 2 7, 4 5 8, etc. The first of these numbers represents feet, the second inches, and the third twelfths of an inch. Hence, 4 2 7 would mean 4 feet 2 7/12 inches.

Example. Find the length of a common rafter for a roof with a run of 12 feet and a pitch of 1/2.

1. Find the pitch in the third column. This is the fifth line down, the one which reads 12 12 1/2.
2. On the upper edge of the square, find the figure that is the same as the run on the building in question. Here it is 12 (Fig. 15-31, *A*).
3. Trace down the table under 12 to the fifth line, where the numbers 16 11 8 are found. This means 16 feet 11 8/12 inches, which is the length of the rafter in question.

Summary. To find the length of a common rafter using a table based on span and pitch:

1. Find the given pitch in the third column of the pitch table. Note the line in which it is located.
2. On the outer edge of the square, find the figure that is the same as the total run of the building.
3. Trace down the table under this number to the line noted in the first step. This gives the length of the rafter in feet, inches, and twelfths of an inch.

Note 1: If the run is over 23 feet, first find the length for a run of 23 feet, then find the length for a run representing the excess over 23 feet, and add the two results.

Note 2: The length of the common rafter given above is from the plumb line or the centerline of the ridgeboard to the outer edge of the plate, measured along the measuring line. When a ridgeboard is used, it is necessary to deduct for 1/2 of the thickness of the ridgeboard. This deduction must be made at *right angles* to the top cut. Thus, if a ridgeboard is 1 3/4" thick, 1/2 of this, or 7/8", must be deducted. This 7/8" must be measured at right angles to the top cut (see Fig. 15-25).

EXERCISES

Obtain the length of common rafter for each of the following:

	Width of bldg.	Run	Pitch	Rise	Length of common rafter
1.		10'-0"	1/6		?
2.	16'-0"		1/3		?
3.	18'-0"			9'-0"	?
4.		18'-0"		27'-0"	?
5.	16'-0"			10'-0"	?
6.		8'-0"		8'-0"	?
7.	12'-6"		1/4		?
8.	15'-0"		1/6		?

Appendix

TABLE 1
Decimal Equivalents of Common Fractions

Fraction		Decimal Equivalent Customary (in.)	Metric (mm)	Fraction		Decimal Equivalent Customary (in.)	Metric (mm)
	1/64	.015625	0.3969		33/64	.515625	13.0969
1/32		.03125	0.7938	17/32		.53125	13.4938
	3/64	.046875	1.1906		35/64	.546875	13.8906
1/16		.0625	1.5875	9/16		.5625	14.2875
	5/64	.078125	1.9844		37/64	.578125	14.6844
3/32		.09375	2.3813	19/32		.59375	15.0813
	7/64	.109375	2.7781		39/64	.609375	15.4781
1/8		.1250	3.1750	5/8		.6250	15.8750
	9/64	.140625	3.5719		41/64	.640625	16.2719
5/32		.15625	3.9688	21/32		.65625	16.6688
	11/64	.171875	4.3656		43/64	.671875	17.0656
3/16		.1875	4.7625	11/16		.6875	17.4625
	13/64	.203125	5.1594		45/64	.703125	17.8594
7/32		.21875	5.5563	23/32		.71875	18.2563
	15/64	.234375	5.9531		47/64	.734375	18.6531
1/4		.250	6.3500	3/4		.750	19.0500
	17/64	.265625	6.7469		49/64	.765625	19.4469
9/32		.28125	7.1438	25/32		.78125	19.8438
	19/64	.296875	7.5406		51/64	.796875	20.2406
5/16		.3125	7.9375	13/16		.8125	20.6375
	21/64	.328125	8.3384		53/64	.828125	21.0344
11/32		.34375	8.7313	27/32		.84375	21.4313
	23/64	.359375	9.1281		55/64	.859375	21.8281
3/8		.3750	9.5250	7/8		.8750	22.2250
	25/64	.390625	9.9219		57/64	.890625	22.6219
13/32		.40625	10.3188	29/32		.90625	23.0188
	27/64	.421875	10.7156		59/64	.921875	23.4156
7/16		.4375	11.1125	15/16		.9375	23.8125
	29/64	.453125	11.5094		61/64	.953125	24.2094
15/32		.46875	11.9063	31/32		.96875	24.6063
	31/64	.484375	12.3031		63/64	.984375	25.0031
1/2		.500	12.7000	1		1.000	25.4000

TABLE 2

Units of Measure

Linear Measure

12 inches = 1 foot	3 feet = 1 yard	5 1/2 yards = 1 rod
40 rods = 1 furlong		8 furlongs = 1 mile

Equivalent Values

Inches		Feet		Yards		Rods		Furlongs		Miles
36	=	3	=	1						
198	=	16.5	=	5.5	=	1				
7,920	=	660	=	220	=	40	=	1		
63,360	=	5,280	=	1,760	=	320	=	8	=	1

Square Measure

144 square inches	=	1 square foot
9 square feet	=	1 square yard
30 1/4 square yards	=	1 square rod
160 square rods	=	1 acre
640 acres	=	1 square mile

1 square mile = 640 acres = 102,400 square rods = 3,097,600 square yards
1 square mile = 27,878,400 square feet = 4,014,489,600 square inches

Cubic Measure

1,728 cubic inches = 1 cubic foot
27 cubic feet = 1 cubic yard
128 cubic feet = 1 cord
24 3/4 cubic feet = 1 perch
1 cubic yard = 27 cubic feet = 46,656 cubic inches

Measure of Angles or Arcs

60 seconds	=	1 minute
60 minutes	=	1 degree
90 degrees	=	1 right angle or quadrant
360 degrees	=	1 circle

1 circle = 360 degrees = 21,600 minutes = 1,296,000 seconds
1 minute of arc on the earth's surface = 1 nautical mile = 1.17 times a land mile or 6,080 feet.

TABLE 2 (Continued)
Units of Measure

Weight — Avoirdupois

437 1/2 grains = 1 ounce   16 ounces = 1 pound

100 pounds = 1 hundredweight (cwt.)

2,000 pounds = 1 ton   2,240 pounds = 1 long ton

1 ton = 20 cwt. = 2,000 pounds = 32,000 ounces = 14,000,000 grains

1 pound avdp. = 7,000 grains

Liquid Measure

4 gills   = 1 pint   4 quarts = 1 gallon

2 pints   = 1 quart   31 1/2 gallons = 1 barrel

63 gallons or 2 barrels = 1 hogshead

1 hogshead = 2 barrels = 63 gallons = 252 quarts = 504 pints = 1,016 gills

The U.S. gallon contains 231 cubic inches = .134 cubic feet.

1 cubic foot of water = 7.481 gallons and weighs 62.425 pounds at 39.2F

1 gallon of water weighs 8.45 pounds.

(For ordinary work 1 cubic foot is considered 7 1/2 gallons; 1 gallon 8 1/3 pounds)

Mariners' Measure

6 feet = 1 fathom   5,280 feet = 1 statute mile

120 fathoms   =   1 cable length

6,085 feet   =   1 nautical mile

7 1/2 cable lengths   =   1 mile

Miscellaneous

3 inches	=	1 palm	18 inches	= 1 cubit
4 inches	=	1 hand	21.8 inches	= 1 Bible cubit
9 inches	=	1 span	2 1/2 feet	= 1 military pace
12 articles	=	1 dozen	1 league	= 3 miles
12 dozen	=	1 gross	20 articles	= 1 score
12 gross	=	1 great gross	24 sheets	= 1 quire
2 articles	=	1 pair	20 quires	= 1 ream

TABLE 3
Trigonometric Functions

Angle	Sine	Cosine	Tangent	Angle	Sine	Cosine	Tangent
1°	.0175	.9998	.0175	46°	.7193	.6947	1.0355
2°	.0349	.9994	.0349	47°	.7314	.6820	1.0724
3°	.0523	.9986	.0524	48°	.7431	.6691	1.1106
4°	.0698	.9976	.0699	49°	.7547	.6561	1.1504
5°	.0872	.9962	.0875	50°	.7660	.6428	1.1918
6°	.1045	.9945	.1051	51°	.7771	.6293	1.2349
7°	.1219	.9925	.1228	52°	.7880	.6157	1.2799
8°	.1392	.9903	.1405	53°	.7986	.6018	1.3270
9°	.1564	.9877	.1584	54°	.8090	.5878	1.3764
10°	.1736	.9848	.1763	55°	.8192	.5736	1.4281
11°	.1908	.9816	.1944	56°	.8290	.5592	1.4826
12°	.2079	.9781	.2126	57°	.8387	.5446	1.5399
13°	.2250	.9744	.2309	58°	.8480	.5299	1.6003
14°	.2419	.9703	.2493	59°	.8572	.5150	1.6643
15°	.2588	.9659	.2679	60°	.8660	.5000	1.7321
16°	.2756	.9613	.2867	61°	.8746	.4848	1.8040
17°	.2924	.9563	.3057	62°	.8829	.4695	1.8807
18°	.3090	.9511	.3249	63°	.8910	.4540	1.9626
19°	.3256	.9455	.3443	64°	.8988	.4384	2.0503
20°	.3420	.9397	.3640	65°	.9063	.4226	2.1445
21°	.3584	.9336	.3839	66°	.9135	.4067	2.2460
22°	.3746	.9272	.4040	67°	.9205	.3907	2.3559
23°	.3907	.9205	.4245	68°	.9272	.3746	2.4751
24°	.4067	.9135	.4452	69°	.9336	.3584	2.6051
25°	.4226	.9063	.4663	70°	.9397	.3420	2.7475
26°	.4384	.8988	.4877	71°	.9455	.3256	2.9042
27°	.4540	.8910	.5095	72°	.9511	.3090	3.0777
28°	.4695	.8829	.5317	73°	.9563	.2924	3.2709
29°	.4848	.8746	.5543	74°	.9613	.2756	3.4874
30°	.5000	.8660	.5774	75°	.9659	.2588	3.7321
31°	.5150	.8572	.6009	76°	.9703	.2419	4.0108
32°	.5299	.8480	.6249	77°	.9744	.2250	4.3315
33°	.5446	.8387	.6494	78°	.9781	.2079	4.7046
34°	.5592	.8290	.6745	79°	.9816	.1908	5.1446
35°	.5736	.8192	.7002	80°	.9848	.1736	5.6713
36°	.5878	.8090	.7265	81°	.9877	.1564	6.3138
37°	.6018	.7986	.7536	82°	.9903	.1392	7.1154
38°	.6157	.7880	.7813	83°	.9925	.1219	8.1443
39°	.6293	.7771	.8098	84°	.9945	.1045	9.5144
40°	.6428	.7660	.8391	85°	.9962	.0872	11.4301
41°	.6561	.7547	.8693	86°	.9976	.0698	14.3007
42°	.6691	.7431	.9004	87°	.9986	.0523	19.0811
43°	.6820	.7314	.9325	88°	.9994	.0349	28.6363
44°	.6947	.7193	.9657	89°	.9998	.0175	57.2900
45°	.7071	.7071	1.0000	90°	1.0000	.0000	

TABLE 4

Powers, Roots, and Reciprocals

Number	Square	Cube	Square root	Cube root	Reciprocal
1	1	1	1.000000	1.000000	1.0000000
2	4	8	1.414214	1.259912	.5000000
3	9	27	1.732051	1.442250	.3333333
4	16	64	2.000000	1.587401	.2500000
5	25	125	2.236068	1.709976	.2000000
6	36	216	2.449490	1.817112	.1666667
7	49	343	2.645751	1.912931	.1428571
8	64	512	2.828427	2.000000	.1250000
9	81	729	3.000000	2.080084	.1111111
10	100	1000	3.162278	2.154435	.1000000
11	121	1331	3.316625	2.223980	.0909091
12	144	1728	3.464102	2.289429	.0833333
13	169	2197	3.605551	2.351335	.0769231
14	196	2744	3.741657	2.410142	.0714286
15	225	3375	3.872983	2.466212	.0666667
16	256	4096	4.000000	2.519842	.0625000
17	289	4913	4.123106	2.571282	.0588235
18	324	5832	4.242641	2.620741	.0555556
19	361	6859	4.358899	2.668402	.0526316
20	400	8000	4.472136	2.714418	.0500000
21	441	9261	4.582576	2.758924	.0476190
22	484	10,648	4.690416	2.802039	.0454545
23	529	12,167	4.795832	2.843867	.0434783
24	576	13,824	4.898980	2.884499	.0416667
25	625	15,625	5.000000	2.924018	.0400000
26	676	17,576	5.099020	2.962496	.0384615
27	729	19,683	5.196152	3.000000	.0370370
28	784	21,952	5.291503	3.036589	.0357143
29	841	24,389	5.385165	3.072317	.0344828
30	900	27,000	5.477226	3.107233	.0333333
31	961	29,791	5.567764	3.141381	.0322581
32	1024	32,768	5.656854	3.174802	.0312500
33	1089	35,937	5.744563	3.207534	.0303030
34	1156	39,304	5.830952	3.239612	.0294118
35	1225	42,875	5.916080	3.271066	.0285714

TABLE 4 (Continued)

Powers, Roots, and Reciprocals

Number	Square	Cube	Square root	Cube root	Reciprocal
36	1296	46,656	6.000000	3.301927	.0277778
37	1369	50,653	6.082763	3.332222	.0270270
38	1444	54,872	6.164414	3.361975	.0263158
39	1521	59,319	6.244998	3.391211	.0256410
40	1600	64,000	6.324555	3.419952	.0250000
41	1681	68,921	6.403124	3.448217	.0243902
42	1764	74,088	6.480741	3.476027	.0238095
43	1849	79,507	6.557439	3.503398	.0232558
44	1936	85,184	6.633250	3.530348	.0227273
45	2025	91,125	6.708204	3.556893	.0222222
46	2116	97,336	6.782330	3.583048	.0217391
47	2209	103,823	6.855655	3.608826	.0212766
48	2304	110,592	6.928203	3.634241	.0208333
49	2401	117,649	7.000000	3.059306	.0204082
50	2500	125,000	7.071068	3.684013	.0200000
51	2601	132,651	7.141428	3.708430	.0196078
52	2704	140,608	7.211103	3.732511	.0192308
53	2809	148,877	7.280110	3.756286	.0188679
54	2916	157,464	7.348469	3.779763	.0185185
55	3025	166,375	7.416199	3.802953	.0181818
56	3136	175,616	7.483315	3.825862	.0178571
57	3249	185,193	7.549834	3.848501	.0175439
58	3364	195,112	7.615773	3.870877	.0172414
59	3481	205,379	7.681146	3.892997	.0169492
60	3600	216,000	7.745967	3.914868	.0166667
61	3721	226,981	7.810250	3.936497	.0163934
62	3844	238,328	7.874008	3.957892	.0161290
63	3969	250,047	7.937254	3.979057	.0158730
64	4096	262,144	8.000000	4.000000	.0156250
65	4225	274,625	8.062258	4.020726	.0153846
66	4356	287,496	8.124038	4.041240	.0151515
67	4489	300,763	8.185353	4.061548	.0149254
68	4624	314,432	8.246211	4.081655	.0147059
69	4761	328,509	8.306624	4.101566	.0144928
70	4900	343,000	8.366600	4.121285	.0142857

TABLE 4 (Continued)

Powers, Roots, and Reciprocals

Number	Square	Cube	Square root	Cube root	Reciprocal
71	5041	357,911	8.426150	4.140818	.0140845
72	5184	373,248	8.485281	4.160168	.0138889
73	5329	389,017	8.544004	4.179339	.0136986
74	5476	405,224	8.602325	4.198336	.0135135
75	5625	421,875	8.660254	4.217163	.0133333
76	5776	438,976	8.717798	4.235824	.0131579
77	5929	456,533	8.774964	4.254321	.0129870
78	6084	474,552	8.831761	4.272659	.0128205
79	6241	493,039	8.888194	4.290840	.0126582
80	6400	512,000	8.944272	4.308870	.0125000
81	6561	531,441	9.000000	4.326749	.0123457
82	6724	551,368	9.055385	4.344482	.0121951
83	6889	571,787	9.110434	4.362071	.0120482
84	7056	592,704	9.165151	4.379519	.0110048
85	7225	614,125	9.219545	4.396830	.0117647
86	7396	636,056	9.273619	4.414005	.0116279
87	7569	658,503	9.327379	4.431048	.0114943
88	7744	681,472	9.380832	4.447960	.0113636
89	7921	704,969	9.433981	4.464745	.0112360
90	8100	729,000	9.486833	4.481405	.0111111
91	8281	753,571	9.539392	4.497941	.0109890
92	8464	778,688	9.591663	4.514357	.0108696
93	8649	804,357	9.643651	4.530655	.0107527
94	8836	830,584	9.695360	4.546836	.0106383
95	9025	857,375	9.746794	4.562903	.0105263
96	9216	884,736	9.797959	4.578857	.0104167
97	9409	912,673	9.848858	4.594701	.0103093
98	9604	941,192	9.899495	4.610436	.0102041
99	9801	970,299	9.949874	4.626065	.0101010
100	10,000	1,000,000	10.000000	4.641589	.0100000

TABLE 5 CONVERSION OF ENGLISH AND METRIC MEASURES

Linear Measure

Unit	Inches to milli- meters	Milli- meters to inches	Feet to meters	Meters to feet	Yards to meters	Meters to yards	Miles to kilo- meters	Kilo- meters to miles
1	25.40	0.03937	0.3048	3.281	0.9144	1.094	1.609	0.6214
2	50.80	0.07874	0.6096	6.562	1.829	2.187	3.219	1.243
3	76.20	0.1181	0.9144	9.842	2.743	3.281	4.828	1.864
4	101.60	0.1575	1.219	13.12	3.658	4.374	6.437	2.485
5	127.00	0.1968	1.524	16.40	4.572	5.468	8.047	3.107
6	152.40	0.2362	1.829	19.68	5.486	6.562	9.656	3.728
7	177.80	0.2756	2.134	22.97	6.401	7.655	11.27	4.350
8	203.20	0.3150	2.438	26.25	7.315	8.749	12.87	4.971
9	228.60	0.3543	2.743	29.53	8.230	9.842	14.48	5.592

Example   1 in. = 2540 mm ,   1 m = 3.281 ft.,   1 km = 0.6214 mi.

Surface Measure

Unit	Square inches to square centi- meters	Square centi- meters to square inches	Square feet to square meters	Square meters to square feet	Square yards to square meters	Square meters to square yards	Acres to hec- tares	Hec- tares to acres	Square miles to square kilo- meters	Square kilo- meters to square miles
1	6.452	0.1550	0.0929	10.76	0.8361	1.196	0.4047	2.471	2.59	0.3861
2	12.90	0.31	0.1859	21.53	1.672	2.392	0.8094	4.942	5.18	0.7722
3	19.356	0.465	0.2787	32.29	2.508	3.588	1.214	7.413	7.77	1.158
4	25.81	0.62	0.3716	43.06	3.345	4.784	1.619	9.884	10.36	1.544
5	32.26	0.775	0.4645	53.82	4.181	5.98	2.023	12.355	12.95	1.931
6	38.71	0.93	0.5574	64.58	5.017	7.176	2.428	14.826	15.54	2.317
7	45.16	1.085	0.6503	75.35	5.853	8.372	2.833	17.297	18.13	2.703
8	51.61	1.24	0.7432	86.11	6.689	9.568	3.237	19.768	20.72	3.089
9	58.08	1.395	0.8361	96.87	7.525	10.764	3.642	22.239	23.31	3.475

Example   1 sq. in. = 6.452 cm^2,   1 m^2 = 1.196 sq. yds., 1 sq. mi. = 2.59 km^2

Cubic Measure

Unit	Cubic inches to cubic centi- meters	Cubic centi- meters to cubic inches	Cubic feet to cubic meters	Cubic meters to cubic feet	Cubic yards to cubic meters	Cubic meters to cubic yards	Gallons to cubic feet	Cubic feet to gallons
1	16.39	0.06102	0.02832	35.31	0.7646	1.308	0.1337	7.481
2	32.77	0.1220	0.05663	70.63	1.529	2.616	0.2674	14.96
3	49.16	0.1831	0.08495	105.9	2.294	3.924	0.4010	22.44
4	65.55	0.2441	0.1133	141.3	3.058	5.232	0.5347	29.92
5	81.94	0.3051	0.1416	176.6	3.823	6.540	0.6684	37.40
6	98.32	0.3661	0.1699	211.9	4.587	7.848	0.8021	44.88
7	114.7	0.4272	0.1982	247.2	5.352	9.156	0.9358	52.36
8	131.1	0.4882	0.2265	282.5	6.116	10.46	1.069	59.84
9	147.5	0.5492	0.2549	371.8	6.881	11.77	1.203	67.32

Example   1 cm^3 = 0.06102 cu. in.,   1 gal. = 0.1337 cu. ft.

Volume or Capacity Measure

Unit	Liquid ounces to cubic centi- meters	Cubic centi- meters to liquid ounces	Pints to liters	Liters to pints	Quarts to liters	Liters to quarts	Gallons to liters	Liters to gallons	Bushels to hecto- liters	Hecto- liters to bushels
1	29.57	0.03381	0.4732	2.113	0.9463	1.057	3.785	0.2642	0.3524	2.838
2	59.15	0.06763	0.9463	4.227	1.893	2.113	7.571	0.5284	0.7048	5.676
3	88.72	0.1014	1.420	6.340	2.839	3.785	11.36	0.7925	1.057	8.513
4	118.3	0.1353	1.893	8.454	3.170	4.227	15.14	1.057	1.410	11.35
5	147.9	0.1691	2.366	10.57	4.732	5.284	18.93	1.321	1.762	14.19
6	177.4	0.2029	2.839	12.68	5.678	6.340	22.71	1.585	2.114	17.03
7	207.0	0.2367	3.312	14.79	6.624	7.397	26.50	1.849	2.467	19.86
8	236.6	0.2705	3.785	16.91	7.571	8.454	30.28	2.113	2.819	22.70
9	266.2	0.3043	4.259	19.02	8.517	9.510	34.07	2.378	3.171	25.54

Example   1 ℓ = 2.113 pts.,   1 gal. = 3.785 ℓ

TABLE 5A STANDARD TABLES OF METRIC MEASURE

Linear Measure		
Unit	Value in Meters	Symbol or Abbreviation
micron	0.000 001	μ
millimeter	0.001	mm
centimeter	0.01	cm
decimeter	0.1	dm
meter (unit)	1.0	m
dekameter	10.0	dam
hectometer	100.0	hm
kilometer	1 000.00	km
myriameter	10 000.00	Mm
megameter	1 000 000.00	

Surface Measure		
Unit	Value in Square Meters	Symbol or Abbreviation
square millimeter	0.000 001	mm^2
square centimeter	0.000 1	cm^2
square decimeter	0.01	dm^2
square meter (centiare)	1.0	m^2
square dekameter (are)	100.0	a^2
hectare	10 000.0	ha^2
square kilometer	1 000 000.0	km^2

Volume		
Unit	Value in Liters	Symbol or Abbreviation
milliliter	0.001	mℓ
centiliter	0.01	cℓ
deciliter	0.1	dℓ
liter (unit)	1.0	ℓ
dekaliter	10.0	daℓ
hectoliter	100.0	hℓ
kiloliter	1 000.0	kℓ

Mass		
Unit	Value in Grams	Symbol or Abbreviation
microgram	0.000 001	μg
milligram	0.001	mg
centigram	0.01	cg
decigram	0.1	dg
gram (unit)	1.0	g
dekagram	10.0	dag
hectogram	100.0	hg
kilogram	1 000.0	kg
myriagram	10 000.0	Mg
quintal	100 000.0	q
ton	1 000 000.0	t

Cubic Measure		
Unit	Value in Cubic Meters	Symbol or Abbreviation
cubic micron	10^{-18}	μ^3
cubic millimeter	10^{-9}	mm^3
cubic centimeter	10^{-6}	cm^3
cubic decimeter	10^{-3}	dm^3
cubic meter	1	m^3
cubic dekameter	10^3	dam^3
cubic hectometer	10^6	hm^3
cubic kilometer	10^9	km^3

TABLE 6

Circumferences and Areas of Circles

Diameter	Circum-ference	Area	Diameter	Circum-ference	Area
1	3.1416	0.7854	51	160.22	2042.82
2	6.2832	3.1416	52	163.36	2123.72
3	9.4248	7.0686	53	166.50	2206.18
4	12.5664	12.5664	54	169.65	2290.22
5	15.7080	19.635	55	172.79	2375.83
6	18.850	28.274	56	175.93	2463.01
7	21.991	38.485	57	179.07	2551.76
8	25.133	50.266	58	182.21	2642.08
9	28.274	63.617	59	185.35	2733.97
10	31.416	78.540	60	188.50	2827.43
11	34.558	95.033	61	191.64	2922.47
12	37.699	113.10	62	194.78	3019.07
13	40.841	132.73	63	197.92	3117.25
14	43.982	153.94	64	201.06	3216.99
15	47.124	176.71	65	204.20	3318.31
16	50.265	201.06	66	207.34	3421.19
17	53.407	226.98	67	210.49	3525.65
18	56.549	254.47	68	213.63	3631.68
19	59.690	283.53	69	216.77	3739.28
20	62.832	314.16	70	219.91	3848.45
21	65.973	346.36	71	223.05	3959.19
22	69.115	380.13	72	226.19	4071.50
23	72.257	415.48	73	229.34	4185.39
24	75.398	452.39	74	232.48	4300.84
25	78.540	490.87	75	235.62	4417.86
26	81.681	530.93	76	238.76	4536.46
27	84.823	572.56	77	241.90	4656.63
28	87.965	615.75	78	245.04	4778.36
29	91.106	660.52	79	248.19	4901.67
30	94.248	706.86	80	251.33	5026.55
31	97.389	754.77	81	254.47	5153.00
32	100.53	804.25	82	257.61	5281.02
33	103.67	855.30	83	260.75	5410.61
34	106.81	907.92	84	263.89	5541.77
35	109.96	962.11	85	267.04	5674.50
36	113.10	1017.88	86	270.18	5808.80
37	116.24	1075.21	87	273.32	5944.68
38	119.38	1134.11	88	276.46	6082.12
39	122.52	1194.59	89	279.60	6221.14
40	125.66	1256.64	90	282.74	6361.73
41	128.81	1320.25	91	285.88	6503.88
42	131.95	1385.44	92	289.03	6647.61
43	135.09	1452.20	93	292.17	6792.91
44	138.23	1520.53	94	295.31	6939.78
45	141.37	1590.43	95	298.45	7088.22
46	144.51	1661.90	96	301.59	7238.23
47	147.65	1734.94	97	304.73	7389.81
48	150.80	1809.56	98	307.88	7542.96
49	153.94	1885.74	99	311.02	7697.69
50	157.08	1963.50	100	314.16	7853.98

Note.—The surface of a sphere of given diameter may be found directly from the above table, since it is equal to the area of a circle of twice the diameter of the sphere.

TABLE 7

Milling Machine Standard Tapers

(as adopted by the Milling Machine Manufacturers of the National Machine Tool Builders' Association)

No of Taper	A	B	*Threaded End of Draw-in Bolt
10	5/8	3/8	
20	7/8	1/2	
30	1 1/4	5/8	
40	1 3/4	1	{7/16" 14 N.C., R.H. {5/8" 11 N.C., R.H.
50	2 3/4	1 9/16	{5/8" 11 N.C., R.H. {1" 8 N.C., R.H.

*End of Draw-In Bolt is threaded with two sizes of thread. The larger size thread is used in threaded hole in Arbors and in Adapters where possible, but limitations on certain Adapters require the use of a threaded hole to fit the smaller Threaded End of the Draw-In Bolt.

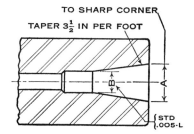

TO SHARP CORNER

TAPER $3\frac{1}{2}$ IN PER FOOT

STD .005-L

TABLE 8

The Jarno Taper

Taper per Foot = 0.6 Inch. Taper per Inch = 0.05 Inch.

$$\text{Diam. Large End} = \frac{\text{No. of Taper}}{8} \qquad \text{Diam. Small End} = \frac{\text{No. of Taper}}{10}$$

$$\text{Length of Taper} = \frac{\text{No. of Taper}}{2}$$

In the Jarno system, the taper of which is 0.6 inch per foot or 1 in 20, the number of the taper is the key by which all the dimensions are immediately determined. That is, the number of the taper is the number of tenths of an inch in diameter at the small end, the number of eighths of an inch at the large end, and the number of halves of an inch in length or depth. For example: the No. 6 taper is six-eighths (3/4) inch diameter at large end, six-tenths (6/10) inch diameter at the small end and six-halves (3) inches in length. Similarly, the No. 16 taper is sixteen-eighths, or 2 inches diameter at the large end; sixteen-tenths or 1.6 inches at the small end and sixteen-halves or 8 inches in length.

TABLE 9

Diameter, Weight, and Resistance of Solid Copper Wire

No. Awg.	Diam- eter, Mils	Area, Circular Mils	Weight, Bare Wire		Resistance at 25°C. (77°F.)		
			Pounds per 1,000 Ft.	Pounds per Mile	Ohms per 1,000 Ft.	Ohms per Mile	Feet per Ohm
0000	460	211,600	641	3,358	0.0499	0.2638	20,040
000	410	167,800	508	2,683	0.0630	0.3325	15,870
00	364.8	133,100	403	2,126	0.0794	0.419	12,590
0	324.9	105,500	319.5	1,687	0.1003	0.529	9,980
1	289.3	83,700	253.3	1,337	0.1262	0.666	7,930
2	257.6	66,400	200.9	1,061	0.1591	0.840	6,290
3	229.4	52,600	159.3	841	0.2008	1.062	4,980
4	204.3	41,700	126.4	668	0.2533	1.338	3,950
5	181.9	33,100	100.2	529	0.3193	1.685	3,134
6	162.0	26,250	79.5	419	0.403	2.127	2,485
7	144.3	20,820	63.0	332.6	0.507	2.682	1,971
8	128.5	16,510	50.0	264.0	0.640	3.382	1,562
9	114.4	13,009	39.63	208.3	0.807	4.26	1,238
10	101.9	10,380	31.43	165.9	1.019	5.37	983
11	90.7	8,230	24.92	131.6	1.284	6.78	779
12	80.8	6,530	19.77	104.3	1.618	8.55	618
13	72.0	5,180	15.68	82.8	2.040	10.77	490
14	64.1	4,110	12.43	65.6	2.575	13.60	388.2
15	57.1	3,257	9.86	52.1	3.244	17.13	308.4
16	50.8	2,583	7.82	41.3	4.09	21.62	244.3
17	45.3	2,048	6.20	32.73	5.16	27.24	193.9
18	40.3	1,624	4.92	26.00	6.51	34.34	153.7
19	35.89	1,288	3.899	20.57	8.20	43.3	121.9
20	31.96	1,022	3.092	16.33	10.34	54.6	96.6
21	28.46	810	2.452	12.93	13.04	68.9	76.6
22	25.35	642	1.945	10.27	16.44	86.9	60.8
23	22.57	509	1.542	8.14	20.75	109.5	48.2
24	20.10	404	1.223	6.46	26.15	138.1	38.25
25	17.90	320.4	0.970	5.12	33.00	174.3	30.30
26	15.94	254.1	0.759	4.06	41.6	219.5	24.04
27	14.20	201.5	0.610	3.220	52.4	276.8	19.07
28	12.64	159.8	0.484	2.556	66.1	349.2	15.13
29	11.26	126.7	0.3836	2.025	83.4	441	11.98
30	10.03	100.5	0.3042	1.606	105.4	556	9.48
31	8.93	79.7	0.2413	1.273	132.6	700	7.55
32	7.95	63.2	0.1913	1.011	167.2	883	5.98
33	7.08	50.1	0.1517	0.807	210.8	1,113	4.74
34	6.30	39.75	0.1203	0.636	265.8	1,403	3.762
35	5.61	31.52	0.0954	0.504	335.5	1,772	2.980
36	5.00	25.00	0.0757	0.400	423	2,232	2.366
37	4.45	19.83	0.0600	0.3168	533	2,814	1.877
38	3.965	15.72	0.0476	0.2514	673	3,553	1.487
39	3.531	12.47	0.03774	0.1991	847	4,470	1.180
40	3.145	9.89	0.02993	0.1579	1,068	5,640	0.936

TABLE 10

Wire and Sheet-Metal Gages

Gage Numbers	United States	American or Brown & Sharpe	Washburn & Moen, Am. Steel &Wire Co. Roebling	Trenton Iron Co.	Birming-ham or Stub's Iron Wire	Stub's Steel Wire	British Imperial	Gage Number
7-0	.500	. . . .	. . . .	. . . . . .	. . . . . .	. . . . .	.500	7-0
6-0	.469	. . . .	.460	. . . .	. . . . . .	. . . . .	.464	6-0
5-0	.438	. . . .	.430	.450	. . . . . .	. . . . .	.432	5-0
4-0	.406	.460	.394	.400	.454	. . . . .	.400	4-0
000	.375	.410	.363	.360	.425	. . . . .	.372	000
00	.344	.365	.331	.330	.380	. . . . .	.348	00
0	.313	.325	.307	.305	.340	. . . . .	.324	0
1	.281	.289	.283	.285	.300	.227	.300	1
2	.266	.258	.263	.265	.284	.219	.276	2
3	.250	.229	.244	.245	.259	.212	.252	3
4	.234	.204	.225	.225	.238	.207	.232	4
5	.219	.182	.207	.205	.220	.204	.212	5
6	.203	.162	.192	.190	.203	.201	.192	6
7	.188	.144	.177	.175	.180	.199	.176	7
8	.172	.128	.162	.160	.165	.197	.160	8
9	.156	.114	.148	.145	.148	.194	.144	9
10	.141	.102	.135	.130	.134	.191	.128	10
11	.125	.0907	.121	.118	.120	.188	.116	11
12	.109	.0808	.106	.105	.109	.185	.104	12
13	.0938	.072	.0915	.0925	.095	.182	.092	13
14	.0781	.0641	.080	.0806	.083	.180	.080	14
15	.0703	.0571	.072	.070	.072	.178	.072	15
16	.0625	.0508	.0625	.061	.065	.175	.064	16
17	.0563	.0453	.054	.0525	.058	.172	.056	17
18	.050	.0403	.0475	.045	.049	.168	.048	18
19	.0438	.0359	.041	.040	.042	.164	.040	19
20	.0375	.032	.0348	.035	.035	.161	.036	20
21	.0344	.0285	.0318	.031	.032	.157	.032	21
22	.0313	.0253	.0286	.028	.028	.155	.028	22
23	.0281	.0226	.0258	.025	.025	.153	.024	23
24	.025	.0201	.023	.0225	.022	.151	.022	24
25	.0219	.0179	.0204	.020	.020	.148	.020	25
26	.0188	.0159	.0181	.018	.018	.146	.018	26
27	.0172	.0142	.0173	.017	.016	.143	.0164	27
28	.0156	.0126	.0162	.016	.014	.139	.0149	28
29	.0141	.0113	.015	.015	.013	.134	.0136	29
30	.0125	.010	.014	.014	.012	.127	.0124	30
31	.0109	.0089	.0132	.013	.010	.120	.0116	31
32	.0102	.008	.0128	.012	.009	.115	.0108	32
33	.0094	.0071	.0118	.011	.008	.112	.010	33
34	.0086	.0063	.0104	.010	.007	.110	.0092	34
35	.0078	.0056	.0095	.0095	.005	.108	.0084	35
36	.007	.005	.009	.009	.004	.106	.0076	36
37	.0066	.0045	.0085	.0085	. . . .	.103	.0068	37
38	.0063	.004	.008	.008	. . . .	.101	.006	38
39	. . . . .	.0035	.0075	.0075	. . . . . .	.099	.0052	39
40	. . . . .	.0031	.007	.007	. . . . . .	.097	.0048	40

TABLE 11

American National Coarse and Fine Threads
Dimensions and Tap Drill Sizes

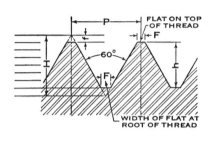

P (Pitch) $= \dfrac{1}{\text{No. threads per inch}}$

h (Depth) $= .649519\,P = \dfrac{.649519}{n}$

$H = .866025\,P$

$\dfrac{H}{8} = f = .108253\,P$

F (flat) $= .125\,P = \dfrac{P}{8}$

$n = $ No. of threads per inch

Nominal Size	Threads per Inch			Major Diameter Inches	Pitch Diameter Inches	Root Diameter Inches	Commercial Tap Drill Approx. 75% Full Thread	Decimal Equivalent of Tap Drill
	N.C.	N.F.	N.S.					
0	....	80	....	.0600	.0519	.0438	3/64	.0469
1	....	....	56	.0730	.0614	.0498	54	.0550
1	64	....	....	.0730	.0629	.0527	53	.0595
1	....	72	....	.0730	.0640	.0550	53	.0595
2	56	....	....	.0860	.0744	.0628	50	.0700
2	....	64	....	.0860	.0759	.0657	50	.0700
3	48	....	....	.0990	.0855	.0719	47	.0785
3	....	56	....	.0990	.0874	.0758	45	.0820
4	....	....	32	.1120	.0917	.0714	45	.0820
4	....	....	36	.1120	.0940	.0759	44	.0860
4	40	....	....	.1120	.0958	.0795	43	.0890
4	....	48	....	.1120	.0985	.0849	42	.0935
5	....	....	36	.1250	.1070	.0889	40	.0980
5	40	....	....	.1250	.1088	.0925	38	.1015
5	....	44	....	.1250	.1102	.0955	37	.1040
6	32	....	....	.1380	.1177	.0974	36	.1065
6	....	....	36	.1380	.1200	.1019	34	.1110
6	....	40	....	.1380	.1218	.1055	33	.1130
8	....	....	30	.1640	.1423	.1207	30	.1285
8	32	....	....	.1640	.1437	.1234	29	.1360
8	....	36	....	.1640	.1460	.1279	29	.1360
8	....	....	40	.1640	.1478	.1315	28	.1405
10	24	....	....	.1900	.1629	.1359	25	.1495
10	....	....	28	.1900	.1668	.1436	23	.1540
10	....	....	30	.1900	.1684	.1467	22	.1570
10	....	32	....	.1900	.1697	.1494	21	.1590
12	24	....	....	.2160	.1889	.1619	16	.1770
12	....	28	....	.2160	.1928	.1696	14	.1820
12	....	....	32	.2160	.1957	.1754	13	.1850
1/4	20	....	....	.2500	.2175	.1850	7	.2010
1/4	....	28	....	.2500	.2268	.2036	3	.2130
5/16	18	....	....	.3125	.2764	.2403	F	.2570
5/16	....	24	....	.3125	.2854	.2584	I	.2720
3/8	16	....	....	.3750	.3344	.2938	5/16	.3125
3/8	....	24	....	.3750	.3479	.3209	Q	.3320
7/16	14	....	....	.4375	.3911	.3447	U	.3680
7/16	....	20	....	.4375	.4050	.3726	25/64	.3906

TABLE 11 (Continued)
American National Coarse and Fine Threads Dimensions and Tap Drill Sizes

Nominal Size	Threads per Inch			Major Diameter Inches	Pitch Diameter Inches	Root Diameter Inches	Commercial Tap Drill Approx. 75% Full Thread	Decimal Equivalent of Tap Drill
	N.C.	N.F.	N.S.					
1/2	13	. . . .	. . . .	.5000	.4500	.4001	27/64	.4219
1/2	. . . .	20	. . . .	.5000	.4675	.4351	29/64	.4531
9/16	12	. . . .	. . . .	.5625	.5084	.4542	31/64	.4844
9/16	. . . .	18	. . . .	.5625	.5264	.4903	33/64	.5156
5/8	11	. . . .	. . . .	.6250	.5660	.5069	17/32	.5312
5/8	. . . .	18	. . . .	.6250	.5889	.5528	37/64	.5781
3/4	10	. . . .	. . . .	.7500	.6850	.6201	21/32	.6562
3/4	. . . .	16	. . . .	.7500	.7094	.6688	11/16	.6875
7/8	9	. . . .	. . . .	.8750	.8028	.7307	49/64	.7656
7/8	. . . .	14	. . . .	.8750	.8286	.7822	13/16	.8125
7/8	. . . .	. . . .	18	.8759	.8389	.8028	53/64	.8281
1	8	. . . .	. . . .	1.0000	.9188	.8376	7/8	.8750
1	. . . .	14	. . . .	1.0000	.9536	.9072	15/16	.9375
1 1/8	7	. . . .	. . . .	1.1250	1.0322	.9394	63/64	.9844
1 1/8	. . . .	12	. . . .	1.1250	1.0709	1.0168	1 3/64	1.0469
1 1/4	7	. . . .	. . . .	1.2500	1.1572	1.0644	1 7/64	1.1094
1 1/4	. . . .	12	. . . .	1.2500	1.1959	1.1418	1 11/64	1.1719
1 3/8	6	. . . .	. . . .	1.3750	1.2667	1.1585	1 7/32	1.2187
1 3/8	. . . .	12	. . . .	1.3750	1.3209	1.2668	1 19/64	1.2969
1 1/2	6	. . . .	. . . .	1.5000	1.3917	1.2835	1 11/32	1.3437
1 1/2	. . . .	12	. . . .	1.5000	1.4459	1.3918	1 27/64	1.4219
1 3/4	5	. . . .	. . . .	1.7500	1.6201	1.4902	1 9/16	1.5625
2	4 1/2	. . . .	. . . .	2.0000	1.8557	1.7113	1 25/32	1.7812
2 1/4	4 1/2	. . . .	. . . .	2.2500	2.1057	1.9613	2 1/32	2.0313
2 1/2	4	. . . .	. . . .	2.5000	2.3376	2.1752	2 1/4	2.2500
2 3/4	4	. . . .	. . . .	2.7500	2.5876	2.4252	2 1/2	2.5000
3	4	. . . .	. . . .	3.0000	2.8376	2.6752	2 3/4	2.7500
3 1/4	4	. . . .	. . . .	3.2500	3.0876	2.9252	3	3.0000
3 1/2	4	. . . .	. . . .	3.5000	3.3376	3.1752	3 1/4	3.2500
3 3/4	4	. . . .	. . . .	3.7500	3.5876	3.4252	3 1/2	3.5000
4	4	. . . .	. . . .	4.0000	3.8376	3.6752	3 3/4	3.7500

TABLE 11A

Unified American Standard Threads Basic Dimensions

Nominal Size	Basic Major Diameter, Inches	Thds. per Inch	Basic Pitch Diameter, Inches	Minor Diameter Ext. Thds. Inches	Minor Diameter Int. Thds. Inches	Lead Angle at Basic Pitch Diameter Deg. Min.		Section at Minor Diameter Sq. In.	Tensile Stress Area Sq. In.
Coarse-Thread Series									
1/4	0.2500	20	0.2175	0.1887	0.1959	4	11	0.0269	0.0317
5/16	0.3125	18	0.2764	0.2443	0.2524	3	40	0.0454	0.0322
3/8	0.3750	16	0.3344	0.2983	0.3073	3	24	0.0678	0.0773
7/16	0.4375	14	0.3911	0.3499	0.3602	3	20	0.0933	0.1060
1/2	0.5000	13	0.4500	0.4056	0.4157	3	7	0.1257	0.1416
9/16	0.5625	12	0.5084	0.4603	0.4723	2	59	0.1620	0.1816
5/8	0.6250	11	0.5660	0.5135	0.5266	2	56	0.2018	0.2256
3/4	0.7500	10	0.6850	0.6273	0.6417	2	40	0.3020	0.3340
7/8	0.8750	9	0.8028	0.7387	0.7547	2	31	0.4193	0.4612
1	1.0000	8	0.9188	0.8466	0.8647	2	29	0.5510	0.6051
1 1/8	1.1250	7	1.0322	0.9497	0.9704	2	31	0.6931	0.7627
1 1/4	1.2500	7	1.1572	1.0747	1.0954	2	15	0.8896	0.9684
1 3/8	1.3750	6	1.2667	1.1705	1.1946	2	24	1.0541	1.1536
1 1/2	1.5000	6	1.3917	1.2955	1.3196	2	11	1.2936	1.4041
1 3/4	1.7500	5	1.6201	1.5046	1.5335	2	15	1.7441	1.8983
2	2.0000	4 1/2	1.8557	1.7274	1.7594	2	11	2.3001	2.4971
2 1/4	2.2500	4 1/2	2.1037	1.9774	2.0094	1	55	3.0212	3.246
2 1/2	2.5000	4	2.3376	2.1933	2.2294	1	57	3.7161	3.9976
2 3/4	2.7500	4	2.5876	2.4433	2.4794	1	46	4.6194	4.9326
3	3.0000	4	2.8376	2.6933	2.7294	1	36	5.6209	5.9659
3 1/4	3.2500	4	3.6876	2.9433	2.9794	1	29	6.7205	7.0992
3 1/2	3.5000	4	3.3376	3.1933	3.2294	1	22	7.9183	8.3268
3 3/4	3.7500	4	3.5876	3.4433	3.4794	1	16	9.2143	9.6546
4	4.0000	4	3.8376	3.6933	3.7294	1	11	10.6084	11.0805
Fine-Thread Series									
1/4	0.2500	28	0.2268	0.2062	0.2113	2	52	0.0326	0.0362
5/16	0.3125	24	0.2854	0.2614	0.2674	2	40	0.0524	0.0579
3/8	0.3750	24	0.3479	0.3239	0.3299	2	11	0.0809	0.0876
7/16	0.4375	20	0.4050	0.3762	0.3834	2	15	0.1090	0.1185
1/2	0.5000	20	0.4675	0.4387	0.4459	1	57	0.1486	0.1597
9/16	0.5623	18	0.5264	0.4943	0.5024	1	55	0.1888	0.2026
5/8	0.6250	18	0.3889	0.5568	0.5649	1	43	0.2400	0.2555
3/4	0.7500	16	0.7094	0.6733	0.6823	1	36	0.3513	0.3724
7/8	0.8750	14	0.8286	0.7874	0.7977	1	34	0.4805	0.5088
1	1.0000	14	0.9536	0.9124	0.9227	1	22	0.6464	0.6791
1	1.0000	12	0.9459	0.8978	0.9098	1	36	0.6245	0.6624
1 1/8	1.1250	12	1.0709	1.0228	1.0348	1	25	0.8118	0.8549
1 1/4	1.2500	12	1.1959	1.1478	1.1598	1	16	1.0237	1.0721
1 3/8	1.3750	12	1.3209	1.2728	1.2848	1	9	1.2602	1.3137
1 1/2	1.5000	12	1.4459	1.3978	1.4098	1	3	1.5212	1.5799

TABLE 12

Formulas for Determining the Dimensions of Gears by Diametral Pitch

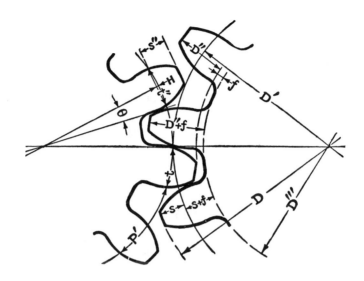

P	=	diametral pitch or the number of teeth to one inch of pitch diameter.
N	=	number of teeth.
D'	=	pitch diameter.
s	=	addendum.
f	=	clearance at bottom of tooth.
s + f	=	dedendum.
t	=	thickness of tooth on pitch line.
t"	=	chordal thickness of tooth.
D"	=	working depth of tooth.
D" + f	=	whole depth of tooth.
D	=	outside diameter.
D'''	=	bottom diameter.
P'	=	circular pitch.
H	=	height of arc.
s"	=	distance from chord to top of tooth.
θ	=	1/4 the angle subtended by circular pitch.
M	=	module (in millimeters).

TABLE 12 (Continued)

Formulas for Determining the Dimensions of Gears by Diametral Pitch

$P \quad = \quad \dfrac{N}{D'}; \text{ or} = \dfrac{N+2}{D}; \text{ or} = \dfrac{\pi}{P'}; \text{ or} = \dfrac{1}{M}$

$N \quad = \quad D'P; \text{ or} = DP - 2; \text{ or} = \dfrac{D'\pi}{P'}$

$D' \quad = \quad \dfrac{N}{P}; \text{ or} = \dfrac{NP'}{\pi}; \text{ or} = .3183\ P'N; \text{ or} = \dfrac{DN}{N+2}; \text{ or} = NM$

$s \quad = \quad \dfrac{1}{P}; \text{ or} = \dfrac{P'}{\pi}; \text{ or} = .3183\ P'; \text{ or} = \dfrac{D'}{N}; \text{ or} = \dfrac{D}{N+2}$

$f \quad = \quad \dfrac{t}{10}; \text{ or} = .1571\ M$

$s + f \quad = \quad \dfrac{1.157}{P}; \text{ or} = .3683\ P'; \text{ or} = 1.1571\ M$

$t \quad = \quad \dfrac{P'}{2}; \text{ or} = \dfrac{\pi}{2P} = \dfrac{1.5708}{P}; \text{ or } 1.5708\ M$

$t'' \quad = \quad D' \sin \theta$

$D'' \quad = \quad 2s; \text{ or} = \dfrac{2}{P}; \text{ or} = .6366\ P'; \text{ or} = 2M$

$D'' + f \quad = \quad \dfrac{2.157}{P}; \text{ or} = .6866\ P'; \text{ or} = 2.1571\ M$

$D \quad = \quad D' + 2s; \text{ or} = \dfrac{N+2}{P}; \text{ or} = \dfrac{P'\ (N+2)}{\pi}$

$D''' \quad = \quad D - 2(D'' + f); \text{ or} \quad \dfrac{N - 2.31416}{P}; \text{ or} = (N - 2.3141)\ M$

$P' \quad = \quad \dfrac{\pi}{P}; \text{ or} = \pi M; \text{ or} = \dfrac{D'\pi}{N}; \text{ or} = \dfrac{D\pi}{N+2}$

$H \quad = \quad \dfrac{D'\ (1 - \cos \theta)}{2}$

$s'' \quad = \quad s + H$

$\theta \quad = \quad \dfrac{90°}{N}$

1 millimeter = 0.03937 inch

TABLE 13 Weights of Materials

Materials	Pounds per Cubic Foot	Pounds per Cubic Inch
Aluminum	168.5	.0975
Brass, cast (20 to 36 percent zinc)	523 to 538	.303 to .311
Brick, pressed	150	
Brick, common, hard	125	
Bronze	545 to 554	.315 to .321
Cement, American, Rosendale.	56	
Cement, Portland	90	
Chromium	443	.256
Clay, loose	63	
Coal, broken, loose, anthracite	54	
Coal, broken, loose, bituminous.	49	
Concrete	154	
Copper	556	.322
Earth, common loan	76	
Earth, packed	95	
Gravel, dry, loose	90 to 106	
Gravel, well shaken	99 to 117	
Gold	1,203	.696
Ice	58.7	
Iron, cast.	490	.284
Iron, wrought	486 to 493	.281 to .285
Lead	707	.409
Lime	53	
Masonry, well dressed	165	
Masonry, dry rubble	138	
Mortar, hardened	103	
Nickel	555	.321
Quartz	165	
Sand, dry, loose.	90 to 106	
Sand, well shaken	99 to 117	
Silver	655	.379
Snow, freshly fallen.	5 to 12	
Snow, wet and compacted	15 to 50	
Steel, cast or rolled	474 to 486	.274 to .281
Stone, gneiss.	168	
Stone, granite	170	
Stone, limestone	168	
Stone, marble	168	
Stone, sandstone	151	
Stone, shale	162	
Stone, slate	175	
Tar	62	
Tin	456	.264
Water (at 4°C.)	62.36	.036
Wood, dry, ash	38	
Wood, dry, cherry.	42	
Wood, dry, chestnut	41	
Wood, dry, elm	35	
Wood, dry, hemlock	25	
Wood, dry, hickory	53	
Wood, dry, lignum vitae	83	
Wood, dry, mahogany	53	
Wood, dry, maple.	49	
Wood, dry, oak, white	50	
Wood, dry, oak, other kinds	32 to 45	
Wood, dry, pine, white.	25	
Wood, dry, pine, yellow	34 to 45	
Wood, dry, spruce.	25	
Wood, dry, sycamore.	37	
Wood, dry, walnut, black	38	
Zinc.	445	.258

TABLE 14

Shortcut Formulas

Circumference of a circle = diameter x 3.1416

Circumference of a circle = radius x 6.2832

Area of a circle = square of the radius x 3.1416

Area of a circle = square of the diameter x .7854

Area of a circle = circumference^2 x .0796

Area of a circle = half the circumference x half its diameter

Radius of a circle = circumference x .1592

Radius of a circle = square root of the area x .5642

Diameter of a circle = circumference x .3183

Diameter of a circle = square root of the area x 1.1284

Side of an inscribed equilateral triangle = diameter of circle x .866

Side of an inscribed square = diameter of circle x .707

Side of an inscribed square = circumference of circle x .225

Side of an equal square = circumference of circle x .282

Side of an equal square = diameter of circle x .886

Area of a triangle = base x half the altitude

Area of an ellipse = product of both axes and .7854

Volume of a sphere = surface x one-sixth of its diameter

Surface of a sphere = circumference x its diameter

Surface of a sphere = square of the diameter x 3.1416

Surface of a sphere = square of the circumference x .3183

Volume of a sphere = cube of the diameter x .5236

Volume of a sphere = cube of the circumference x .0169

Side of an inscribed cube = radius of sphere x 1.1547

Volume of a cone or pyramid = area of its base x one-third of its altitude

Surface of a cube = area of one of its sides x 6

Area of a trapezoid = altitude x one-half the sum of its parallel sides

TABLE 15
Weights of Flat Sizes of Steel in Pounds per Linear Foot

Thickness	Width of Stock

Thickness	1/2	3/4	1	1 1/4	1 1/2	1 3/4	2
1/8	.213	.320	.426	.530	.640	.745	.850
3/16	.319	.480	.639	.790	.960	1.12	1.28
1/4	.425	.640	.852	1.06	1.28	1.49	1.70
5/16	.531	.800	1.06	1.33	1.60	1.86	2.13
3/8	.638	.960	1.28	1.59	1.91	2.23	2.55
7/16	.744	1.12	1.49	1.86	2.23	2.60	2.98
1/2	....	1.28	1.70	2.13	2.55	2.98	3.40
9/16	....	1.44	1.91	2.39	2.87	3.35	3.83
5/8	....	1.60	2.12	2.66	3.19	3.72	4.26
11/16	....	1.76	2.34	2.92	3.51	4.09	4.68
3/4	....	....	2.55	3.19	3.83	4.46	5.10
13/16	....	....	2.76	3.45	4.14	4.83	5.53
7/8	....	....	2.98	3.72	4.46	5.21	5.96
15/16	....	....	3.19	3.98	4.78	5.58	6.38
1	....	....	....	4.25	5.10	5.96	6.80

TABLE 16
Weights of Sheet Steel and Iron

Number of Gauge	Approx. Thickness in Inches	Weight per Square Foot		
		Galvanized Iron	Black Iron	Black Steel
10	.138	5.781	5.625	5.737
11	.123	5.156	5.000	5.100
12	.107	4.531	4.375	4.462
13	.092	3.906	3.750	3.825
14	.077	3.281	3.125	3.156
15	.069	2.969	2.812	2.869
16	.061	2.656	2.50	2.550
17	.055	2.406	2.25	2.295
18	.049	2.156	2.00	2.040
19	.044	1.906	1.75	1.785
20	.037	1.656	1.50	1.530
21	.034	1.531	1.375	1.402
22	.031	1.406	1.25	1.275
23	.028	1.281	1.125	1.147
24	.025	1.156	1.00	1.020
25	.022	1.031	.875	.892
26	.019	.906	.75	.765
27	.017	.844	.687	.701
28	.016	.781	.625	.637
29	.014	.719	.562	.574
30	.012	.656	.5	.510

TABLE 17
Weights in Square and Round Bars of Steel in Pounds per Linear Foot
Based on 489.6 Lb. per Cubic Foot*

Thickness or Diam., Inches	Weight of Square Bar 1 Foot Long	Weight of Round Bar 1 Foot Long	Thickness or Diam., Inches	Weight of Square Bar 1 Foot Long	Weight of Round Bar 1 Foot Long
1/32	.0033	.0026	3	30.60	24.03
1/16	.0133	.0104	3 1/8	33.20	26.08
1/8	.0531	.0417	3 1/4	35.92	28.20
3/16	.1195	.0938	3 3/8	38.73	30.42
1/4	.2123	.1669	3 1/2	41.65	32.71
5/16	.3333	.2608	3 5/8	44.68	35.09
3/8	.4782	.3756	3 3/4	47.82	37.56
7/16	.6508	.5111	3 7/8	51.05	40.10
1/2	.8500	.6676	4	54.40	42.73
9/16	1.076	.8449	4 1/4	61.41	48.24
5/8	1.328	1.043	4 1/2	68.85	54.07
11/16	1.608	1.262	4 3/4	76.71	60.25
3/4	1.913	1.502	5	85.00	66.76
13/16	2.245	1.763	5 1/4	93.72	73.60
7/8	2.603	2.044	5 1/2	102.8	80.77
15/16	2.989	2.347	5 3/4	112.4	88.29
1	3.400	2.670	6	122.4	96.14
1 1/16	3.838	3.014	6 1/4	132.8	104.3
1 1/8	4.303	3.379	6 1/2	143.6	112.8
1 3/16	4.795	3.766	6 3/4	154.9	121.7
1 1/4	5.312	4.173	7	166.6	130.9
1 5/16	5.857	4.600	7 1/4	178.7	140.4
1 3/8	6.428	5.019	7 1/2	191.3	150.2
1 7/16	7.026	5.518	7 3/4	204.2	160.3
1 1/2	7.650	6.008	8	217.6	171.0
1 9/16	8.301	6.520	8 1/4	231.4	181.8
1 5/8	8.978	7.051	8 1/2	245.6	193.0
1 11/16	9.682	7.604	8 3/4	260.3	204.4
1 3/4	10.41	8.178	9	275.4	216.3
1 13/16	11.17	8.773	9 1/4	291.1	228.5
1 7/8	11.95	9.388	9 1/2	306.8	241.0
1 15/16	12.76	10.02	9 3/4	323.2	253.9
2	13.60	10.68	10	340.0	267.0
2 1/8	15.35	12.06	10 1/4	357.2	280.6
2 1/4	17.22	13.52	10 1/2	374.9	294.4
2 3/8	19.18	15.07	10 3/4	392.9	308.6
2 1/2	21.25	16.69	11	411.4	323.1
2 5/8	23.43	18.40	11 1/4	430.3	337.9
2 3/4	25.00	20.20	11 1/2	449.6	353.1
2 7/8	28.10	22.07	11 3/4	469.4	368.6

To compute the weight of sheet steel:
Multiply the thickness by 40.8; the result is the weight in pounds per square foot.
Example. If a piece of sheet steel is .005" thick, its weight is .005 x 40.8 = 204 lb. per square foot.
To compute the weight of sheet iron:
Multiply the thickness by 40; the result is the weight in pounds per square foot.
Example. If a piece of sheet iron is .005" thick, its weight is .005 x 40 = .200 lb. per square foot.

*For wrought iron, deduct 2 percent; for high-speed steel, add 11 percent.

TABLE 18 Electrical Symbols

Symbol	Description	Symbol	Description
	Fluorescent Fixture-120V.		Surface Wall or Ceiling Mounted P.A. System Horn
	Fluorescent Fixture-277 V.		Combination Clock and Speaker
	Incandescent Fixture		Microphone Outlet, Wall Mounted
	Incandescent Wall Bracket		Microphone Outlet in Floor
	Exit Light, Ceiling Mounted		Microphone Suspended from Ceiling
	Exit Light, Wall Mounted		Clock
	Mercury Vapor Fixture		Timer with Control Panel
	Fluorescent Fixture on Emergency Circuit		Clock, Skeleton Dial
	Incandescent or Mercury Fixture on Emergency Circuit		Master Clock in General Office
M^{Sp}	Single Pole Switch		Program Bell
S_P	Switch and Pilot Light		Yard Program Bell Mounted on Exterior of Building
S_K	Key-Operated Switch		Public Telephone Outlet, Wall Mounted
S_3	Three-Way Switch		Public Telephone Floor Outlet
S_4	Four-Way Switch		Public Telephone Pay Station
3(4) SK	Three-Way, or Four-Way Key-Operated Switch		Unit Ventilator
S_M	Momentary Contact Contractor Controlled Push Button		Motor
S	Manual Motor Starter	UPo	Conduit Turned Up
	Dual Floodlight Unit, Wall Mounted	DN●	Conduit Turned Down
	Television Antenna Outlet, Wall Mounted	OFB OR ●FB	Floor Box
	Duplex Receptacle		Lighting Panel 277/480 V. or 120/208 V. (As Indicated)
	Duplex Receptacle Above Counter Height		Power Panel 277/480 V. or 120/208 V. (As Indicated)
	110 V. Floor Receptacle		Sound System Amplifier
	Heavy Duty Receptacle		Telephone Strip Cabinet
D.S.	Disconnect Switch, Unfused		Conduit Run Concealed in Ceiling
	Disconnect Switch, Fused		Conduit Run Concealed in Floor
	Magnetic Motor Starter	E.P.	Explosion Proof
	Fire Alarm Station	W.P.	Weather Proof
	Fire Alarm Gong		Electric Equipment Cutoff Switch
	Fire Alarm City Connection		Gas Only Cutoff Switch
	Thermal Detector		Gas, Water, Vacuum and/or Air Cutoff Switch
	Fire Alarm Disarrangement Station and City Box Disconnect		120 V.A.C. Solenoid Valve
	Junction Box	o	Pole-Top Mounted M.V. Globe
	Recessed Ceiling Mounted P.A. System Speaker		Pole-Top Mounted M.V. Area Light
			Pole-Arm Mounted Twin M.V. Area Light Each End
			Oil Burner Emergency Cutoff Switch
			Electric Thermostat
			Single Receptacle

TABLE 19

Shrinkage of Castings for Each Foot of Length

In large cylinders. . . . . 3/32 in.

In small cylinders . . . . 1/16 in.

In beams and girders . . . . 1/10 in.

In thick brass . . . . 5/32 in.

In thin brass . . . . 3/16 in.

In cast iron pipe . . . . 1/8 in.

In steel . . . . 1/4 in.

In zinc. . . . . 5/16 in.

In lead. . . . . 5/16 in.

In tin . . . . 1/4 in.

In copper. . . . . 3/16 in.

In bismuth . . . . 5/32 in.

In malleable iron . . . . 1/8 in.

In aluminum. . . . . 3/16 in.

Index